Chapman & Hall/CRC Mathematical and Computational Biology Series

SPATIAL ECOLOGY

EDITED BY

STEPHEN CANTRELL
THE UNIVERSITY OF MIAMI
CORAL GABLES, FLORIDA, U.S.A.

CHRIS COSNER
THE UNIVERSITY OF MIAMI
CORAL GABLES, FLORIDA, U.S.A.

SHIGUI RUAN
THE UNIVERSITY OF MIAMI
CORAL GABLES, FLORIDA, U.S.A.

CRC Press is an imprint of the
Taylor & Francis Group an **informa** business
A CHAPMAN & HALL BOOK

CHAPMAN & HALL/CRC
Mathematical and Computational Biology Series

Aims and scope:
This series aims to capture new developments and summarize what is known over the whole spectrum of mathematical and computational biology and medicine. It seeks to encourage the integration of mathematical, statistical and computational methods into biology by publishing a broad range of textbooks, reference works and handbooks. The titles included in the series are meant to appeal to students, researchers and professionals in the mathematical, statistical and computational sciences, fundamental biology and bioengineering, as well as interdisciplinary researchers involved in the field. The inclusion of concrete examples and applications, and programming techniques and examples, is highly encouraged.

Proposals for the series should be submitted to one of the series editors above or directly to:
CRC Press, Taylor & Francis Group
4th, Floor, Albert House
1-4 Singer Street
London EC2A 4BQ
UK

Published Titles

Bioinformatics: A Practical Approach
Shui Qing Ye

Cancer Modelling and Simulation
Luigi Preziosi

Combinatorial Pattern Matching Algorithms in Computational Biology Using Perl and R
Gabriel Valiente

Computational Biology: A Statistical Mechanics Perspective
Ralf Blossey

Computational Neuroscience: A Comprehensive Approach
Jianfeng Feng

Data Analysis Tools for DNA Microarrays
Sorin Draghici

Differential Equations and Mathematical Biology
D.S. Jones and B.D. Sleeman

Engineering Genetic Circuits
Chris J. Myers

Exactly Solvable Models of Biological Invasion
Sergei V. Petrovskii and Bai-Lian Li

Gene Expression Studies Using Affymetrix Microarrays
Hinrich Göhlmann and Willem Talloen

Handbook of Hidden Markov Models in Bioinformatics
Martin Gollery

Introduction to Bioinformatics
Anna Tramontano

An Introduction to Systems Biology: Design Principles of Biological Circuits
Uri Alon

Kinetic Modelling in Systems Biology
Oleg Demin and Igor Goryanin

Knowledge Discovery in Proteomics
Igor Jurisica and Dennis Wigle

Meta-analysis and Combining Information in Genetics and Genomics
Rudy Guerra and Darlene R. Goldstein

Modeling and Simulation of Capsules and Biological Cells
C. Pozrikidis

Niche Modeling: Predictions from Statistical Distributions
David Stockwell

Normal Mode Analysis: Theory and Applications to Biological and Chemical Systems
Qiang Cui and Ivet Bahar

Optimal Control Applied to Biological Models
Suzanne Lenhart and John T. Workman

Pattern Discovery in Bioinformatics: Theory & Algorithms
Laxmi Parida

Python for Bioinformatics
Sebastian Bassi

Spatial Ecology
Stephen Cantrell, Chris Cosner, and Shigui Ruan

Spatiotemporal Patterns in Ecology and Epidemiology: Theory, Models, and Simulation
Horst Malchow, Sergei V. Petrovskii, and Ezio Venturino

Stochastic Modelling for Systems Biology
Darren J. Wilkinson

Structural Bioinformatics: An Algorithmic Approach
Forbes J. Burkowski

The Ten Most Wanted Solutions in Protein Bioinformatics
Anna Tramontano

Chapman & Hall/CRC
Taylor & Francis Group
6000 Broken Sound Parkway NW, Suite 300
Boca Raton, FL 33487-2742

Printed in the United States of America on acid-free paper
10 9 8 7 6 5 4 3 2 1

International Standard Book Number: 978-1-4200-5985-4 (Hardback)

Library of Congress Cataloging-in-Publication Data

Spatial ecology / [edited by] Stephen Cantrell, Chris Cosner, and Shigui Ruan.
p. cm. -- (Chapman & Hall/CRC mathematical and computational biology series)
Includes bibliographical references and index.
ISBN 978-1-4200-5985-4 (alk. paper)
1. Spatial ecology. 2. Spatial ecology--Mathematical models. I. Cantrell, Robert Stephen. II. Cosner, Chris. III. Ruan, Shigui, 1963- IV. Title. V. Series.

QH541.15.S62S615 2010
577.015'118--dc22 2009021500

Visit the Taylor & Francis Web site at
http://www.taylorandfrancis.com

and the CRC Press Web site at
http://www.crcpress.com

Contents

Preface

Mathematical models and concepts have been important in ecology since its inception as a discipline. However, symbiosis between the disciplines of mathematics and ecology on any appreciable scale is a far more recent phenomenon. Fortunately, in the last few decades, there has been a growing recognition among many theoretical and empirical ecologists and mathematical scientists that they can and should work together to the benefit of both disciplines, science more broadly, and society at large. Promoting this kind of interaction and integration was the theme of a conference we helped organize here at the University of Miami in January 2005, along with colleagues from the Department of Biology and the Rosenstiel School of Marine and Atmospheric Sciences. The title of the meeting was "Workshop in Spatial Ecology." The choice of topic was deliberate and two-fold. First of all, space and spatial features are now solidly established as essential considerations in ecology, both in terms of theory and practice. Second, the mathematical challenges in advancing understanding of the role of space in ecology are substantial and mathematically seductive. We believed that the benefits of bringing together a select group of top-flight ecologists and mathematicians, many of whom would not have heretofore met each other, would be enormous, and if the atmosphere at the meeting is any indication, we were correct. Not long into our interactions, the suggestion arose for some kind of a follow-up volume to the workshop; not a conference proceedings per se, but something more substantial, more thoughtful, that would promote the kind of interplay between mathematics and ecology, and between theory and data, that we so enjoyed during the workshop. We immediately thought of two volumes of essays on ecological theory that have greatly influenced our development as mathematical scientists interested in serious ecological questions: *Perspectives in Ecological Theory*, edited by J. Roughgarden, R. M. May, and S. A. Levin, Princeton University Press, 1989; and *Spatial Ecology: The Role of Space in Population Dynamics and Interspecific Interactions*, edited by D. Tilman and P. Kareiva, Princeton University Press, 1997. We thought that a volume along those lines that considered emerging challenges in spatial ecology could be highly valuable to a new generation of mathematical scientists and ecologists, especially if the choice of contributors to the volume reflected the current trend toward increased interaction of mathematical and ecological scientists and the resulting trend toward integration of the two disciplines. It was in that spirit that we arrived at the current volume.

We have identified emergent challenges in spatial ecology: understanding the impact of space on community structure, incorporating the scale and structure of landscapes into mathematical models, and developing the connections between spatial ecology and the three other disciplines of evolutionary theory, epidemiology, and economics. This volume is divided into sections focused on those topics. Many of the authors of essays in this volume spoke at the Workshop in Spatial Ecology, but quite a number did not attend. Nevertheless, all of them share a commitment to the advancement of ecology as a truly quantitative science, particularly as it touches upon the role of space.

One of the fundamental problems in spatial ecology is to understand how spatial effects influence the dynamics of populations and the structure of communities. There has been significant progress in recent years on developing and analyzing spatial models for a single population in a temporally constant environment, and at least some on models for two competitors or a predator and its prey, but there has been much less work on models for spatial effects in communities involving several species or trophic levels or environmental variability in both space and time. On the mathematical side, much of the progress on understanding spatial models has been related to the development of a theory that can give criteria for unconditional persistence or extinction, that is, determining when a model has some sort of globally attracting set with certain species present and others absent. There has been some progress, but not as much, on methods for treating models that have multiple attracting sets so that their predictions are conditional on factors such as the initial state of the system. The chapter by DeAngelis et al. describes how models and simulations can provide insight into community and food chain structure in assemblages of fish species in wetland environments where the area of fish habitat is seasonally fluctuating. The chapter by Amarasekare presents results from models that illuminate how dispersal and spatial heterogeneity influence the mechanisms and patterns of species coexistence in multi-trophic communities with intraguild predation or predator-mediated coexistence. The chapter by Jiang and Shi describes recent progress on the mathematical theory for treating models where Allee effects, strong competition, or other mechanisms give rise to "bistability," that is, to multiple stable equilibria.

Classical modeling approaches in spatial ecology typically treat space as homogeneous and isotropic. For example, most spatially explicit models based on partial differential equations envision that organisms disperse through a uniform environment via simple diffusion. However, many ecological processes occur in spatial structures that display various sorts of heterogeneity and/or directionality at various scales, and the nature of the spatial structure of populations themselves is not always obvious. Organisms may disperse via nonrandom mechanisms that arise directly from the physical environment, for example by advection, and may decide whether or how to move in nonrandom ways based on environmental cues. The idea of connecting spatial scale and structure and dispersal behavior to phenomena in population dynamics, evolution, epidemiology, and economics is a recurring theme that is present in many of the chapters in this collection, and it is the specific focus of several of them. The chapter by Ovaskainen and Crone discusses how diffusion models can be extended and refined to describe dispersal in heterogeneous landscapes consisting of patches and corridors of various types, and to account for dispersal behavior that may involve effects such as habitat preferences. The next three chapters are motivated or partly motivated by the problem of understanding ecological processes in river systems. Those systems present a branching spatial structure that is different from that of typical terrestrial environments, and dispersal in them is influenced by physical advection. The chapter by Fagan et al. addresses the issue of formulating metapopulation models for river networks and examines the effects of the "branchiness" of the network on metapopulations inhabiting it. The chapter by Hadeler et al. treats a variety of effects related to models for populations with quiescent phases. One particular problem that is discussed in that chapter is the "drift paradox," that is, the problem of understanding how populations in streams can resist being washed out by advection. Shifting into a quiescent phase is one possible mechanism by which a population can resist washout. The chapter by Nisbet et al. treats population dynamics in advective media, reviews the drift paradox, and identifies characteristic length scales related to population dynamics in such media. The final chapter in this section, by Hinrichsen and Holmes, addresses the problem of determining the spatial structure (or absence thereof) of a population from measurements at different sites. Specifically, it treats the application of state-space models to the problem of determining whether multi-site data correspond to independent populations with independent environmental drivers, independent populations with a shared environmental driver, a collection of populations with the same growth rate but independent environmental drivers, or a single population. Each of those cases would call for a distinct modeling approach, so determining which one represents the actual situation is important for connecting models with data.

The remaining chapters in the collection treat topics related to space and ecology, but do so relative to the perspectives of evolutionary theory, epidemiology, or economics. These areas are related to ecology by both direct connections among the phenomena they examine and philosophical similarities in the issues they address. Ecology describes the framework in which the natural selection that drives evolution occurs. In a sense, epidemiology describes the population interactions between microbes and other organisms, and may involve other aspects of ecology in the contexts of vector-borne or zoonotic diseases. The economics of harvesting resources such as fish or forests are tied to the ecology of those resources. More broadly, all of these disciplines aim to describe the large-scale emergent behavior of systems consisting of many interacting independent agents that may cooperate, compete, or exploit each other. For that reason modeling ideas and approaches that have worked well in the context of one of them may be relevant to others. Finding unifying approaches to these disciplines may be one of the grand intellectual challenges of current scientific thought.

At our present level of understanding, the conclusions about spatial aspects of evolution that can be drawn from models seem to depend to a considerable extent on the detailed assumptions built into the models. The chapters on topics related to evolution in this volume provide a guided tour through a number of scenarios and modeling approaches that represent active areas of current research, and suggest some paths toward conceptual unification. The chapter by Hanski discusses how realistic metapopulation models may provide a unifying approach to ecological and evolutionary theory in fragmented habitats. Those models account for the areas of patches and the connectivity among patches. Hanski discusses how metapopulation models can be used to study the evolution of migration rates in fragmented environments. The chapter by Holt and Barfield also connects metapopulation theory to evolution. It specifically addresses the problem of understanding how environmental heterogeneity influences the evolution of species' niches. The chapters by Cantrell et al. and Bolker, in contrast, treat problems related to the evolution of dispersal in continuous environments. Both of those chapters take the viewpoint of adaptive dynamics as a starting point. The key assumption behind adaptive dynamics is that the strategies which can be expected to be successful in an evolutionarily sense are those that allow populations using them to successfully invade resident populations using other strategies. Although Cantrell et al. and Bolker use similar philosophical approaches based on adaptive dynamics, they use different types of models to study different scenarios. Cantrell et al. use reaction-advection-diffusion models in spatially heterogeneous environments to examine the evolution of dispersal mechanisms arising from local movement behavior that may be responsive to environmental conditions. Bolker uses spatial moment equations to examine how the nature and scale of spatial autocorrelation in environmental suitability influence the evolution of the shape of nonlocal dispersal kernels. Taken together the chapters treating spatial aspects of evolutionary theory show how strongly assumptions about the nature of dispersal and the scale and structure of the environment influence the conclusions that can be drawn about the evolutionary causes and effects of dispersal.

Recent concerns about the emergence or resurgence and global spread of infectious diseases have motivated renewed interest in epidemiology. Many potentially dangerous pathogens are zoonotic or vector-borne and thus have aspects that are directly related to ecology. Some similar problems arise in both disciplines, and often these can be addressed by similar modeling approaches. That point is well illustrated by the chapters on epidemiology. The chapter by Lloyd and Sattenspiel uses a metapopulation approach to examine how nonlinear disease dynamics interact with seasonal forcing to determine spatiotemporal patterns in disease dynamics. The chapter by Potts and Kimbrell describes how simulation models can be used to compare different control strategies for vector-borne diseases. The chapter by Ruan and Wu reviews a selection of reaction-diffusion models for the spread of diseases with animal hosts. In each case the modeling approach is reminiscent of ideas that are widely used in ecology, but is modified by the specific features of the epidemiological system it describes.

As human populations increase they put increased pressure on natural resources, which makes it crucial that we learn how to use them in sustainable ways and if possible to optimize the benefits derived from them. To do that it is necessary to understand how economics interacts with ecology and then apply ideas from optimal control theory. The chapter by Sanchirico and Wilen addresses the problem of optimal fisheries management from the viewpoint of metapopulation modeling. The chapter by Olson explores the general issue of constructing models that describe the dynamics of resources, human factors in harvesting them, and the flow of capital investment needed to support the harvesting. The chapter by Herrera and Lenhart reviews some results on optimal control in metapopulation models and shows how to extend the approach to reaction-diffusion models and related models based on partial differential equations.

Our friend and colleague Alan Lazer once remarked: “It is better to open up an area of research than to close one down.” As the chapters in this volume show, there is a great deal more to be done before the discipline of spatial ecology is ready to be closed down. We hope that this volume will inspire readers to open up new areas of research in the mathematical theory of spatial ecology and its connections with evolutionary theory, epidemiology, and economics.

Robert Stephen Cantrell
Chris Cosner
Shigui Ruan

Coral Gables, FL

List of Contributors

Priyanga Amarasekare
Department of Ecology and
Evolutionary Biology
University of California at Los Angeles
Los Angeles, CA
USA
Email: *amarasek@eeb.ucla.edu*

Kurt E. Anderson
Department of Biology
University of California at Riverside
Riverside, CA
USA

Michael Barfield
Department of Zoology
University of Florida
111 Bartram Hall, PO Box 118525
Gainesville, FL
USA
Email: *mjb01@ufl.edu*

Benjamin M. Bolker
Department of Zoology
University of Florida
620B Bartram Hall, Box 118525
Gainesville, FL
USA
Email: *bolker@zoo.ufl.edu*

Robert Stephen Cantrell
Department of Mathematics
University of Miami
Coral Gables, FL
USA
Email: *rsc@math.miami.edu*

Chris Cosner
Department of Mathematics
University of Miami
Coral Gables, FL
USA
Email: *gcc@math.miami.edu*

Elizabeth E. Crone
Department of Ecosystem and
Conservation Sciences
University of Montana
Missoula, MT
USA
Email: *elizabeth.crone@umontana.edu*

Don L. DeAngelis
U. S. Geological Survey
Florida Integrated Science Centers
and Department of Biology
University of Miami
P. O. Box 249118
Coral Gables, FL
USA
Email: *ddeangelis@bio.miami.edu*

Douglas D. Donalson
Everglades National Park
40001 State Road 9336
Homestead, FL
USA

William F. Fagan
Department of Biology
University of Maryland
College Park, MD
USA
Email: *bfagan@umd.edu*

Ulrike Feudel
Carl von Ossietzky University
Oldenburg
Institute for Chemistry and Biology
of the Marine Environment (ICBM)
Theoretical Physics/Complex Systems
PF 2503, D - 26111 Oldenburg
Germany

Evan H. Campbell Grant
Graduate Program in Marine, Estuarine
and Environmental Sciences
University of Maryland
College Park, MD
USA

Karl P. Hadeler
Department of Mathematics
University of Tübingen
Auf der Morgenstelle 10
D-72076 Tübingen
Germany
Email: *k.p.hadeler@uni-tuebingen.de*

Ilkka Hanski
Department of Biological and
Environmental Sciences
P.O. Box 65
Viikinkaari 1, FI-00014
University of Helsinki
Finland
Email: *ilkka.hanski@helsinki.fi*

Guillermo E. Herrera
Department of Economics
Bowdoin College
Brunswick, ME
USA

Thomas Hillen
Department of Mathematical
and Statistical Sciences
University of Alberta
Edmonton, Canada
Email: *thillen@ualberta.ca*

Richard A. Hinrichsen
Hinrichsen Environmental Consulting
901 NE 43rd ST #101
Seattle, WA
USA
Email: *hinrich@seanet.com*

Elizabeth E. Holmes
National Marine Fisheries Service
Northwest Fisheries Science Center
2725 Montlake Blvd. E.
Seattle, WA
USA
Email: *eli.holmes@noaa.gov*

Robert D. Holt
Department of Zoology
University of Florida
111 Bartram Hall, PO Box 118525
Gainesville, FL
USA
Email: *rdholt@zoology.ufl.edu*

Jifa Jiang
Department of Mathematics
Shanghai Normal University
Shanghai
China
Email: *jiangjf@mail.tongji.edu.cn*

Tristan Kimbrell
Beasley School of Law
Temple University
Philadelphian, PA
USA

Suzanne Lenhart
Department of Mathematics
University of Tennessee
Knoxville, TN
USA
Email: *lenhart@math.utk.edu*

Mark A. Lewis
Department of Mathematical
and Statistical Sciences
and Department of Biological Sciences
University of Alberta
Edmonton, Canada
Email: *mlewis@math.ualberta.ca*

Alun L. Lloyd
Department of Mathematics and
Biomathematics Graduate Program
North Carolina State University
Raleigh, NC
USA
Email: *alun_lloyd@ncsu.edu*

Yuan Lou
Department of Mathematics
Ohio State University
Columbus, OH
USA
Email: *lou@math.ohio-state.edu*

Heather J. Lynch
Department of Biology
University of Maryland
College Park, MD
USA

Edward McCauley
Department of Biological Sciences
University of Calgary
Calgary, Canada

Roger M. Nisbet
Department of Ecology, Evolution,
and Marine Biology
University of California at Santa Barbara
Santa Barbara, CA
Email: *nisbet@lifesci.ucsb.edu*

Donald B. Olson
RSMAS/MPO
University of Miami
4600 Rickenbacker Causeway
Miami, FL
USA
Email: *dolson@rsmas.miami.edu*

Otso Ovaskainen
Department of Biological
and Environmental Sciences
PO Box 65, FI-00014
University of Helsinki
Finland
Email: *otso.ovaskainen@helsinki.fi*

Matthew D. Potts
Department of Environmental Science,
Policy and Management
University of California at Berkeley
Berkeley, CA
USA
Email: *mdpotts@nature.berkeley.edu*

Shigui Ruan
Department of Mathematics
University of Miami
Coral Gables, FL
USA
Email: *ruan@math.miami.edu*

James N. Sanchirico
Department of Environmental Science and Policy
University of California at Davis
Davis, CA
and
University Fellow
Resources for the Future
Washington, DC
USA
Email: *jsanchirico@ucdavis.edu*

Lisa Sattenspiel
Department of Anthropology
University of Missouri
107 Swallow Hall
Columbia, MO
USA

Junping Shi
Department of Mathematics
College of William and Mary
Williamsburg, VA
USA
and
School of Mathematics
Harbin Normal University
Harbin, Helongjiang
China
Email: *shij@math.wm.edu*

Joel C. Trexler
Department of Biological Sciences
Florida International University
Miami, FL
USA
E-mail: *trexlerj@fiu.edu*

Peter J. Unmack
Department of Biology
Brigham Young University
Provo, UT
USA

James E. Wilen
Department of Agricultural and Resource Economics
University of California at Davis
Davis, CA
USA

Jianhong Wu
Department of Mathematics and Statistics
York University
Toronto, Canada
Email: *wujh@mathstat.yorku.ca*

CHAPTER 1

Competition dynamics in a seasonally varying wetland

Don L. DeAngelis
U. S. Geological Survey and University of Miami

Joel C. Trexler
Florida International University

Douglas D. Donalson
Everglades National Park

Abstract. We have used one- and two-dimensional, spatially explicit models to simulate fish communities in freshwater wetlands in which the seasonality of rainfall in these wetlands causes annual fluctuations in the amount of flooded area, or fish habitat. We have modeled the competition between small fish species that differ from each other in efficiency of resource utilization and dispersal ability. The simulations showed that these tradeoffs, along with the spatial and temporal variability of the environment, allow coexistence of several species competing exploitatively for a common resource type. This mechanism, while sharing some characteristics with other mechanisms proposed for coexistence of competing species, is novel in detail. Simulated fish densities resembled patterns observed in Everglades empirical data. We are also modeling trophic chains and how these chains respond to the annual fluctuations in available habitat. These studies are a step towards understanding the community and food chain structure of fishes in seasonally fluctuating environments. They raise many theoretical questions that we plan to discuss in our essay.

1.1 Introduction

Models used in applied aspects of ecology, such as dealing with specific questions of conservation, assessment, and restoration, are usually far different from models used to elucidate theoretical issues. The former tend to include details that may be important to the particular applied question, while the latter are kept as simple as possible to reveal theoretical insights. However, theory should and can play a more prominent role in influencing the way that ecosystems are managed. The concept of trophic cascades from food web theory and metapopulation theory from spatial ecology are examples where theoretical models are beginning to make inputs to management plans.

As ecological theory is extended to more and more complex phenomena in which spatial heterogeneity and temporal fluctuations play a role, its potential application to real ecosystems and to specific applied issues is increasing. Practical models, even though necessarily more detailed and specific than those of theoretical ecology, may contain kernels of simpler theoretical concepts and models. Here we consider such a case of application of theory to a key component of the Everglades ecosystem.

The Everglades is a large freshwater marsh, characterized by the strong seasonal rainfall pattern of the region, which creates a cycle of wet and dry seasons. Water depths vary seasonally, but are seldom greater than one meter in this hydroscape of thousands of square kilometers. Because of the flat landscape, relatively small differences in mean water level amplify into large differences in the amount of wetted area and flooding duration, which affect the plant and animal communities. A community of small-bodied fishes, along with macroinvertebrates like crayfish, is a crucial component of the Everglades ecosystem (Kushlan 1990), as it is in many other seasonal wetlands, such as the Pantanal (Heckman 1998). These fishes are important connections that link both the small herbivorous fauna that feed on periphyton and the detritivores with the higher trophic level species, such as wading birds.

A question of great practical importance is how water levels in the Everglades should be regulated to maintain a system that is as close as possible to the natural ecosystem. The wetland small-fish community is strongly influenced by seasonal hydrologic fluctuations (Loftus and Kushlan 1987, Trexler et al. 2002). Human-induced changes in hydrology over the last several decades have altered hydroperiods in most wetland areas, thereby diminishing this fish forage-base or changing the pattern of its availability. Lack of sufficient biomass and availability of prey is hypothesized to have been a major factor in the decline of wading bird nesting at traditional Everglades' rookeries (Ogden 1994).

The species richness of the fish community is deemed to be important, both for its intrinsic value and for the contribution of species richness to biomass productivity of the community. The coexistence of many fish species of similar small body size and resource use also poses interesting questions for ecological theory. Numerous hypothesized mechanisms have been proposed for the maintenance of species richness in communities and the maintenance of the diverse Everglades freshwater fish community may be related to some current ecological theory on nonequilibrium communities. Environmental fluctuations are often proposed as means for maintaining richness in a dynamic community by preventing competitively dominant species from eliminating others. Chesson (2000) reviewed mathematical models showing that environmental fluctuations could promote diversity in nonequilibrium communities, when the fluctuations effectively provide distinct niches for the competing species. These circumstances may occur when the competing species have tradeoffs in key physiological and/or behavioral traits that allow the relative advantages to alternate among species in a fluctuating environment.

It is possible that some of the tradeoffs involve differences in the ability to move quickly into newly flooded areas and in the competitive ability in the permanently flooded areas. As vast areas of wetland are re-flooded each year, opportunistic fish species can disperse into and exploit those areas first; while other species appear better at dominating more permanently inundated areas of marsh. Species better at exploiting more stable areas should have higher reproductive and/or survival rates in long-hydroperiod areas, and they should be slower to disperse.

This idea is related to some current theoretical ideas developed for other communities. For example, Litchman and Klausmeier (2001) developed a model based on tradeoffs in coexisting species, phytoplankton species in their case, competing under seasonally periodic light availability. One species ('opportunist') was able to grow faster under initially high levels of light, but, when phytoplankton biomass increased to the point that self-shading occurred, the advantage shifted to the other ('gleaner') species. Both species declined during the period of the year when external solar radiation was low. For certain ratios of light to dark period, coexistence was possible.

The model of Litchman and Klausmeier (2001) relies on periodic temporal variations for coexistence. Other theoretical ideas emphasize spatial movement, as in "successional mosaic" models (Armstrong 1976, Tilman 1994, Holmes and Wilson 1998). In that hypothesis, disturbances occur asynchronously across the landscape, creating new habitats ready to be recolonized. If some members of the regional species pool have traits that allow invasion of newly available gaps where they increase rapidly, while others invade slowly but are better competitors and eventually displace the pioneers, species diversity can be maintained. Areas within this dynamic landscape offer a range of successional stages at a given time, allowing niches for many different life-history traits. Other models of this class assume that all patches are continuously occupied by all the species, but differences in dispersal rates, along with differences in resource growth rates on different patches, can maintain more than one species on a given resource (Abrams and Wilson 2004, Namba and Hashimoto 2004).

Our conceptual model, which attempts to account for at least some aspects of coexistence within the South Florida wetland fish community, contains elements of the above nonequilibrium hypotheses. However, the mechanism we propose differs slightly from each of those. As in the "successional mosaic" hypothesis, fish species populations move at different rates into newly opened (flooded) habitat, with the more competitive species moving more slowly than the more opportunistic ones. But this re-colonization process does not occur in randomly and asynchronously opened habitat patches, as in gap creation in forest systems. As in the Litchman and Klausmeier (2001) model, rather than random disturbances, deterministic periodic temporal variation is assumed, here as large annual pulses during the seasonal flooding period. In addition, during the dry season, the recession of water forces all populations together into permanent or semi-permanent waterbodies, so that all species may be squeezed together for part of an annual cycle. The gradual opening of new habitat by the rising water gives the more effectively dispersing fish species a temporary advantage, during which they can build in numbers before being subjected to competition by the other invaders. When the waters recede, the opportunistic fish are subjected again to heavy competition, but if they have built up high enough numbers, the species may persist.

1.2 Model

The mechanism for small fish coexistence described above was incorporated into a detailed spatial simulation model of competing fish species described by DeAngelis et al. (2005). However, the mechanism can be transparently illustrated by a more abstract model. We first describe it conceptually and then show that it is plausible by showing model output for a particular parameterization.

Table 1.1. X_1, X_2, and X_3 represent the three species. $\uparrow$ represents increasing population size, $\downarrow$ represents decreasing population size, $\rightarrow$ represents emigration from a region, $\leftarrow$ represent immigration to a region, and c stands for constant. There are 6 time periods denoted in the table, and 6 transitions between time periods, which may be very short.

	Period of Time During the Year											
Region	I	I→II	II	II→III	III	III→IV	IV	IV→V	V	V→VI	VI	VI→I
A	X_1 c $X_2 \downarrow$ $X_3 \downarrow$	X_1 c X_2 $X_3 \rightarrow$	X_1 c $X_2 \downarrow$ $X_3 \downarrow$	X_1 c $X_2 \rightarrow$ X_3	X_1 c $X_2 \downarrow$ $X_3 \downarrow$	X_1 c X_2 X_3	X_1 c $X_2 \downarrow$ $X_3 \downarrow$	X_1 c X_2 X_3	X_1 c $X_2 \downarrow$ $X_3 \downarrow$	X_1 c $X_2 \leftarrow$ X_3	X_1 c $X_2 \downarrow$ $X_3 \downarrow$	X_1 c X_2 $X_3 \leftarrow$
B		$X_3 \leftarrow$	$X_2 \leftarrow$ $X_3 \uparrow$	$X_2 \uparrow$ X_3	X_2 $X_3 \downarrow$	$X_2 \uparrow$ $X_3 \rightarrow$	X_2 $X_3 \downarrow$	$X_2 \uparrow$ $X_3 \leftarrow$	$X_2 \rightarrow$ $X_3 \downarrow$	X_3	$X_3 \uparrow$	$X_3 \rightarrow$
C						$X_3 \leftarrow$	$X_3 \uparrow$	$X_3 \rightarrow$				

The conceptual model considers three fish populations, each of which has a tradeoff in its competitive ability and ability to disperse into newly flooded areas. Instead of considering a smooth elevation gradient, we assume a step-wise gradient of three elevations. The first region, Region A, is low elevation and permanently flooded. Region B is flooded for a fraction of the year and Region C is flooded for a smaller fraction of the year. Fish Species 1 can only survive in Region A; the water is too shallow for it in Regions B and C. Both Species 2 and 3 can invade Regions B when it floods, but Species 3 can invade sooner and stay longer. When Region C, the highest elevation region is flooded, only Species 3 can invade.

The competitive dynamics are simplified in a crucial way by making competition asymmetrical. Species 1 has a negative effect on Species 2 and 3, and Species 2 has a negative effect on Species 3, but the reverse does not occur. We assume further that each population grows logistically in the absence of competition, but when in the presence of a competitively superior species, a population (e.g., Species i) is affected via Lotka-Volterra competition (i.e., $-c_{ij}X_iX_j$) by the competitively superior species j. Because the population of Species 1 does not move out of Region A and because it is assumed to suffer no negative effects of competition, it remains constant at its carrying capacity.

Imagine a yearly cycle in which water level rises and falls in a smooth, relatively deterministic manner. The dynamics of the community can be described by considering the year divided into 12 intervals, as shown in Table 1.1. The simplifications made above allow the model to be solved analytically. The equations and parameters for a particular quantitative realization of this conceptual model are shown in the Appendix. Conceptually, the temporal sequence of dynamics through a year should be as follows.

Time period I. The water level is low, so all three fish species are confined to Region A. Species 1 is the best competitor. It is assumed to remain constant during this and all other period. The other two species are declining.

Time period I→II. This is an interval during which the water depth in Region B reaches a level that some part of the population of Species 3 can invade. (This, and all other transition intervals, is considered to be very brief in the model.)

Time period II. The population of Species 3 increases in density in Region B, following logistic growth. Both Species 2 and 3 continue to decline in Region A.

Time period II→III. With rising water level, during this brief interval part of the population of Species 2 invades Region B. The remaining populations of both Species 2 and 3 in Region A continue to decline.

Time period III. The population of Species 2 increases in Region B, and Species 3 may either grow, or decline, depending on the balance between its own growth rate and the Lotka-Volterra competition from Species 2.

Time period III→IV. Water level continues to increase, such that part of the population of Species 3 invades Region C.

Time period IV. The population of Species 3 increases in Region C according to logistic growth. The dynamics in the other regions remain the same.

Time period IV→V. The water level is now falling and is shallow enough in Region C that some of Species 3 migrates back to Region B, though some fraction of the population is stranded in Region C and dies.

Time period V. The dynamics in Regions A and B continue as before.

Time period V→VI. The water level is now falling and is shallow enough in Region B that some of Species 2 migrates back to Region A, though some fraction of the population is stranded in Region B and dies.

Time period VI. The population of Species 3 is able to increase again in Region B without competition from Species 2.

Time period VI→I. The water level is now falling and is shallow enough in Region B that some of Species 3 migrates back to Region A, though some fraction of the population is stranded in Region B and dies. The cycle now repeats.

1.3 Results

A quantitative evaluation of the model can be made using a hypothetical set of parameter values. At the beginning of the year, water level is assumed to be low and all populations are squeezed together in Region A. Populations of Species 2 and 3 decline at first in Region A (Figure 1.1a), a decline that is sharpened by a migration of parts of these populations (Species 3 first, and then Species 2) to Region B as water levels rise (Figure 1.1b). In Region B, Species 3 is first able to increase, but after Species 2 invades and starts to increase, the population of Species 3 declines. Species 3 is then able to invade Region C with further increase in water levels, where it grows logistically until falling water level causes that region to dry out (Figure 1.1c). Part of the population of Species 3 is able to migrate back to Region B, where competition with Species 2 continues. Finally, falling water levels compress both Species 2 and 3 back into Region A and the cycle begins again. This yearly cycle is stable and the system will return to it if perturbed.

This simple model illustrates the role that periodic fluctuations in the environment, in this case in water level, can play in biodiversity. Species 3 cannot exist if the amplitude of the regular flooding is decreased. If this amplitude is decreased such that the period of time that Region C is flooded decreases sufficiently from the 110 day period shown in Figure 1.1, Species 3 will disappear from the system (Figure 1.2).

1.4 Discussion

The model displays a highly simplified version of the actual dynamics of fish species along an elevation gradient subject to temporal fluctuations in water level. However, this is a first building block onto which more complexities can be added. The ability of this mechanism to operate in more realistic models has been demonstrated in a multi-species simulation model in which as many as five fish species with different competitive and dispersal abilities were able to coexist along an elevation gradient (DeAngelis et al. 2005). A surprising outcome of that model was that a species that was both a poorer competitor and had less dispersal ability than at least one other species in the model was still able to coexist. That result illustrates the emergent complexities that multiple competing species in a spatially and temporally varying environment can create. The simple model here, with highly asymmetric competition, cannot produce such complex phenomena as that. However, even the simple model of this paper demonstrates the importance of amplitude of annual fluctuations in water level. A decrease in amplitude may lead to the loss of a population that requires sufficient time in an area without competition to maintain population size.

The real Everglades ecosystem contains further complexities that must be encompassed by any model that aims at realism. One such complexity is the existence of microscale elevation heterogeneity, which leads to the existence of small permanent and temporary ponds in areas that have otherwise dried out. These can serve as

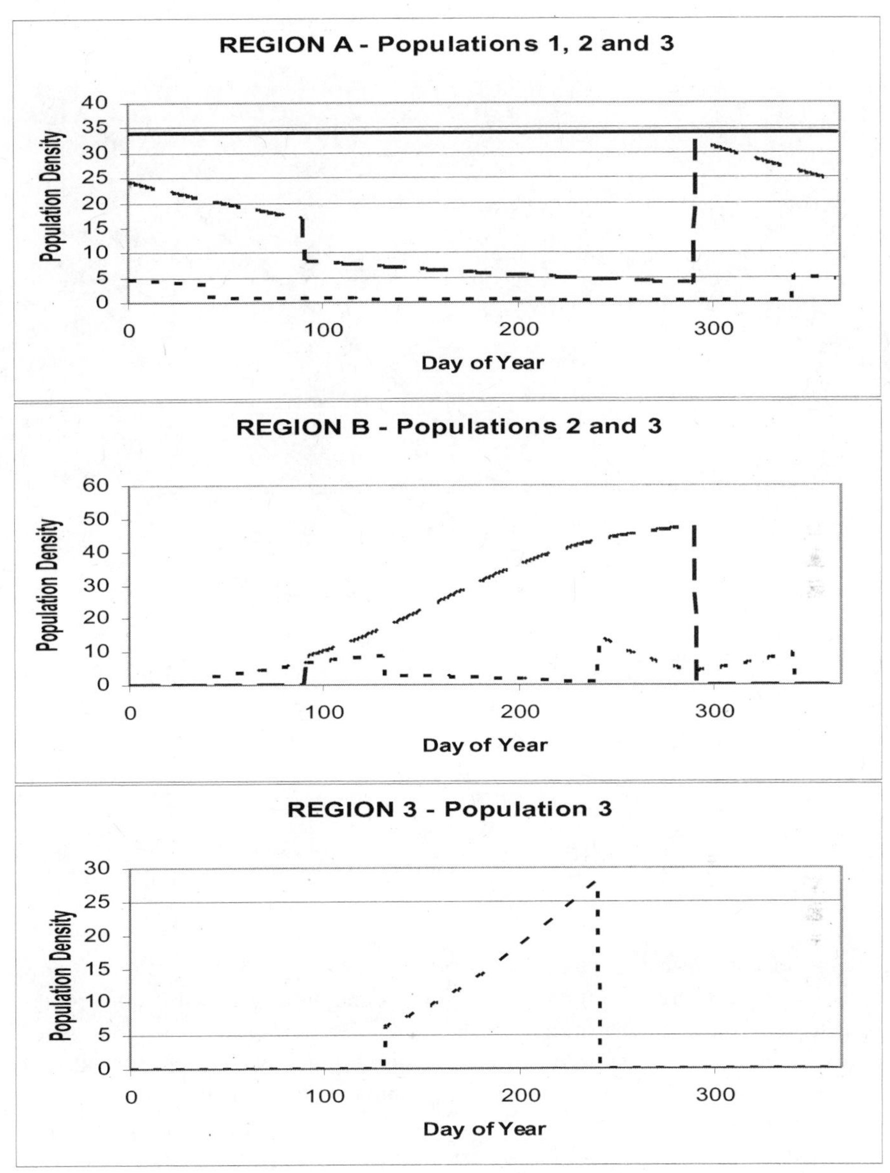

Figure 1.1 This shows the dynamics of three populations in three discrete regions of increasing elevation in a wetland, subject to regular seasonal fluctuations in water level that result in Regions B and C being flooded only part of the year. (a) Species 1 (solid line) exists on in Region A, and has a negative effect on the two other species. (b) Species 2 (dashed line) and 3 (dotted line) can migrate instantaneously to Region B when water becomes sufficiently deep. (c) Species 3 can briefly occupy the highest elevation area, Region C. The parameter values used are as follows. $T_1 = 40.$, $T_2 = 90.$, $T_3 = 130.$, $T_4 = 240.$, $T_5 = 290.$, $T_6 = 340.$, $r_2 = 0.012$, $r_3 = 0.02$, $k_2 = 50.$, $k_3 = 50.$, $c_{12} = 0.004$, $c_{13} = 0.005$, $c_{23} = 0.0002$, $f_{2wet} = 0.5$, $f_{2dry} = 0.6$, $f_{3wet} = 0.7$, $f_{3dry} = 0.5$.

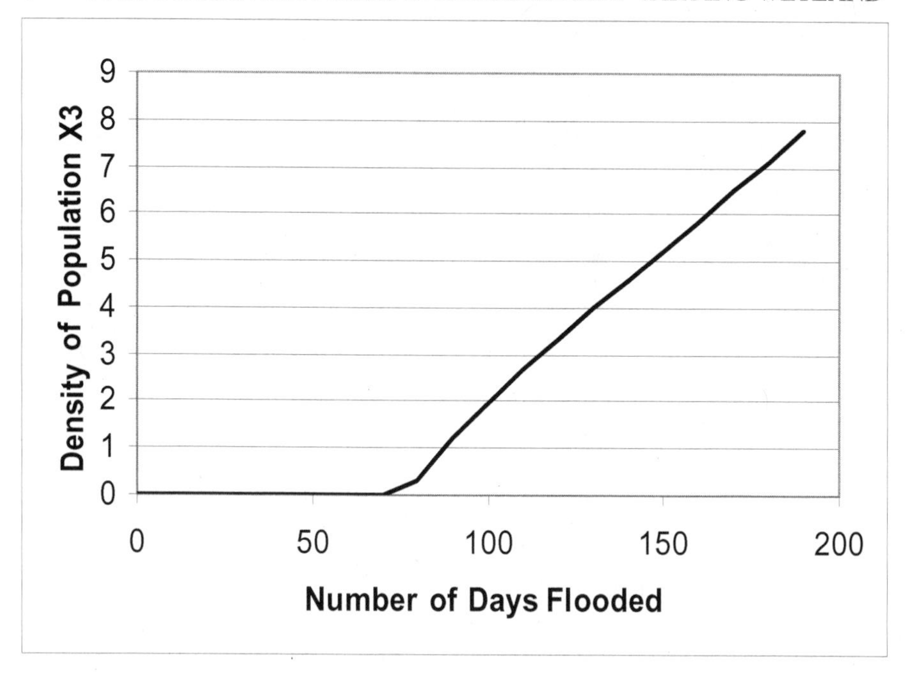

Figure 1.2 Size of the population of Species 3 in Region A at the end of the year, as a function of the length of the period that Region C is flooded.

refuges for fish, so that population recovery in a new flooded region does not have to depend on the arrival of immigrants from distant larger permanent waterbodies. Another complexity is that of the total food web. Predator-prey interactions generate oscillations, and the movement of pulses of migrating fish across the landscape creates spatially varying concentrations of periphyton, detritus, and nutrients. These dynamics are now being studied using a large, spatially explicit simulation model. This model, by using a 100×100 cell grid, also allows us to extend the analysis beyond the simple topography of the model described here, and also to more complex temporal changes in water levels, which may be highly irregular in the Everglades. These all may be expected to contribute to novel emergent qualities in the community dynamics. However, the new model still contains at its heart, though in far more elaborated form, the mechanism of species coexistence illustrated in the Appendix. As expected, it can produce results of coexistence that reflect those of the simpler model. Importantly, both the simple model and the more complex one demonstrate the importance of environmental fluctuations in maintaining species richness.

1.5 Appendix

A number of simplifying assumptions are made so that the mechanism behind coexistence of competing fish along an elevation gradient can be explained analytically. The equations for the three fish species, where the elevation gradient is divided into three regions of different elevation, are as follows.

Region A - Lowest elevation

This region is always flooded and always occupied by all three species. It is assumed that Species 1 is the dominant competitor, whose biomass density stays close to its carrying capacity, k_1. Species 2 and 3 always decline in this region due to asymmetric or one-sided competition, but are reinforced by immigration from the Region B when it dries, which prevents these populations from going to zero. The simplified equations for the three species are always

$$X_1^* = k_1$$

$$\frac{dX_2}{dt} = -c_{12}X_1^*X_2 \tag{1.1}$$

$$\frac{dX_3}{dt} = -c_{13}X_1^*X_3$$

Region B - Intermediate elevation

Both Species 2 and 3 can invade this region when it floods, though Species 3 invades first, at time T_1, and leaves at time T_6, while Species 2 invades at time T_2 and leaves at time T_5. Species 2 is competitively dominant and is always described by the equation

$$\frac{dX_2}{dt} = r_2\left(1 - \frac{X_2}{k_2}\right)X_2 \tag{1.2}$$

When Species 3 is alone, during the time intervals $T_1 < t < T_2$ and $T_5 < t < T_6$, it is described by

$$\frac{dX_3}{dt} = r_3\left(1 - \frac{X_3}{k_3}\right)X_3 \tag{1.3}$$

However, when both Species 2 and 3 are present, Species 3 is described as having the negative effect of one-sided competition from Species 2, as follows, where, for simplicity, we ignore the carrying capacity effect on Species 3:

$$\begin{aligned} \frac{dX_3}{dt} &= (r_3 - c_{23}X_2)X_3 \\ &= \left(r_3 - \frac{c_{23}f_{2wet}X_2^*(T_2)k_2e^{r_2(t-T_2)}}{f_{2wet}X_2^*(T_2)e^{r_2(t-T_2)} + (k_2 - f_{2wet}X_2^*(T_2))}\right)X_3 \end{aligned} \tag{1.4}$$

Here f_{2wet} is the fraction of population of Species 2 that migrates from Region A to Region B when it floods.

Region C - Highest elevation

Only Species 3 can invade this region, during the interval $T_3 < t < T_4$. Its growth is described by

$$\frac{dX_3}{dt} = r_3 \left(1 - \frac{X_3}{k_3}\right) X_3 \tag{1.5}$$

When these equations are integrated over each of the time intervals, with appropriate initial conditions at the start of each interval, the following mathematical expressions are obtained in each time period and region (see Table 1.1):

Time Period I ($0 < t < T_1$)

Region A:

$$\begin{aligned} X_{2A}(t) &= X_{2A}(0)e^{-c_{12}X_1^* t} \\ X_{3A}(t) &= X_{3A}(0)e^{-c_{13}X_1^* t} \end{aligned} \tag{1.6}$$

Time Period II ($T_1 < t < T_2$)

Region A: Here f_{3wet} is the fraction of population of Species 3 that migrates to Region B from Region A when it floods.

$$\begin{aligned} X_{2A}(t) &= X_{2A}(T_1)e^{-c_{12}X_1^*(t-T_1)} \\ X_{3A}(t) &= (1 - f_{3wet})X_{3A}(T_1)e^{-c_{13}X_1^*(t-T_1)} \end{aligned} \tag{1.7}$$

Region B:

$$X_{3B}(t) = \frac{f_{3wet}X_{3A}(T_1)k_3 e^{r_3(t-T_1)}}{f_{3wet}X_{3A}(T_1)e^{r_3(t-T_1)} + (k_3 - f_{3wet}X_{3A}(T_1))} \tag{1.8}$$

Time Period III ($T_2 < t < T_3$)

Region A:

$$\begin{aligned} X_{2A}(t) &= (1 - f_{2wet})X_{2A}(T_2)e^{-c_{12}X_1^*(t-T_2)} \\ X_{3A}(t) &= X_{3A}(T_2)e^{-c_{13}X_1^*(t-T_2)} \end{aligned} \tag{1.9}$$

Region B:

$$X_{2B}(t) = \frac{f_{2wet}X_{2A}(T_2)k_2e^{r_2(t-T_2)}}{f_{2wet}X_{2A}(T_2)e^{r_2(t-T_2)} + (k_2 - f_{2wet}X_{2A}(T_2))}$$

$$X_{3B}(t) = X_{3b}(T_2)e^{Q_{3B}} \tag{1.10}$$

$$Q_{3B} = r_3(t - T_2) + \frac{c_{23}k_2}{r_2}\ln(R(t)/k_2)$$

$$R(t) = f_{2wet}X_{2A}(T_2)e^{r_2(t-T_2)} + (k_2 - f_{2wet}X_{2A}(T_2))$$

Time Period IV ($T_3 < t < T_4$)

Region A:

$$X_{2A}(t) = X_{2A}(T_3)e^{-c_{12}X_1^*(t-T_3)} \tag{1.11}$$

$$X_{3A}(t) = X_{3A}(T_3)e^{-c_{13}X_1^*(t-T_3)}$$

Region B: Here f_{3wet} is the fraction of population of Species 3 that migrates from Region B to Region C when it floods

$$X_{2B}(t) = \frac{X_{2B}(T_3)k_2e^{r_2(t-T_3)}}{X_{2B}(T_3)e^{r_2(t-T_3)} + (k_2 - X_{2B}(T_3))}$$

$$X_{3B}(t) = (1 - f_{3wet})X_{3B}(T_3)e^{Q_{3B}} \tag{1.12}$$

$$Q_{3B} = r_3(t - T_3) + \frac{c_{23}k_2}{r_2}\ln(R(t)/k_2)$$

$$R(t) = X_{2A}(T_3)e^{r_2(t-T_3)} + (k_2 - X_{2A}(T_3))$$

Region C:

$$X_{3C}(t) = \frac{f_{3wet}X_{3B}(T_3)k_3e^{r_3(t-T_3)}}{f_{3wet}X_{3B}(T_3)e^{r_3(t-T_3)} + (k_3 - f_{3wet}X_{3B}(T_3))} \tag{1.13}$$

Time Period V ($T_4 < t < T_5$)

Region A:

$$X_{2A}(t) = X_{2A}(T_4)e^{-c_{12}X_1^*(t-T_4)} \tag{1.14}$$

$$X_{3A}(t) = X_{3A}(T_4)e^{-c_{13}X_1^*(t-T_4)}$$

Region B: Here f_{3dry} is the fraction of population of Species 3 that migrates from

Region C to Region B when the former is too shallow.

$$X_{2B}(t) = \frac{X_{2A}(T_4)k_2e^{r_2(t-T_4)}}{X_{2B}(T_4)e^{r_2(t-T_4)} + (k_2 - X_{2B}(T_4))}$$

$$X_{3B}(t) = (X_{3B}(T_4) + f_{3dry}X_{3C}(T_4))e^{Q_{3B}} \tag{1.15}$$

$$Q_{3B} = r_3(t - T_4) + \frac{c_{23}k_2}{r_2}\ln(R(t)/k_2)$$

$$R(t) = X_{2B}(T_4)e^{r_2(t-T_4)} + (k_2 - X_{2B}(T_4))$$

Time Period VI ($T_5 < t < T_6$)

Region A: Here f_{2dry} is the fraction of population of Species 2 that migrates from Region B to Region A when the former is too shallow.

$$X_{2A}(t) = (X_{2A}(T_5) + f_{2dry}X_{2B}(T_5))e^{-c_{12}X_1^*(t-T_5)}$$

$$X_{3A}(t) = X_{3A}(T_5)e^{-c_{13}X_1^*(t-T_5)} \tag{1.16}$$

Region B:

$$X_{3B}(t) = \frac{X_{3B}(T_5)k_3e^{r_3(t-T_3)}}{X_{3B}(T_5)e^{r_3(t-T_3)} + (k_3 - X_{3B}(T_5))} \tag{1.17}$$

Time Period VII ($T_6 < t < 365$)

Region A: Here f_{3wet} is the fraction of population of Species 3 that migrates from Region B to Region A when the former is too shallow.

$$X_{2A}(t) = (X_{2A}(T_5)e^{-c_{12}X_1^*(t-T_6)}$$

$$X_{3A}(t) = (X_{3A}(T_5) + f_{3dry}X_{3B}(T_6))e^{-c_{13}X_1^*(t-T_5)} \tag{1.18}$$

Then set

$$X_2(0) = X_2(365)$$

$$X_3(0) = X_3(365) \tag{1.19}$$

and begin a new annual cycle.

1.6 References

P.A. Abrams and W.G. Wilson (2004), Coexistence of competitors in metacommunities due to spatial variation in resource growth rates; does R^* predict the outcome of competition? *Ecology Letters* **7**:929-940.

R.M. Armstrong (1976), Fugitive species: Experiments with fungi and some theoretical considerations, *Ecology* **57**:953-963.

P. Chesson (2000), Mechanisms of maintenance of species diversity, *Ann. Rev. Ecol. Syst.* **31**:343-366.

D.L. DeAngelis, J.C. Trexler, and W.F. Loftus (2005), Life history trade-offs and community dynamics of small fishes in a seasonally pulsed wetland, *Canadian Journal of Fisheries and Aquatic Sciences* **62**:781-790.

C.W. Heckman (1998), *The Pantanal of Poconé,* Kluwer Academic Publishers, Dordrecht.

E.E. Holmes and H.B. Wilson (1998), Running from trouble: Long-distance dispersal and the competitive coexistence of inferior species, *Amer. Natur.* **151**:578-586.

J.A. Kushlan (1990), Freshwater marshes, in *"Ecosystems of Florida,"* ed. by R. L. Meyers and J. J. Ewel, University of Central Florida Press, Orlando, FL, pp. 324-363.

E. Litchman and C.a. Klausmeier (2001), Competition of phytoplankton under fluctuating light, *Amer. Natur.* **157**:170-187.

W.F. Loftus and J.A. Kushlan (1987), Freshwater fishes of southern Florida, *Bull. Florida State Museum, Biol. Sci.* **31**:147-344.

T. Namba and C. Hashimoto (2004), Dispersal-mediated coexistence of competing predators, *Theor. Pop. Biol.* **66(1)**:53-70.

J.C. Ogden (1994), A comparison of wading bird nesting colony dynamics (1931-1946 and 1974-1989) as an indication of ecosystem conditions in the southern Everglades, in *"Everglades: The System and its Restoration,"* ed. by S.M. Davis and J.C. Ogden, St. Lucie Press, Delray Beach, FL, pp. 533-570.

D. Tilman (1994), Competition and biodiversity in spatially structured habitats, *Ecology* **75**:2-16.

J.C. Trexler, W.F. Loftus, F. Jordan, J.H. Chick, K.L. Kandl, T.C. McElroy and O.L. Bass Jr. (2002), Ecological scale and its implications for freshwater fishes in the Florida Everglades, in *"The Everglades, Florida Bay, and Coral Reefs of the Florida Keys: An Ecosystem Sourcebook,"* ed. by J. W. Porter and K. G. Porter, CRC Press, Boca Raton, FL, pp. 153-181.

CHAPTER 2

Spatial dynamics of multitrophic communities

Priyanga Amarasekare
University of California at Los Angeles

Abstract. I discuss the influence of dispersal on two multitrophic communities: intraguild predation and keystone predation. The key finding is an asymmetry between species in their dispersal effects and responses. In both intraguild predation and keystone predation, dispersal of the predator-resistant inferior competitor has a large effect, but dispersal of the predator-susceptible superior competitor has little or no effect, on coexistence and species' distributions. In the case of keystone predation, the inferior competitor's dispersal also mediates the predator's dispersal effects: predator dispersal has no effect when the inferior competitor is immobile, and a large effect when it is mobile. The direct and indirect effects of the inferior competitor's dispersal changes species' distributions from inter-specific segregation in resource-poor and resource-rich habitats to inter-specific aggregation in resource-rich habitats. The important point is that the interaction between competition and predation creates asymmetries between species that lead to unexpected effects of dispersal. These asymmetries suggest the existence of keystone dispersers, species that, through their dispersal, have disproportionately large effects on species distributions and diversity in multitrophic communities.

2.1 Introduction

The interplay between species interactions and dispersal is the key determinant of diversity in spatially structured environments (Leibold et al. 2004, 2005). A great deal is known about this interplay in communities with one or two trophic levels (e.g., resource, consumer; Levin 1974; Holt 1985; Murdoch et al. 1992; Amarasekare and Nisbet 2001; Jansen 2001; Abrams and Wilson 2004) but relatively little is known about it in communities with multiple trophic levels (e.g., resource, consumer, natural enemy).

Most theory on spatial coexistence focuses on nontrophic or pairwise trophic interactions where species cannot coexist in the absence of dispersal (e.g., competitive dominance, predator overexploitation, Allee effects induced by the absence of a mutualistic partner). In such situations, dispersal can allow coexistence given spatial variation in species' traits (Levin 1974; Holt 1985, 1993; Amarasekare and Nisbet 2001; Codeco and Grover 2001; Amarasekare 2004; Leibold et al. 2004). Two aspects of multitrophic communities suggest the need for a different framework for understanding the interplay between dispersal and species interactions. First, multitrophic communities are characterized by two types of interactions (trophic and nontrophic interactions) that are dynamically quite different. Second, in multitrophic communities species occupying a particular trophic level can coexist in the absence of dispersal, but the operation of such coexistence mechanisms is variable in space and time. Thus, local and spatial coexistence mechanisms can operate simultaneously, and their interaction can lead to emergent properties (Amarasekare 2006, 2007). Dispersal effects on multitrophic communities are therefore likely to be quite different from dispersal effects on communities with only one type of species interaction.

Two examples of multitrophic community modules illustrate these differences. Intraguild predation (IGP) occurs when species competing for a common resource also prey on or parasitize one another (e.g., Polis et al. 1989; Arim and Marquet 2004); keystone predation (KP) occurs when species competing for a common resource also share a natural enemy (e.g., Sih et al. 1985; Navarette and Menge 1996). In both cases the two consumer species can coexist via a trade-off that allows for local niche partitioning. In intraguild predation local niche partitioning is possible because the inferior resource competitor can prey on or parasitize its competitor; in keystone predation it occurs because the inferior competitor gains more of the resource by being less susceptible to the predator. A key feature of these trade-offs is that their expression depends on traits of species occupying other trophic levels within the community (Amarasekare 2007, 2008). In intraguild predation it is the common resource; in keystone predation it is the common resource and/or natural enemy. In the absence of dispersal or other ameliorating factors, spatial variation in resource productivity or predator mortality can shift the advantage to one consumer species and cause the other's exclusion. For instance, when resource productivity is low (predator mortality is high), exploitative competition dominates and the inferior resource competitor is excluded; when resource productivity is high (predator mortality is low), predation dominates and the species more susceptible to predation is excluded (Holt and Polis 1997; Diehl and Feissel 2000; Noonberg and Abrams 2005). Thus, the trade-off between competition and predation allows coexistence only at intermediate productivity/mortality levels. This illustrates another feature that distinguishes multitrophic interactions. In nontrophic or pairwise trophic interactions, spatial variation in species' traits typically facilitates coexistence (Leibold et al. 2004). In multitrophic interactions, spatial variation in resource or predator traits can constrain the coexistence of intermediate consumers. Thus, diversity maintenance in multitrophic communities depends crucially on whether dispersal by intermediate consumers can counteract the diversity reducing effects of spatial variation that act through a shared resource or natural enemy (Amarasekare 2007).

Here I present some theoretical insights on the spatial dynamics of multitrophic communities characterized by competition and predation. These insights are based on a comparative analysis of dispersal effects on intraguild predation and keystone predation (Fig. 2.1). I focus on these two multitrophic interactions because they are widespread in nature, occurring in a wide variety of taxa from microbes to mammals (Polis et al. 1989; Chase and Leibold 2003; Arim and Marquet 2004).

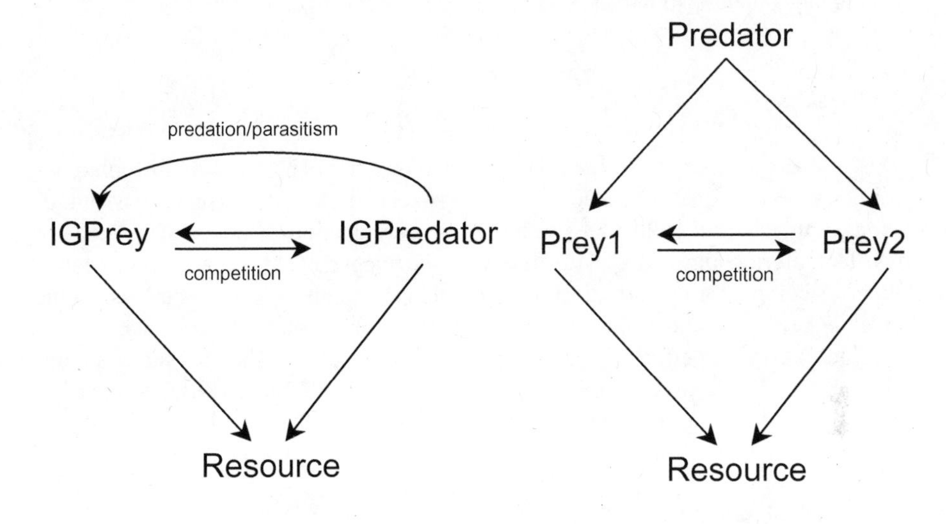

Figure 2.1 Multitrophic (resource-consumer-predator) interactions with competition and predation. On the left is intraguild predation where two consumers compete for a common resource but one consumer can prey on or parasitize the other. On the right is keystone predation where two consumers compete for a common resource but also share a common natural enemy.

2.2 Theoretical framework

When studying the interplay between dispersal and the local multitrophic dynamics, it is instructive to consider a metacommunity that experiences spatial variation in resource or predator traits (e.g., resource productivity and predator mortality; Noonberg and Abrams 2005; Amarasekare 2006) but no spatial variation in the traits of the consumers themselves. The consumer species experience spatial variation in resource/predator traits via dispersal. The minimal model that can incorporate local dynamics, dispersal, and spatial heterogeneity is a three-patch metacommunity with each patch exhibiting a level of resource productivity/predator mortality that leads to the three outcomes observed in the absence of dispersal (Leibold 1996; Holt and Polis 1997; Amarasekare 2006, 2007, 2008): (i) resource productivity (predator mortality) is too low (too high) for the predator-resistant inferior competitor to invade when rare, (ii) resource productivity (predator mortality) is too high (too low) for the predator-susceptible superior competitor to invade when rare, and (iii) resource productivity (predator mortality) is within the range that allows both species to invade and coexist via a trade-off between competition and predation. Here I focus on the key insights that emerge from analysis of such three-patch models. I defer the mathematical details to the Appendix.

2.3 Results

The crucial outcome of the interplay between dispersal and local multitrophic dynamics is an asymmetry between consumer species in their dispersal effects and responses. In both intraguild predation and keystone predation, dispersal of the predator-resistant inferior competitor has a large effect, but dispersal of the predator-susceptible superior competitor has no effect, on coexistence and species' distributions. In the case of keystone predation, the inferior competitor's dispersal also mediates the predator's dispersal effects: predator dispersal has no effect when the inferior competitor is immobile, and a large effect when it is mobile (Table 2.1). Below I explain the biological mechanisms underlying this dispersal asymmetry.

2.3.1 Intraguild predation

The dispersal asymmetry between the superior competitor (the IGPrey) and the inferior competitor (IGPredator) can be understood by considering the way each species' abundance changes with changes in resource productivity. The IGPrey's abundance-productivity relationship is strongly affected by the IGPredator's dispersal rate, while the IGPredator's abundance-productivity relationship is qualitatively unaffected by the dispersal rates of either the IGPrey or the IGPredator (Fig. 2.2; Amarasekare 2006). For instance, the IGPredator's abundance increases monotonically with increasing productivity regardless of its dispersal rate (Fig. 2.3). In contrast, the IGPrey's abundance-productivity relationship is highly sensitive to the IGPredator's dispersal rate (Fig. 2.3). When the IGPredator's dispersal rate is low relative to the within-patch mortality rate, the IGPrey's abundance declines monotonically with increasing productivity, which is the same pattern observed in the absence of dispersal (Fig. 2.3). When the IGPredator's dispersal rate is moderate relative to the within-patch mortality rate, the IGPrey's abundance-productivity relationship becomes hump-shaped with the highest abundance at intermediate productivity. When the IGPredator's dispersal rate is high relative to the within-patch mortality rate, the IGPrey's abundance increases monotonically with increasing productivity (Fig. 2.3; Amarasekare 2006). Thus, low IGPredator dispersal leads to inter-specific segregation with the IGPrey being concentrated in areas of low resource productivity and the IGPredator in areas of high resource productivity. In contrast, high IGPredator dispersal leads to inter-specific aggregation with both species being concentrated in areas of high productivity (Fig. 2.3). These differences arise solely because of the IGPrey's differential response to the magnitude of IGPredator's dispersal. The IG-

Predator's abundance-productivity relationship is qualitatively insensitive to changes in its own dispersal rate as well as that of the IGPrey's (Figs. 2.2 and 2.3).

The mechanisms underlying these abundance-productivity relationships can be understood by examining the change in the IGPrey's abundance as a function of the IG-Predator's dispersal rate (Fig. 2.2). The important point is the existence of a threshold value of the IGPredator's dispersal rate, below which the IGPrey's abundance is the lowest in the high productivity patch, and above which the IGPrey's abundance is the lowest in the low productivity patch (Fig. 2.2c). Below this threshold the high productivity patch is a sink for the IGPrey (i.e., it cannot maintain a positive per capita growth rate when rare in the absence of dispersal). It is rescued from extinction by immigration from the intermediate productivity patch where local coexistence occurs, in the absence of dispersal, via a trade-off between competition and IGP. Thus at low dispersal rates of the IGPredator we have within-patch coexistence via a combination of source-sink dynamics in the IGPrey, and a competition-IGP trade-off. Within-patch coexistence is still possible once the IGPredator's dispersal rate exceeds the threshold, but it occurs via a combination of mechanisms that do not require the IGPrey's dispersal (Amarasekare 2006). Coexistence in the intermediate productivity patch occurs via the competition-IGP trade-off while coexistence in the high productivity patch occurs via high emigration of the IGPredator. The latter mechanism comes about because random dispersal leads to a net movement of individuals from areas of higher to lower abundance. The IGPredator's emigration out of the high productivity patch, where it has the highest abundance, far exceeds immigration into it. This reduces the strength of IGP in the high productivity patch and allows the IGPrey to invade when rare, even in the absence of its own dispersal. The key point is that spatial coexistence of the IGPrey and IGPredator, and the resulting abundance-productivity relationships, arise from two different combinations of mechanisms. Which combination operates depends on the magnitude of the IGPredator's dispersal rate (Table 2.1; Amarasekare 2006). With low dispersal, when the competition-predation trade-off operates in the intermediate patch, local coexistence in the low productivity patch occurs via source-sink dynamics in the IGPredator, and coexistence in the high productivity patch occurs via source-sink dynamics in the IGPrey; with moderate to high dispersal, local coexistence in the low productivity patch still occurs via source-sink dynamics of the IGPredator, while local coexistence in the high producitivity patch occurs via emigration of the IGPredator, which operates in the absence of any spatial dynamics in the IGPrey. Thus, the IGPredator's dispersal determines which type of spatial coexistence mechanism operates in the metacommunity.

2.3.2 *Keystone predation*

The dispersal asymmetry is such that the predator-resistant inferior competitor's dispersal has a much stronger effect on coexistence than dispersal of the predator or the predator-susceptible superior competitor (Amarasekare 2008). The inferior competitor's dispersal enables consumer coexistence in high productivity habitats by allow-

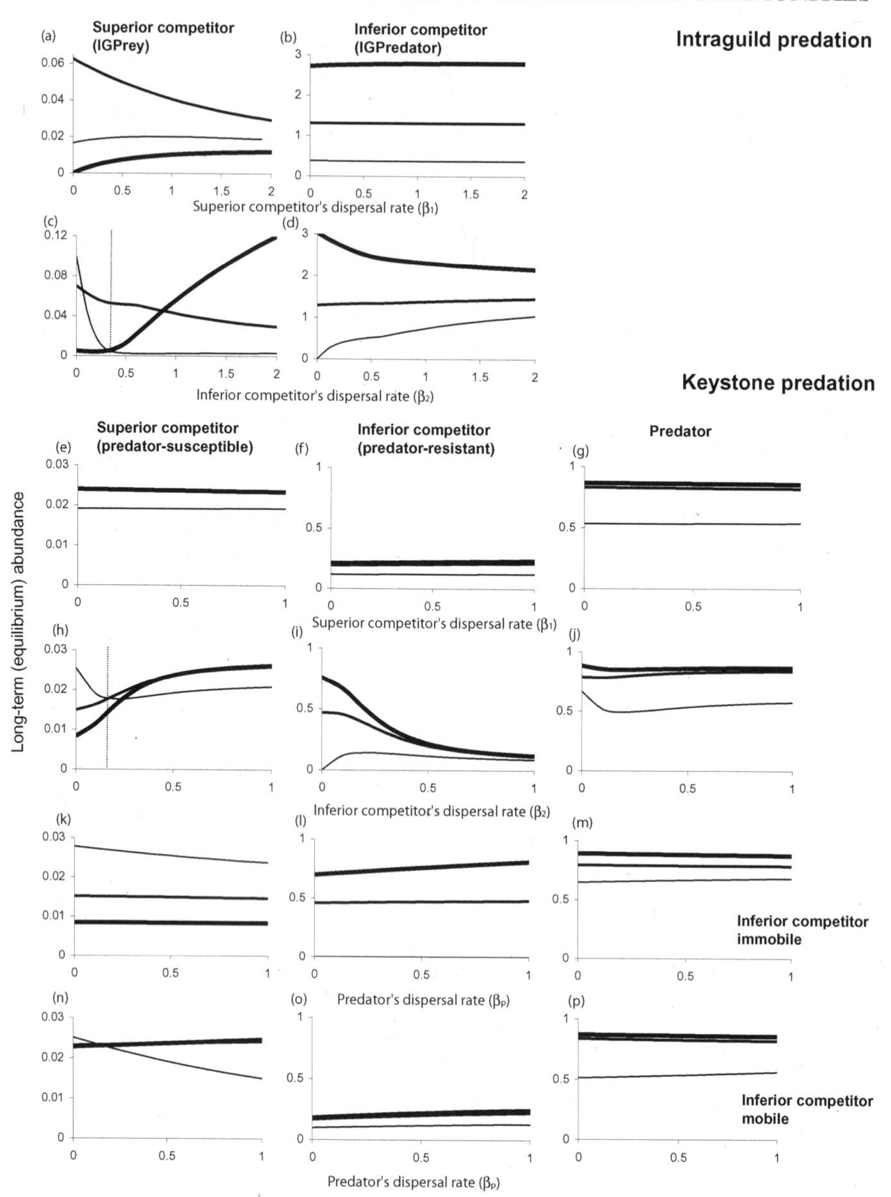

Figure 2.2 Long-term (equilibrium) abundances under variable dispersal rates for intraguild predation (panels (a)-(d)) and keystone predation (panels ((e)-(p)). In each graph, the three lines with increasing thickness depict, respectively, abundances in the low, intermediate, and high productivity patches. The vertical dashed line in panels (c) and (h) depicts the threshold dispersal rate below which source-sink dynamics drive coexistence in the low and high productivity patches, and above which source-sink dynamics of the inferior competitor drives coexistence in the low productivity patch while emigration of the inferior competitor drives coexistence in the high productivity patch. Note that in keystone predation, the predator's dispersal has no effect on the superior competitor when the inferior competitor is immobile, and a large effect when the inferior competitor is mobile. Parameter values used for intraguild predation: $a_1 = 10, a_2 = 2, \alpha = 2, f = 2, \beta_1 = 0.2$ (panels c and d) and $\beta_2 = 0.2$ (panels (a) and (b)); parameter values for keystone predation are: $a_1 = 9, \alpha_1 = 9, a_2 = 2, \alpha_2 = 1, e_1 = e_2 = 1, \delta = 1, f = 2$ and $d_P = 0.5$.

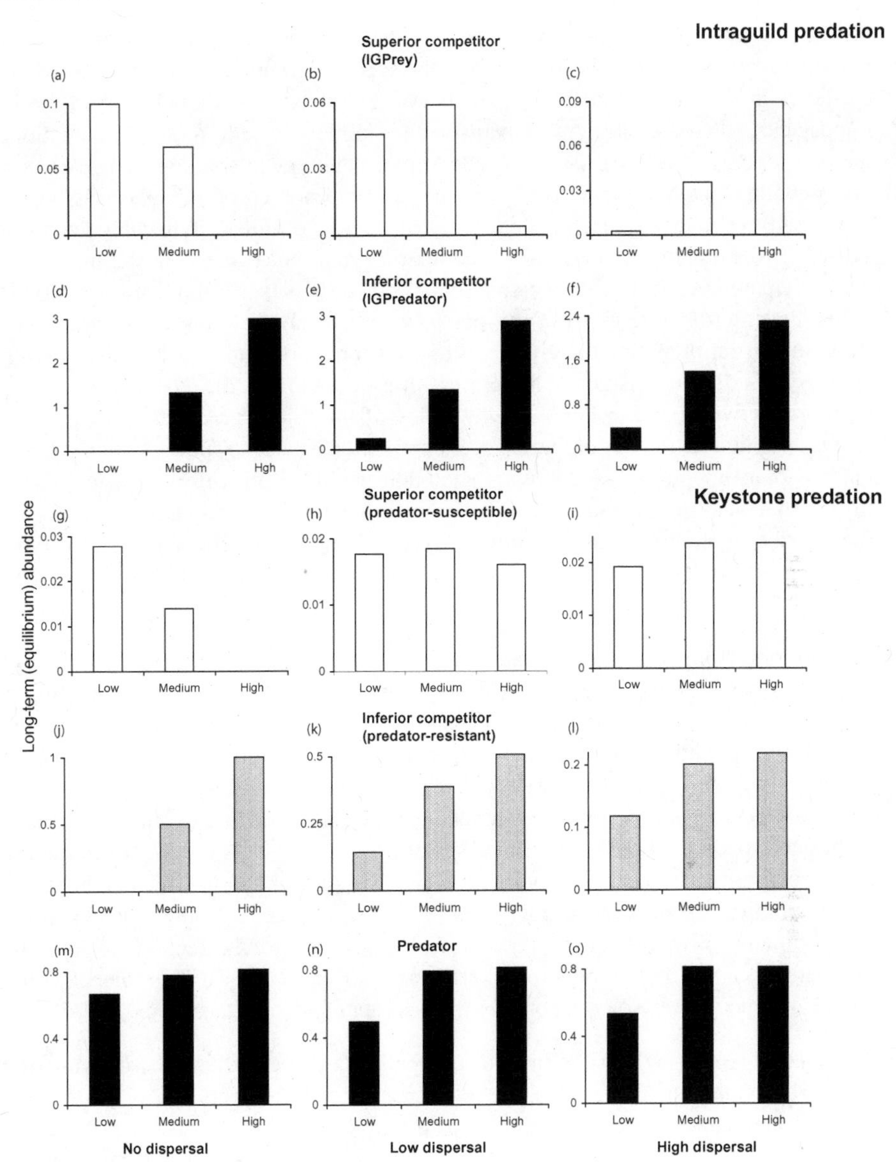

Figure 2.3 The abundance-productivity relationships of the superior and inferior competitors and predator (in the case of keystone predation) as a function of the inferior competitor's dispersal when all species (except the resource) are mobile. In each panel, the x-axis is the resource productivity (r_j) and the y-axis is long-term abundance. In each row, panels from left to right depict the abundance-productivity relationships for a particular species under increasing levels of the inferior competitor's dispersal. The key point is that the superior competitor's abundance-productivity relationship is highly sensitive to the inferior competitor's dispersal rate but the inferior competitor's and the predator's (in the case of keystone predation) abundance-productivity relationships are insensitive to such dispersal. Parameter values used for intraguild predation: $a_1 = 10, a_2 = 2, \alpha = 2, f = 2, \beta_1 = 0.2$; parameter values for keystone predation: $a_1 = 9, a_2 = 2, \alpha_1 = 9, \alpha_2 = 1, e_1 = e_2 = 1, \delta = 1, f = 2, d_P = 0.5, \beta_1 = 0.5, \beta_P = 0.5$.

ing the superior competitor to invade such habitats even when the superior competitor does not disperse. In contrast, dispersal of the superior competitor does not allow coexistence in the low productivity habitats (where predator-mediated coexistence is impossible) when the inferior competitor is immobile. As with intraguild predation, the inferior competitor's dispersal rate determines the type of spatial mechanism that allows within-patch coexistence. When the inferior competitor's dispersal is below a critical threshold (Fig. 2.2h), coexistence in the low productivity patch occurs via source-sink dynamics of the inferior competitor, and coexistence in the high productivity patch occurs via source-sink dynamics of the superior competitor. In both cases, the intermediate productivity patch, where local coexistence can occur via the competition-predation trade-off, acts as the source population. When the inferior competitor's dispersal is above the critical threshold (Fig. 2.2h), coexistence in the high productivity patch occurs via emigration of the inferior competitor, which requires no spatial dynamics of the superior competitor. Since the superior competitor's emigration from the low productivity patch does not allow the inferior competitor to invade that patch, within-patch coexistence still requires source-sink dynamics of the inferior competitor. Thus, the inferior competitor's dispersal determines the type of spatial coexistence mechanism that operates in the metacommunity (Table 2.1; Amarasekare 2008).

The inferior competitor's dispersal also drives species' distributions. Specifically, the inferior competitor induces a qualitative change in the superior competitor's abundance-productivity relationship, causing it to increase rather than decrease with increasing productivity. This change comes about directly through the inferior competitor's own dispersal and indirectly through the predator's dispersal. As a consequence, species' distributions across the landscape also undergo a qualitative change, from inter-specific segregation to inter-specific aggregation (Fig. 2.3). When the inferior competitor does not disperse the predator's dispersal has no effect on the superior competitor's abundance-productivity relationship (Figs. 2.2k-m); when the inferior competitor disperses the predator's dispersal has a strong effect on the superior competitor's abundance-productivity relationship (Figs. 2.2m-o). The superior competitor's dispersal has no effect on any species' abundance pattern (Figs. 2.2h-j).

The mechanism by which the inferior competitor's dispersal changes the superior competitor's abundance-productivity relationship is as follows. The inferior competitor's abundance increases with increasing productivity in the absence of dispersal, and hence its net movement occurs in the direction of decreasing productivity. The inferior competitor's net emigration from the high productivity patch allows the superior competitor to invade that patch. As the inferior competitor's dispersal rate increases the superior competitor is able to establish in the high productivity patch with increasingly higher abundances. When the predator does not disperse, the inferior competitor's net immigration into the low productivity patch decreases the superior competitor's abundance via a mass effect. This is, however, a relatively weak effect. It becomes much stronger when the predator also disperses. Then, the predator's net movement into the low productivity patch imposes high mortality on the superior competitor and decreases its abundance. Thus, the inferior competitor's dispersal has the direct effect, through emigration, of increasing the superior competitor's abundance in the high productivity patch, and the indirect effect, through the predator's dispersal, of reducing the superior competitor's abundance in the low productivity patch.

Table 2.1. Effects of the predator-resistant inferior competitor's dispersal on coexistence and species distributions (reproduced from Amarasekare 2008).

Inferior competitor's dispersal[1]	Coexistence mechanisms[2]	Abundance-productivity relationship		Species distributions
		Superior competitor	Inferior competitor & predator	
Low	Source-sink in both species	↑ with ↑ productivity	↑ with ↑ productivity	Inter-specific segregation
Medium	Emigration-mediated, Source-sink in inf. competitor	Hump-shaped, highest abundance at intermediate productivity	↑ with ↑ productivity	Partial inter-specific segregation
High	Emigration-mediated, Source-sink in inf. competitor	↑ with ↑ productivity	↑ with ↑ productivity	Inter-specific aggregation

[1] relative to the within-patch mortality rate (Amarasekare 2006, 2007, 2008)

[2] in the consumer species when trade-off mediated local niche partitioning also operates

The mechanism by which the inferior competitor's dispersal mediates the effects of the predator's dispersal is as follows. When the inferior competitor does not disperse, the predator's abundance is only weakly related to productivity. This is because the predator's abundance depends on consumers' traits rather than the basal resource's traits. When the inferior competitor disperses, its net movement into the low productivity patch causes an overabundance of the prey less palatable to the predator. This causes a decrease in the predator's abundance in the low productivity patch, thus creating an abundance gradient in the direction of increasing productivity. The resulting influx of predators into the low productivity patch causes a further decline in the superior competitor's abundance in that patch.

Why does the superior competitor's dispersal have no effect on the predator's or the inferior competitor's abundance? The superior competitor's abundance decreases with increasing productivity in the absence of dispersal, and hence its net movement is in the direction of increasing productivity. However, net emigration of the superior competitor from the low productivity patch does not allow the inferior competitor to invade that patch. This is because the inferior competitor's poor resource exploitation ability prevents it from maintaining self-sustaining populations in resource-poor areas; it can only maintain a small sink population, via dispersal, in such habitats. The superior competitor's net immigration into the high productivity patch has no effect on the inferior competitor, because the latter's abundance is much higher than the former's in the presence of the predator. Because of these two constraints, the superior competitor's dispersal also cannot alter the predator's spatial distribution.

2.4 Discussion and conclusions

Multitrophic communities are the basic units of biodiversity. Yet, our knowledge of their spatial dynamics is sketchy at best. Here I have provided a summary of recent work that illustrates the unexpected emergent properties arising from the interplay between dispersal and local multitrophic dynamics.

The most important result is an asymmetry between species in their dispersal effects and responses. In both intraguild predation and keystone predation, dispersal of the predator-resistant inferior competitor has a large effect, but dispersal of the predator-susceptible superior competitor has no effect, on coexistence and species' distributions. In the case of keystone predation, the inferior competitor's dispersal also mediates the predator's dispersal effects: the predator's dispersal has no effect when the inferior competitor is immobile, and a large effect when it is mobile. Together, the direct and indirect effects of the inferior competitor's dispersal change species' distributions from inter-specific segregation in resource-poor and resource-rich habitats to inter-specific aggregation in resource-rich habitats (Amarasekare 2007, 2008).

In communities characterized by intraguild predation, the inferior competitor is also the intraguild predator. Hence it is difficult to determine whether the dispersal asymmetry results from the intraguild predator's role as a competitor *or* predator. A comparative analysis of intraguild predation and keystone predation, as illustrated here, allows one to separate the competitive interactions that occur within a trophic level from the predator-prey interactions that operate between trophic levels. It thus establishes that it is not the predator but the predator-resistant intermediate consumer whose dispersal drives coexistence and species distributions.

The observed dispersal asymmetry arises from the different ways in which the two consumer species solve the conflicting problems of resource acquisition vs. predator avoidance. One species is better at resource acquisition and worse at predator resistance/tolerance, while the other species has the opposite trade-off. This difference leads to species-specific responses to spatial variation in resource productivity (or predator mortality) such that the superior competitor's abundance decreases with increasing productivity, while the inferior competitor's abundance increases with increasing productivity. It is these abundance gradients occurring in opposite directions that drive the dispersal asymmetry. Random dispersal in the face of opposing abundance gradients leads to a net movement of the superior competitor from areas of low to high productivity, and a net movement of the inferior competitor from areas of high to low productivity. The inferior competitor is more limited by resources than predation, and hence net movement of the superior competitor from low to high productivity habitats does not allow the inferior competitor to invade the low productivity habitats in the absence of its own dispersal; it can only maintain a small sink population, via dispersal, in such habitats. Thus, the superior competitor's dispersal induces no qualitative change in the inferior competitor's abundance gradient. The superior competitor is limited less by resources than by predation, and hence net movement of the inferior competitor from high to low productivity habitats allows the superior competitor to invade the high productivity habitat and attain high abundances even in the absence of its own dispersal. The inferior competitor's net movement from high to low productivity areas also reduces predator abundance in the low productivity habitat, thus creating an abundance gradient in the direction of increasing productivity for the predator. The resulting net movement of the predator from high to low productivity habitats inflicts high mortality on the superior competitor in the low productivity habitat. This decreases the superior competitor's abundance in the low productivity habitat below that in the absence of dispersal. The net result is a qualitative change in the superior competitor's abundance gradient, with abundances now increasing with increasing productivity. This in turn induces a qualitative change in the species' distributions across the landscape, from inter-specific segregation to inter-specific aggregation.

The important implication of these results is that the interaction between competition and predation creates asymmetries between species that lead to keystone effects in dispersal. Just as preferential consumption by a top predator plays a keystone role in maintaining local diversity of intermediate consumers, dispersal by the species less susceptible to predation plays a keystone role in the diversity and distribution of intermediate consumers.

The results discussed here reflect multitrophic communities that are interconnected by random dispersal. The keystone disperser effect also emerges when species in such communities interact via nonrandom dispersal (Amarasekare 2007; P. Amarasekare, unpublished manuscript). However, establishing the generality of the keystone disperser phenomenon requires investigations of larger metacommunities containing more complex multitrophic communities. This next step is crucial in developing a theoretical framework for spatial community ecology that can stimulate empirical investigations.

2.5 Acknowledgments

This research was supported by NSF grant DEB-0717350. I thank an anonymous reviewer for many constructive comments on the chapter.

2.6 Appendix: Spatial models

Consider a spatially structured environment consisting of a number of patches of suitable habitat embedded in an inhospitable matrix. There is permanent spatial heterogeneity in habitat quality as would occur if there were differences in soil, nutrient availability, or moisture content that would make some host plant patches or ponds more productive than others. These spatial differences are assumed to occur within a spatial scale that can be traversed by the organisms occupying these habitats.

Within each habitat patch we have a multitrophic interaction characterized by intraguild predation or keystone predation. In both cases coexistence can occur within a habitat patch if there is an interspecific trade-off between competitive ability and susceptibility to predation. The expression of this trade-off, however, depends on on variability in resource and/or predator traits. At very low or very high resource productivity/predator mortality one species gains an overall advantage and excludes the other (Leibold 1996; Holt and Polis 1997; Noonberg and Abrams 2005). Coexistence in variable environments thus requires additional mechanisms besides the competition-predation trade-off.

I consider the simplest mathematical representation of a metacommunity with intraguild predation or keystone predation: a three patch model with each patch exhibiting a level of resource productivity that leads to a qualitatively different outcome: (i) resource productivity is too low for the predator-resistant inferior competitor to invade when rare, (ii) resource productivity is too high for the predator-susceptible superior competitor to invade when rare, and (iii) resource productivity is within the range that allows both species to invade and coexist via a competition-predation trade-off (Holt and Polis 1997; Noonberg and Abrams 2005; Amarasekare 2006, 2007).

I consider a situation in which the resource species is sedentary. The two consumer species do disperse, as does the predator in the case of keystone predation.

2.6.1 Intraguild predation

The spatial dynamics of a community with IGP are given by:

$$\frac{dR_j}{dt} = r_j R_j\Big(1 - \frac{R_j}{K}\Big) - a_1 R_j C_{1j} - a_2 R_j C_{2j} \tag{2.1}$$

$$\frac{dC_{1j}}{dt} = e_1 a_1 R_j C_{1j} - d_1 C_{1j} - \alpha C_{1j} C_{2j} - \beta_1 C_{1j} + \frac{\beta_1}{3} \sum_{j=1}^{3} C_{1j} \tag{2.2}$$

$$\frac{dC_{2j}}{dt} = e_2 a_2 R_j C_{2j} - d_2 C_{2j} + f\alpha C_{1j} C_{2j} - \beta_2 C_{2j} + \frac{\beta_2}{3} \sum_{j=1}^{3} C_{2j} \qquad (2.3)$$

where R_j is the resource abundance in patch j, and C_{ij} is the abundance of Consumer species i in patch j ($i = 1, 2, j = 1, 2, 3$; Amarasekare 2006). The parameter r_j is the per capita rate of resource production in patch j and K is the resource carrying capacity. Resource productivity varies spatially while the resource carrying capacity remains invariant across patches. The parameter a_i is the Consumer i's attack rate, e_i is the number of its offspring resulting from resource consumption, and d_i is its background mortality rate. The parameter α is the Consumer 2's attack rate on Consumer 1, and f is the number of Consumer 2 offspring resulting from intraguild predation. Consumer 1 therefore is the IGPrey, and Consumer 2 is the IGPredator. The parameter β_i is the per capita emigration rate of Consumer i.

I nondimensionalize Equations (2.1)-(2.3) using scaled quantities. Nondimensional analysis not only reduces the number of parameters but also highlights the biologically significant scaling relations between parameters (Nisbet and Gurney 1982; Murray 1993).

I use the substitutions

$$\hat{R_j} = \frac{R_j}{K},\ \hat{C_{ij}} = \frac{C_{ij}}{e_i K},\ \hat{r_j} = \frac{r_j}{d_1},\ \hat{a_i} = \frac{e_i a_i K}{d_i},\ \hat{\alpha} = \frac{e_2 \alpha K}{d_2},\ \hat{\beta_i} = \frac{\beta_i}{d_i},\ \hat{f} = \frac{e_1 f}{e_2},$$

$$\delta = \frac{d_2}{d_1},\ \tau = d_1 t\ \ (d_i \neq 0, i = 1, 2, j = 1, 2, 3)$$

to transform the original variables into nondimensional quantities. The dimensionless time metric τ expresses time in terms of the IGPrey's death rate. This time scaling allows for comparing systems that differ in their natural time scales. Resource abundance is expressed as a fraction of the resource carrying capacity, and varies from 0 to 1. The consumers' abundances are scaled by their respective conversion efficiencies and the resource carrying capacity. The scaled attack rates ($\hat{a_i}$) depend on the resource carrying capacity and the consumer death rate (d_i); the scaled interference parameter $\hat{\alpha}$ shows that the per capita inhibitory effect of the IGPredator on the IGPrey depends on the IGPredator's conversion efficiency and mortality rate as well as the resource carrying capacity. The parameter δ is the ratio of the consumers' mortality rates, and $\hat{\beta_i}$ is the per capita emigration rate of Consumer i relative to its within-patch mortality rate. The other important parameter is the efficiency metric $\hat{f}$. On their own, the efficiency parameters e_i and f have little meaning; as a composite they reveal important scaling relationships between conversion efficiencies for resource consumption and IGP. For instance, large values of $\hat{f}$ imply that for any value of e_1, $f >> e_2$, i.e., the IGPredator obtains a greater benefit from the IGPrey than from the basal resource.

I substitute the nondimensional quantities into Equations (2.1)-(2.3) and drop the hats for convenience. This yields the nondimensional system:

$$\frac{dR_j}{d\tau} = r_j R_j (1 - R_j) - a_1 R_j C_{1j} - \delta a_2 R_j C_{2j} \qquad (2.4)$$

$$\frac{dC_{1j}}{d\tau} = a_1 R_j C_{1j} - C_{1j} - \delta\alpha C_{1j} C_{2j} - \beta_1 C_{1j} + \frac{\beta_1}{3}\sum_{j=1}^{3} C_{1j} \quad (2.5)$$

$$\frac{dC_{2j}}{d\tau} = \delta\Big(a_2 R_j C_{2j} - C_{2j} + f\alpha C_{1j} C_{2j} - \beta_2 C_{2j} + \frac{\beta_2}{3}\sum_{j=1}^{3} C_{2j}\Big). \quad (2.6)$$

Unless otherwise noted, all variables and parameters from this point on are expressed as nondimensional quantities.

Because the goal is to understand the possible interplay between local coexistence mechanisms and those mediated by dispersal, I restrict attention to the situation where local coexistence via a competition-IGP trade-off is possible in at least one patch. The trade-off is such that the IGPrey is the superior resource competitor (i.e., it has a lower $R^{\star}$; Tilman 1982), but the IGPredator can prey on the IGPrey ($\alpha > 0$). From Equations (2.4)-(2.6) $R^{\star}$ in the absence of dispersal is $\frac{1}{a_i}$, and hence competitive superiority of the IGPrey translates into having a higher attack rate than the IGPredator. Throughtout, I use a_i as the measure of competitive ability and α as a measure of the strength of IGP while keeping the mortality ratio (δ) and conversion efficiency (f) fixed.

I introduce spatial variation by setting the resource productivity in each patch to a level that leads to one of the three outcomes observed in the absence of dispersal: (i) IGPrey only, (ii) Coexistence, (iii) IGPredator only. Adopting the convention that patches 1, 2, and 3 represent increasing levels of resource productivity we have $r_1 = (0, r_{C_2})$, $r_2 = (r_{C_2}, r_{C_1})$, and $r_3 = (r_{C_1}, r_{max})$ where r_{C_1} and r_{C_2} are, respectively, the threshold resource productivities required for the IGPrey and IGPredator to invade when rare, and r_{max} is the maximum resource productivity.

Because the focus is on the interplay between IGP and dispersal, I assume that the two consumer species differ in their attack rates (a_i) and dispersal propensities (β_i) but have similar background mortality rates (i.e., $\delta = 1$). Further details of model analyses are given in Amarasekare (2006).

2.6.2 *Keystone predation*

The spatial dynamics of a community with keystone predation are given by:

$$\frac{dR_j}{dt} = r_j R_j\Big(1 - \frac{R_j}{K}\Big) - a_1 R_j C_{1j} - a_2 R_j C_{2j} \quad (2.7)$$

$$\frac{dC_{1j}}{dt} = e_1 a_1 R_j C_{1j} - d_1 C_{1j} - \alpha_1 C_{1j} P_j - \beta_1 C_{1j} + \frac{\beta_1}{3}\sum_{j=1}^{3} C_{1j} \quad (2.8)$$

$$\frac{dC_{2j}}{dt} = e_2 a_2 R_j C_{2j} - d_2 C_{2j} - \alpha_2 C_{2j} P_j - \beta_2 C_{2j} + \frac{\beta_2}{3}\sum_{j=1}^{3} C_{2j} \quad (2.9)$$

$$\frac{dP_j}{dt} = f_1\alpha_1 C_{1j}P_j + f_2\alpha_2 C_{2j}P_j - d_P P_j - \beta_P P_j + \frac{\beta_P}{3}\sum_{j=1}^{3} P_j \tag{2.10}$$

where R_j is the resource abundance, C_{ij} is the abundance of consumer species i, and P_j is the predator abundance in patch j ($i = 1, 2, j = 1, 2, 3$; Amarasekare 2008). The parameter r_j is the per capita rate of resource production in patch j, and K is the spatially invariant resource carrying capacity; a_i is consumer species i's attack rate on the resource, e_i is the number of its offspring resulting from resource consumption, and d_i is its background mortality rate. The parameter α_i is the predator's attack rate on consumer i, and f_i is the number of resulting predator offspring, and d_P is the predator's background mortality rate. The parameters β_i and β_P are, respectively, the per capita emigration rates of consumer i and the predator.

I nondimensionalize Equations (2.7)-(2.10) using the scaled quantities

$$\hat{R_j} = \frac{R_j}{K},\ \hat{C_{ij}} = \frac{C_{ij}}{e_i K},\ \hat{P_j} = \frac{P_j}{f_1 f_2 K},\ \hat{r_j} = \frac{r_j}{d_1},$$

$$\hat{a_i} = \frac{e_i a_i K}{d_i},\ \hat{\alpha_i} = \frac{f_1 f_2 \alpha_i K}{d_i},\ \hat{f} = \frac{f_1}{f_2},\ \hat{e_i} = \frac{e_i}{f_i},\ \delta = \frac{d_2}{d_1},\ \hat{d_p} = \frac{d_P}{d_1},$$

$$\tau = d_1 t,\ \hat{\beta_i} = \frac{\beta_i}{d_i},\ \hat{\beta_P} = \frac{\beta_P}{d_P}\ (d_i \neq 0, i = 1, 2, j = 1, 2, 3).$$

(Note that I have separated the nondimensional parameter $\hat{f} = \frac{e_i}{f_k}, (i, k = 1, 2, i \neq k)$ into $\hat{e_i} = \frac{e_i}{f_i}$ and $\hat{f} = \frac{f_1}{f_2}$ because it allows for a more biologically meaningful scaling relationship; Amarasekare 2008) The nondimensional time metric τ expresses time in terms of the superior competitor's death rate. The nondimensionalized attack rate of the predator on consumer i ($\hat{\alpha_i}$) depends on the predator's conversion efficiencies (f_i), consumer i's mortality rate, and the resource carrying capacity. The parameter $\hat{d_P}$ is the predator's mortality rate relative to that of the superior competitor, and $\hat{\beta_P}$ is the predator's per capita emigration rate scaled by the predator's death rate. Other nondimensional quantities have the same meaning as in the IGP model (Equations (2.4)-(2.6)). Substituting these quantities into Equations (2.7)-(2.10) yields the following nondimensional system:

$$\frac{dR_j}{dt} = r_j R_j(1 - R_j) - a_1 R_j C_{1j} - a_2 R_j C_{2j} \tag{2.11}$$

$$\frac{dC_{1j}}{dt} = e_1 a_1 R_j C_{1j} - C_{1j} - \alpha_1 C_{1j} P_j - \beta_1 C_{1j} + \frac{\beta_1}{3}\sum_{j=1}^{3} C_{1j} \tag{2.12}$$

$$\frac{dC_{2j}}{dt} = \delta\Big(a_2 R_j C_{2j} - C_{2j} - \alpha_2 C_{2j} P_j - \beta_2 C_{2j} + \frac{\beta_2}{3}\sum_{j=1}^{3} C_{2j}\Big) \tag{2.13}$$

$$\frac{dP_j}{dt} = e_1 f \alpha_1 C_{1j} P_j + \frac{\delta e_2}{f}\alpha_2 C_{2j} P_j - d_P P_j - d_P \beta_P P_j + \frac{d_P \beta_P}{3}\sum_{j=1}^{3} P_j. \tag{2.14}$$

Further details of model analyses are given in Amarasekare (2008).

2.7 References

P. A. Abrams and W.G. Wilson (2004), Coexistence of competitors in metacommunities due to spatial variation in resource growth rates: Does R^* predict the outcome of competition? *Ecology Letters* **7**: 929-940.

P. Amarasekare (2006), Productivity, dispersal and the coexistence of intraguild predators and prey, *Journal of Theoretical Biology* **243**: 121-133.

P. Amarasekare (2007), Spatial dynamics of communities with intraguild predation: The role of dispersal strategies, *American Naturalist* **170**: 819-831.

P. Amarasekare (2008), Spatial dynamics of keystone predation, *Journal of Animal Ecology* **77**: 1306-1315.

P. Amarasekare and R. Nisbet (2001), Spatial heterogeneity, source-sink dynamics and the local coexistence of competing species, *American Naturalist* **158**:572-584.

P. Amarasekare, M. Hoopes, N. Mouquet and M. Holyoak (2004), Mechanisms of coexistence in competitive metacommunities, *American Naturalist* **164**: 310-326.

P. R. Armsworth and J.E. Roughgarden (2005a), The impact of directed versus random movement in population dynamics and biodiversity patterns, *American Naturalist* **165**: 449-465.

P. R. Armsworth and J.E. Roughgarden (2005b), Disturbance induces the contrasting evolution of reinforcement and dispersiveness in directed and random movers, *Evolution* **59**(10):2083–2096.

M. Arim and P. Marquet (2004), Intraguild predation: a widespread interaction related to species biology, *Ecology Letters* **7**: 557-564.

B. M. Bolker and S. W. Pacala (1999), Spatial moment equations for plant competition: Understanding spatial strategies and the advantages of short dispersal, *American Naturalist* **153**:575-602.

M. B. Bonsall and R.D. Holt (2003), The effects of enrichment on the dynamics of apparent competitive interactions in stage-structured systems, *American Naturalist* **162**: 780-795.

J. M. Chase (2003), Strong and weak trophic cascades along a productivity gradient, *Oikos* **101**: 187-195.

J. M. Chase and M. A. Leibold (2002), Spatial scale dictates the productivity-biodiversity relationship. Nature, 416, 427-430.

J. M. Chase and M. Leibold (2003), *Ecological Niches: Linking Classical and Contemporary Approaches*, University of Chicago Press, Chicago.

J. M. Chase and W.A. Ryberg (2004), Connectivity, scale-dependence, and the productivity-diversity relationship, *Ecology Letters* **7**: 676-683.

C. T. Codeco and J.P. Grover (2001), Competition along a spatial gradient of resource supply: A microbial experimental model, *American Naturalist* **157**: 300-315.

S. Diehl and M. Feissel (2000), Effects of enrichment on three-level food chains with omnivory, *American Naturalist* **155**: 200-218.

R. D. Holt and G.A. Polis (1997), A theoretical framework for intraguild predation, *American Naturalist* **149**:745-764.

R. D. Holt, J. Grover and D. Tilman (1994), Simple rules for interspecific dominance in systems with exploitative and apparent competition, *American Naturalist* **144**: 741-771.

V.A.A. Jansen (2001), The dynamics of two diffusively coupled predator-prey populations, *Theoretical Population Biology* **59**:119-131.

V. Krivan (1996), Optimal foraging and predator-prey dynamics, *Theoretical Population Biology* **49**: 265-290.

M. A. Leibold, M. Holyoak, N. Mouquet, P. Amarasekare, J.M. Chase, M.F. Hoopes, R.D. Hold, J.B. Shurin, R. Law, D. Tilman, M. Loreau and A. Gonzalez (2004), The metacommunity concept: A framework for multi-scale community ecology, *Ecology Letters* **7**: 601-613.

M. A. Leibold, M. Holyoak and R.D. Holt (2005), *Metacommunities*, University of Chicago Press, Chicago.

S. A. Levin (1974), Dispersion and population interactions, *American Naturalist* **108**:207-228.

W. W. Murdoch, C. J. Briggs, R. M. Nisbet, W.S.C. Gurney and A. Stewart-Oaten (1992), Aggregation and stability in metapopulation models, *American Naturalist* **140**: 41–58.

J. D. Murray (1993), *Mathematical Biology*, Springer-Verlag, New York.

S. A. Navarrete and B.A. Menge (1996), Keystone predation and interaction strength: Interactive effects of predators on their main prey, *Ecological Monographs* **66**: 409-429.

E. Noonburg and P. A. Abrams (2005), Transient dynamics and prey species coexistence with shared resources and predator, *American Naturalist* **165**:322-335.

G. A. Polis, C.A. Myers and R.D. Holt (1989), The ecology and evolution of intraguild predation: Potential competitors that eat each other, *Annual Review of Ecology and Systematics* **20**: 297-330.

A. Sih, P. Crowley, M. McPeek, J. Petranka and K. Strohmeier (1985), Predation, competition and prey communities: A review of field experiments, *Annual Review of Ecology and Systematics* **16**: 269-311.

CHAPTER 3

Bistability Dynamics in Structured Ecological Models

Jifa Jiang[1]
and
Junping Shi[2,3]

[1]Department of Mathematics, Shanghai Normal University,
Shanghai 200092, P.R.China
jiangjf@mail.tongji.edu.cn

[2]Department of Mathematics, College of William and Mary,
Williamsburg, VA 23185, USA
shij@math.wm.edu

[3]School of Mathematics, Harbin Normal University,
Harbin, Heilongjiang 150080, P.R.China

Abstract. Alternative stable states exist in many important ecosystems, and gradual change of the environment can lead to dramatic regime shift in these systems (Beisner et.al. (2003), May (1977), Klausmeier (1999), Rietkerk et.al. (2004), and Scheffer et.al. (2001)). Examples have been observed in the desertification of Sahara region, shift in Caribbean coral reefs, and the shallow lake eutrophication (Carpenter et.al. (1999), Scheffer et.al. (2003), and Scheffer et.al. (2001)). It is well-known that a social-economical system is sustainable if the life-support ecosystem is resilient (Holling (1973) and Folke et.al. (2004)). Here resilience is a measure of the magnitude of disturbances that can be absorbed before a system centered at one locally stable equilibrium flips to another. Mathematical models have been established to explain the phenomena of bistability and hysteresis, which provide qualitative and quantitative information for ecosystem managements and policy making (Carpenter et.al. (1999) and Peters et.al. (2004)). However most of these models of catastrophic shifts are non-spatial ones. A theory for spatially extensive, heterogeneous ecosystems is needed for sustainable management and recovery strategies, which requires a good understanding of the relation between system feedback and spatial scales (Folke et.al. (2004), Walker et.al. (2004) and Rietkerk et.al. (2004)). In this chapter, we survey some recent results on structured evolutionary dynamics including reaction-diffusion equations and systems, and discuss their applications to structured ecological models which display bistability and hysteresis. In Section 1, we review several classical non-spatial models with bistability; we discuss their counterpart reaction-diffusion models in Section 2, and especially diffusion-induced bistability and hysteresis. In Section 3, we introduce some abstract results and concrete examples of threshold manifolds (separatrix) in the bistable dynamics.

3.1 Non-structured models

The logistic model was first proposed by Belgian mathematician Pierre Verhulst (Verhulst (1838)):

$$\frac{dP}{dt} = aP\left(1 - \frac{P}{N}\right), \quad a, N > 0. \tag{3.1}$$

Here a is the maximum growth rate per capita, and N is the carrying capacity. A more general logistic growth type can be characterized by a declining growth rate per capita function. However it has been increasingly recognized by population ecologists that the growth rate per capita may achieve its peak at a positive density, which is called an *Allee effect* (see Allee (1938), Dennis (1989) and Lewis and Kareiva (1993)). An Allee effect can be caused by shortage of mates (Hopf and Hopf (1985), Veit and Lewis (1996)), lack of effective pollination (Groom (1998)), predator saturation (de Roos et.al. (1998)), and cooperative behaviors (Wilson and Nisbet (1997)).

If the growth rate per capita is negative when the population is small, we call such a growth pattern a *strong Allee effect* (see Fig.3.1-c); if $f(u)$ is smaller than the maximum but still positive for small u, we call it a *weak Allee effect* (see Fig.3.1-b). In Clark (1991), a strong Allee effect is called a *critical depensation* and a weak Allee effect is called a *noncritical depensation*. A population with a strong Allee effect is also called *asocial* by Philip (1957). Most people regard the strong Allee effect as the Allee effect, but population ecologists have started to realize that an Allee effect may be weak or strong (see Wang and Kot (2001), Wang, Kot and Neubert (2002)). Some possible growth rate per capita functions were also discussed in Conway (1983,1984). A prototypical model with Allee effect is

$$\frac{dP}{dt} = aP\left(1 - \frac{P}{N}\right) \cdot \frac{P - M}{|M|}, \quad a, N > 0. \tag{3.2}$$

If $0 < M < N$, then the equation is of strong Allee effect type, and if $-N < M < 0$, then it is of weak Allee effect type. At least in the strong Allee effect case, M is called the sparsity constant.

The dynamics of the logistic equation is monostable with one globally asymptotically stable equilibrium, and that of strong Allee effect is bistable with two stable equilibria. A weak Allee effect is also monostable, although the growth is slower at lower density. Another example of a weak Allee effect is the equation of higher order autocatalytic chemical reaction of Gray and Scott (1990):

$$\frac{da}{dt} = -kab^p, \quad \frac{db}{dt} = kab^p, \quad k > 0, \ p \geq 1. \tag{3.3}$$

Here $a(t)$ and $b(t)$ are the concentrations of the reactant A and the autocatalyst B, k is the reaction rate, and $p \geq 1$ is the order of the reaction with respect to the

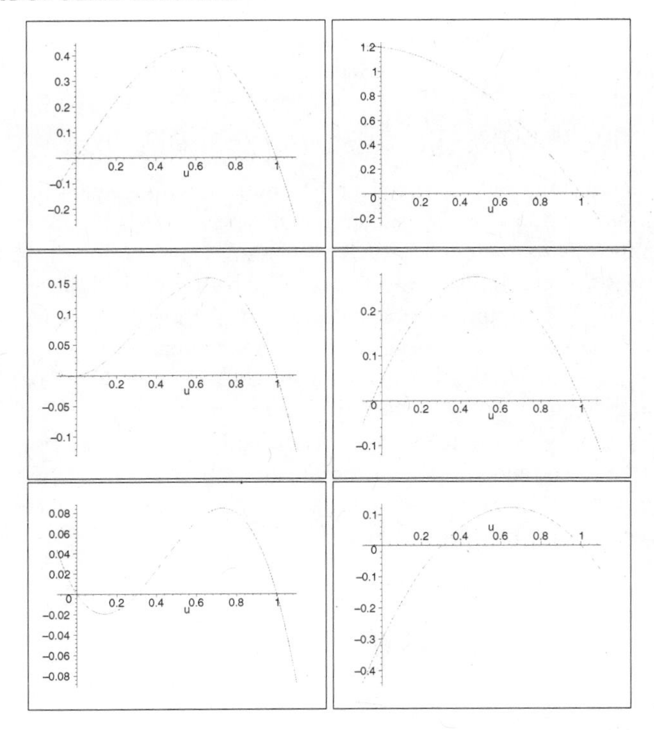

Figure 3.1 (a) logistic (top); (b) weak Allee effect (middle); (c) strong Allee effect (bottom); the graphs on the left are growth rate $uf(u)$, and the ones on the right are growth rate per capita $f(u)$.

autocatalytic species. Notice that $a(t) + b(t) \equiv a_0 + b_0$ is invariant, so that (3.3) can be reduced to

$$\frac{db}{dt} = k(a_0 + b_0 - b)b^p, \quad k, a_0 + b_0 > 0, \; p \geq 1, \tag{3.4}$$

which is of weak Allee effect type if $p > 1$, and of logistic type if $p = 1$. An autocatalytic chemical reaction has been suggested as a possible mechanism of various biological feedback controls (Murray (2003)), and the similarity between chemical reactions and ecological interactions has been observed since Lotka (1920) in his pioneer work.

The cubic nonlinearity in (3.2) has also appeared in other biological models. One prominent example is the FitzHugh-Nagumo model of neural conduction (FitzHugh (1961) and Nagumo et.al. (1962)), which simplifies the classical Hodgkin-Huxley model:

$$\epsilon\frac{dv}{dt} = v(v-a)(1-v) - w, \quad \frac{dw}{dt} = cv - bw, \quad \epsilon, a, b, c > 0, \tag{3.5}$$

where $v(t)$ is the excitability of the system (voltage), and $w(t)$ is a recovery variable representing the force that tends to return the resting state. When c is zero and $w = 0$,

(3.5) becomes (3.2). Another example is a model of the evolution of fecally-orally transmitted diseases by Capasso and Maddalena (1981/82, 1982):

$$\frac{dz_1}{dt} = -a_{11}z_1 + a_{12}z_2, \quad \frac{dz_2}{dt} = -a_{22}z_2 + g(z_1), \quad a_{11}, a_{12}, a_{22} > 0. \tag{3.6}$$

Here $z_1(t)$ denotes the (average) concentration of infectious agent in the environment; $z_2(t)$ denotes the infective human population; $1/a_{11}$ is the mean lifetime of the agent in the environment; $1/a_{22}$ is the mean infectious period of the human infectives; a_{12} is the multiplicative factor of the infectious agent due to the human population; and $g(z_1)$ is the force of infection on the human population due to a concentration z_1 of the infectious agent. If $g(z_1)$ is a monotone increasing concave function, then it is known that the system is monostable with the global asymptotical limit being either an extinction steady state or a nontrivial endemic steady state. However if $g(z_1)$ is a monotone sigmoid function, *i.e.* a monotone convex-concave function with S-shape and saturating to a finite limit, then the system (3.6) possesses two nontrivial endemic steady states and the dynamics of (3.6) is bistable, which can be easily seen from the phase plane analysis.

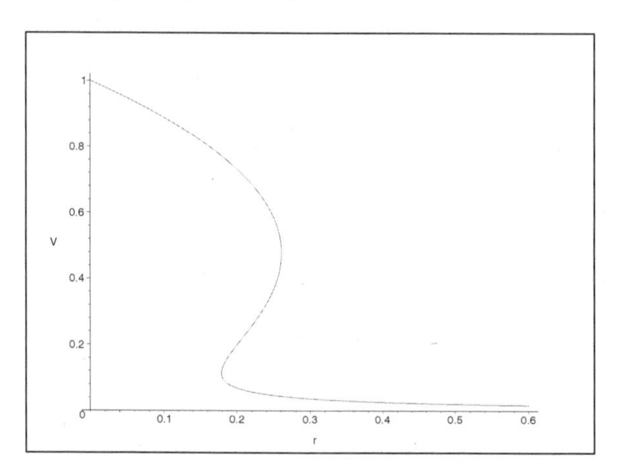

Figure 3.2 Equilibrium bifurcation diagram of (3.8) with $h = 0.1$, where the horizontal axis is r and the vertical axis is V.

Now we turn to some existing models which could lead to catastrophic shifts in ecosystems. In 1960-70s, theoretical predator-prey systems are proposed to demonstrate various stability properties in systems of populations at two or more trophic levels (Rosenzweig and MacArthur (1963) and Rosenzweig (1971)). A simplified model with such a predator-prey feature is that of a grazing system of herbivore-plant interaction as in Noy-Meir (1975), see also May (1977). Here $V(t)$ is the vegetation biomass, and its quantity changes following the differential equation:

$$\frac{dV}{dt} = G(V) - Hc(V), \tag{3.7}$$

where $G(V)$ is the growth rate of vegetation in absence of grazing, H is the herbivore population density, and $c(V)$ is the per capita consumption rate of vegetation by the herbivore. If $G(V)$ is given by the familiar logistic equation, and $c(V)$ is the Holling type II ($p = 1$) or III ($p > 1$) functional response function (Holling (1959)), then (3.7) has the form (after nondimensionalization):

$$\frac{dV}{dt} = V(1-V) - \frac{rV^p}{h^p + V^p}, \quad h, r > 0, \ \ p \geq 1. \tag{3.8}$$

This equation (with $p = 2$) also appears as the model of insect pests such as the spruce budworm (*Choristoneura fumiferana*) in Canada and northern USA (Ludwig et. al. (1978)), in which $V(t)$ is the budworm population. In either situation, the harvesting effort is assumed to be constant as the change of the predator population occurs at a much slower time scale compared to that of the prey. The function $c(V) = \dfrac{\gamma V^p}{h^p + V^p}$ with $p \geq 1$ is called the Hill function in some references. We notice that a Hill function is one of sigmoid functions which is defined in the epidemic model (3.6).

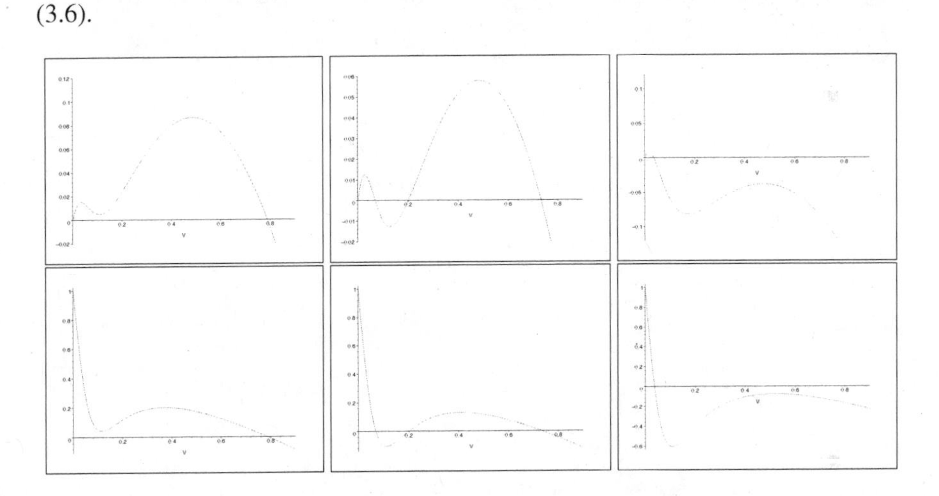

Figure 3.3 (top) Graph of the growth rate function $f(V) = V(1-V) - \dfrac{rV^p}{h^p + V^p}$ with $h = 0.1$; (bottom) Graph of the growth rate per capita $f(V)/V$. (a) $r = 0.17$ (left); (b) $r = 0.2$ (middle); (c) $r = 0.3$ (right).

To describe the catastrophic regime shifts between alternative stable states in ecosystems, a minimal mathematical model

$$\frac{dx}{dt} = a - bx + \frac{rx^p}{h^p + x^p}, \quad a, b, r, h > 0, \tag{3.9}$$

is proposed in Carpenter et.al. (1999), see also Scheffer et.al. (2001). (3.9) can be used in ecosystems such as lakes, deserts, or woodlands. For lakes, $x(t)$ is the level of nutrients suspended in phytoplankton causing turbidity, a is the nutrient loading, b is the nutrient removal rate, and r is the rate of internal nutrient recycling.

The equations (3.8) and (3.9) are examples of differential equation models which ex-

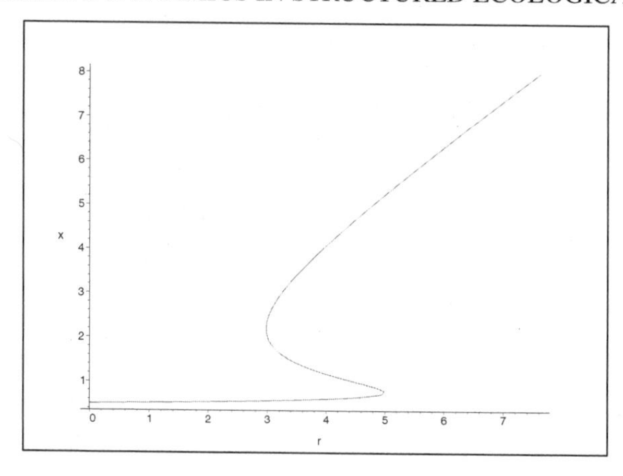

Figure 3.4 Equilibrium bifurcation diagram of (3.9) with $a = 0.5$, $b = 1$, where the horizontal axis is r and the vertical axis is x.

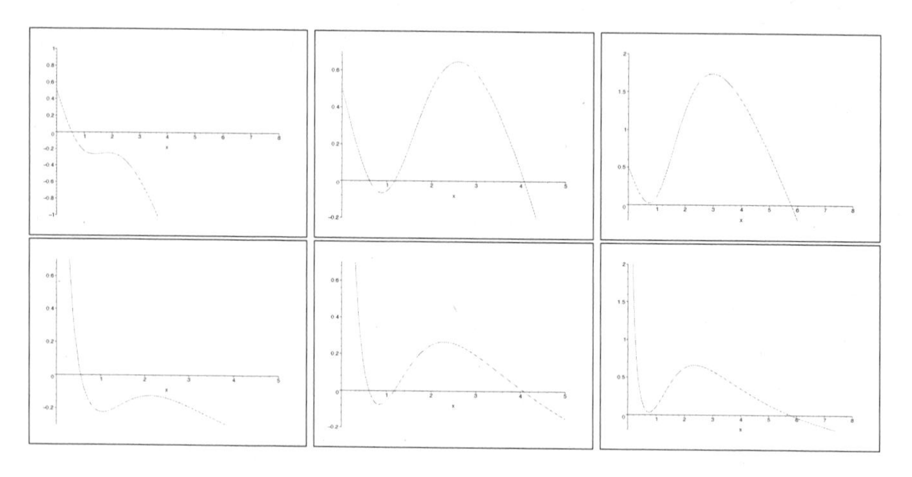

Figure 3.5 (top) Graph of the growth rate function $g(x) = a - bx + \dfrac{rx^p}{h^p + x^p}$ with $a = 0.5$, $b = 1$; (bottom) Graph of the growth rate per capita $f(x)/x$. (a) $r = 2.5$ (left); (b) $r = 4$ (middle); (c) $r = 5.5$ (right).

hibit the existence of multiple stable states and the phenomenon of hysteresis. From the bifurcation diagrams (Fig. 3.2 for (3.8), and Fig. 3.4 for (3.9)), the system has three positive equilibrium points when $r \in (r_1, r_2)$ for some $\infty > r_2 > r_1 > 0$, and the largest and smallest positive equilibrium points are stable. For the grazing system (3.8), the number of stable equilibrium points changes with the herbivore density r. For low r, the vegetation biomass tends to a unique equilibrium slightly lower than 1 (the rescaled carrying capacity); as r increases over r_1, a second stable equilibrium appears through a supercritical saddle-node bifurcation, and it represents a much lower vegetation biomass; as r continues to increases to another parameter threshold

$r_2 > r_1$, the larger stable equilibrium suddenly vanishes through a subcritical saddle-node bifurcation, and the lower stable equilibrium becomes the unique attracting one. As h increases gradually, the vegetation biomass first settles at a higher level for low h, but it collapses to a lower lever as h passes r_2; after this catastrophic shift, even if h is restored slightly, the biomass remains at the low level unless h decreases beyond r_1. This irreversibility of the hysteresis loop gives raise to a serious management problem for the grazing systems, see Noy-Meir (1975) and May (1977). Similar discussions hold for (3.9) as well as r decreases, see Scheffer et.al. (2001), where the drop from high density stable equilibrium to the low one is called "forward shift", and the recovery from the low one to high one is a "backward shift".

It is worth pointing out that the S-shaped bifurcation curve in Fig. 3.2 and Fig. 3.4 can also be viewed as a result of bifurcation with respect to conditions such as nutrient loading, exploitation or temperature rise (Scheffer et.al. (2001)). That is a transition from a monostable system to a bistable one, or mathematically, a cusp bifurcation from a monotone curve to a S-shaped one with two turning points (see Fig. 3.6). Such fold bifurcations have been discussed in much more general settings in Shi (1999), and Liu, Shi and Wang (2007). In general it is hard to rigorously prove the exact transition from monostable to bistable dynamics, especially for higher (including infinite) dimensional problems. In (3.8) with $p = 2$, one can show the cusp bifurcation occurs when h crosses $h_0 = \sqrt{3}/27 \approx 0.1925$. A mathematical survey on the fold and cusp type mappings (especially in infinite dimensional spaces) can be found in Church and Timourian (1997).

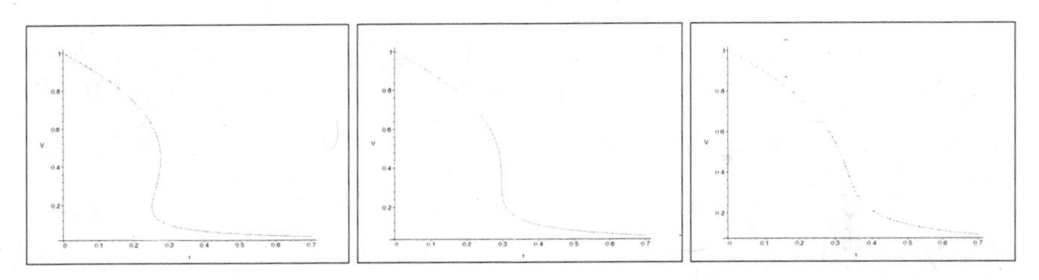

Figure 3.6 Cusp bifurcation in (3.8) with $p = 2$, where the horizontal axis is r and the vertical axis is V. (a) $h = 0.15$ (left); (b) $h = \sqrt{3}/27 \approx 0.1925$ (middle); (c) $h = 0.25$ (right).

We note that in Fig. 3.3-a and Fig. 3.5-c, the system is monostable with only one stable equilibrium point, yet the graph of "growth rate per capita"(see the lower graphs in Fig. 3.3-a and Fig. 3.5-c) has two fluctuations before turning to negative. This is similar to the weak Allee effect defined earlier where the growth rate per capita changes the monotonicity once. These geometric properties of the growth rate per capita functions motivate us to classify all growth rate patterns according to the monotonicity of the function $f(u)/u$ if $f(u)$ is the gross growth rate in a model $u' = f(u)$:

1. $f(u)$ is of *logistic* type, if $f(u)/u$ is strictly decreasing;
2. $f(u)$ is of *Allee effect* type, if $f(u)/u$ changes from increasing to decreasing when u increases;

3. $f(u)$ is of *hysteresis* type, if $f(u)/u$ changes from decreasing to increasing then to decreasing again when u increases.

In all cases, we assume that $f(u)$ is negative when u is large, thus $f(u)$ has at least one zero $u_1 > 0$. In the Allee effect case, if $f(u)$ has another zero in $(0, u_1)$, then it is a strong Allee effect, otherwise it is a weak one; in the hysteresis case, if $f(u)$ has two more zeros in $(0, u_1)$, then it is strong hysteresis, otherwise it is weak. Here we exclude the degenerate cases when $f(u_0) = f'(u_0) = 0$ (double zeros). Considering the ODE model $u' = f(u)$, the weak Allee effect or hysteresis dynamics appears to be no different from the logistic case in terms of the asymptotic behavior, since $f(u) > 0$ for $u \in (0, u_1)$ and $f(u) < 0$ for $u > u_1$. The definitions here are not only for mathematical interest. In the next section, we shall show that the addition of diffusion to the equation can dramatically change the dynamics for the weak Allee effect or hysteresis.

3.2 Diffusion induced bistability and hysteresis

Dispersal of the state variable in a continuous space can be modeled by a partial differential equation with diffusion (see Okubo and Levin (2001), Murray (2003), Cantrell and Cosner (2003)):

$$\frac{\partial u}{\partial t} = d\Delta u + f(u), \quad t > 0, \quad x \in \Omega. \tag{3.10}$$

Here $u(x,t)$ is the density function of the state variable at spatial location x and time t, $d > 0$ is the diffusion coefficient, the habitat Ω is a bounded region in $\mathbf{R}^n$ for $n \geq 1$, $\Delta u = \sum_{i=1}^{n} \frac{\partial^2 u}{\partial x_i^2}$ is the Laplace operator, and $f(u)$ represents the non-spatial growth pattern. We assume that the habitat Ω is surrounded by a completely hostile environment, thus it satisfies an absorbing boundary condition:

$$u(x) = 0, \quad x \in \partial\Omega. \tag{3.11}$$

It is known (Henry (1981)) that for equation (3.10) with boundary condition (3.11), there is a unique solution $u(x,t)$ of the initial value problem with an initial condition $u(x,0) = u_0(x) \geq 0$, provided that $f(u), u_0(x)$ are reasonably smooth. Moreover, if the solution $u(x,t)$ is bounded, then it tends to a steady state solution as $t \to \infty$ if one of the following conditions is satisfied: (i) $f(u)$ is analytic; (ii) if all steady state solutions of (3.10) and (3.11) are non-degenerate (see for example, Polácik (2002) and references therein). Hence the asymptotical behavior of the reaction-diffusion equation can be reduced to a discussion of the structure of the set of steady state solutions and related dynamical behaviors. The steady state solutions of (3.10) and (3.11) satisfy a semilinear elliptic type partial differential equation:

$$d\Delta u(x) + f(u(x)) = 0, \quad x \in \Omega, \quad u(x) = 0, \quad x \in \partial\Omega. \tag{3.12}$$

Since we are interested in the impact of diffusion on the extinction/persistence of population, we use the diffusion coefficient d as the bifurcation parameter. One can also use the size of the domain Ω as an equivalent parameter. To be more precise, we use the change of variable $y = x/\sqrt{d}$ to convert the equation (3.12) to:

$$\Delta u(y) + f(u(y)) = 0, \quad y \in \Omega_d, \quad u(y) = 0, \quad y \in \partial\Omega_d, \tag{3.13}$$

where $\Omega_d = \{y : \sqrt{d}y \in \Omega\}$. This point of view fits the classic concept of critical patch size introduced by Skellam (1951). When $\Omega = (0, l)$, the one-dimensional region, the size of the domain is simply the length of the interval. In higher dimension, Ω_d is a family of domains which have the same shape but "size" proportional to $d^{-1/2}$. Here "size" can be defined as the one-dimensional scale of the domain. Size can also be defined through the principal eigenvalue of $-\Delta$ on the domain Ω with zero boundary condition, which is the smallest positive number $\lambda_1(\Omega)$ such that

$$\Delta\phi(x) + \lambda_1\phi(x) = 0, \quad x \in \Omega, \quad \phi(x) = 0, \quad x \in \partial\Omega, \tag{3.14}$$

has a positive solution ϕ. Apparently $\lambda_1(\Omega_d) = \lambda_1(\Omega)/d$. In application a habitat slowly eroded by external influence can be approximated by such a family of domains Ω_d with similar shape but shrinking size. This is a special case of habitat fragmentation. In the following we use d as bifurcation parameter, and when d increases, the size (or the principal eigenvalue) of the domain Ω_d decreases.

The multiplicity and global bifurcation of solutions of (3.12) have been considered by many mathematicians over the last half century. Several survey papers and monographs can be consulted, see for example (Amann (1976), Cantrell and Cosner (2003), Lions (1981), and Shi (2009)) and the references therein. In this section we review some related results on that subject for the nonlinearity $f(u)$ discussed in Section 1 and their connection to ecosystem persistence/extinction.

For the Verhurst logistic model, the corresponding reaction-diffusion model was introduced by Fisher (1937) and Kolmogoroff, Petrovsky, and Piscounoff (1937) in studying the propagation of an advantageous gene over a spatial region, and the traveling wave solution was considered. The boundary value problem

$$d\Delta u + u\left(1 - \frac{u}{N}\right) = 0, \quad x \in \Omega, \quad u = 0, \quad x \in \partial\Omega, \tag{3.15}$$

was studied by Skellam (1951) when $\Omega = (0, L)$. Indeed in this case an explicit solution and dependence of L on D can be obtained via an elliptic integral (Skellam (1951)). When Ω is a general bounded domain, it was shown (see Cohen and Laetsch (1970), Cantrell and Cosner (1989), Shi and Shivaji (2006)) that when $0 < d^{-1} < \lambda_1(\Omega) \equiv \lambda_1$, the only nonnegative solution of (3.15) is $u = 0$, and it is globally asymptotically stable; when $d^{-1} > \lambda_1$, (3.15) has a unique positive solution u_d which is globally asymptotically stable. It is also known that $u_d(x)$ is is an decreasing function of d for $d < \lambda_1^{-1}$, and $u_d(x) \to 0$ as $d^{-1} \to \lambda_1^+$. Hence the critical number λ_1 represents the critical patch size. When the size of habitat gradually decreases, the biomass decreases too, and when it passes the critical patch size, the biomass becomes zero through a continuous change. Hence the bifurcation diagram of (3.15) is a continuous monotone curve as shown in Fig.3.7 (a).

The bifurcation diagram in Fig.3.7 (a) changes when an Allee effect exists in the growth function $f(u)$. For the boundary value problem

$$d\Delta u + u\left(1 - \frac{u}{N}\right) \cdot \frac{u - M}{|M|} = 0, \quad x \in \Omega, \quad u = 0, \quad x \in \partial\Omega, \tag{3.16}$$

one can use M as a parameter of the bifurcation in the bifurcation diagrams. We always assume $M < N$. When $M \leq -N$, the growth rate per capita is decreasing as in logistic case, thus the bifurcation diagram is monotone as in Fig 3.7 (a). When $-N < M < 0$, the growth rate per capita is of weak Allee effect type, and a new type of bifurcation diagram appears (Fig 3.7 (b)). We notice that the nonlinearity in (3.16) is normalized so that the growth rate per capita at $u = 0$ is always 1 when $M < 0$. Rigorous mathematical results about exact multiplicity of steady state solutions and global bifurcation diagram Fig 3.7 (b) are obtained in Korman and Shi (2001), and Shi and Shivaji (2006) for a more general nonlinearity and the domain being a ball in $\mathbf{R}^n$. We also mention that if the dispersal does not satisfy a linear diffusion law but a nonlinear one, then a weak Allee effect can also occur, and the bifurcation diagram of steady state solutions is like Fig. 3.7-b, see Cantrell and Cosner (2002), and Lee et.al. (2006).

Compared to the logistic case, a backward (subcritical) bifurcation occurs at $(d^{-1}, u) = (\lambda_1, 0)$, and a new threshold parameter value $0 < \lambda_* < \lambda_1$ exists. For $d^{-1} < \lambda_*$ (*extinction regime*), the population is destined to extinction no matter what the initial population is; for $d^{-1} > \lambda_1$ (*unconditional persistence regime*), the population always survive with a positive steady state. However in the intermediate *conditional persistence regime*, $\lambda_* < d^{-1} < \lambda_1$, there are exactly two positive steady state solutions $u_{1,d}$ and $u_{2,d}$. In fact, it can be shown that the three steady state solutions (including 0) can be ordered so that $u_{1,d}(x) > u_{2,d}(x) > 0$. Here $u_{1,d}$ and 0 are both locally stable. Hence the diffusion effect induces a bistability for a monostable model of weak Allee effect. A sudden collapse of the population occurs if d increases (or the domain size decreases) when d^{-1} crosses λ_*, and the system shifts abruptly from $u_{1,d}$ to 0 and it is not recoverable. This may explain that in some ecosystems with weak Allee effect, a catastrophic shift could still occur although the corresponding ODE model predicts unconditional persistence.

For $0 < M < N$ in (3.16), a strong Allee effect means that bistability occurs even for the small diffusion case (d small). If $N/2 \leq M < N$, $u = 0$ is the unique non-negative solution of (3.16) thus extinction is the only possibility. If $0 < M < N/2$, there exist at least two positive steady state solutions of (3.16) following a classical result of variational methods due to Rabinowitz (1973/74). When the domain is a ball in $\mathbf{R}^n$, it was shown by Ouyang and Shi (1998, 1999) that (3.16) has at most two positive solutions and the bifurcation diagram is exactly like Fig.3.7-c. Earlier the exact bifurcation diagram for the one-dimensional problem was obtained by Smoller and Wasserman (1981). It is well-known that in this case that a small initial population always leads to extinction, thus a single threshold value λ_* exists to separate the extinction and conditional persistence regimes. Earlier work on the dynamics of (3.10) and (3.11) with strong Allee effect was considered in Bradford and Philip (1970a, 1970b) and Yoshizawa (1970).

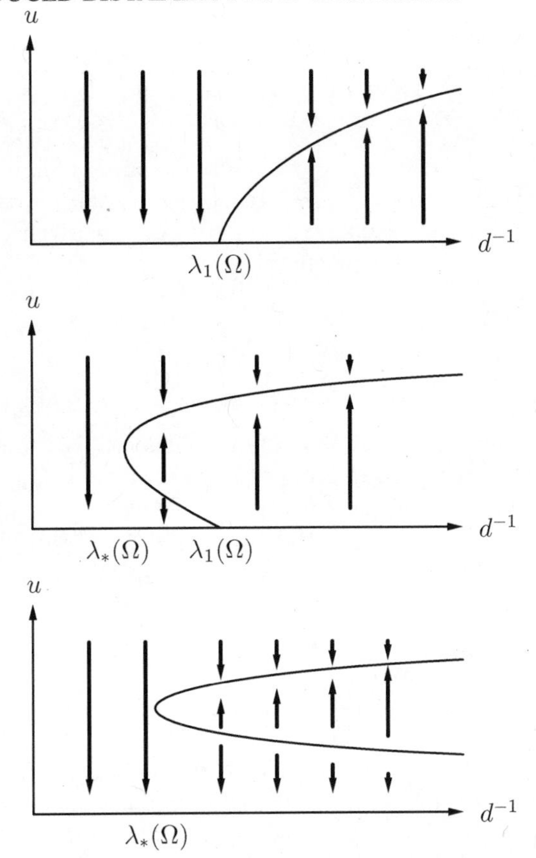

Figure 3.7 Bifurcation diagrams for (3.16): (a) logistic (upper); (b) weak Allee effect (middle); (c) strong Allee effect (lower).

The exact multiplicity results proved in Ouyang and Shi (1998, 1999) (see also Shi (2009)) hold for more general nonlinearities $f(u)$, and the criterion on $f(u)$ for the exact multiplicity are given by the shape of the function $f(u)/u$ and the convexity of $f(u)$. Another example is the border line case for (3.16) between the weak ($M < 0$) and strong Allee effect ($M > 0$), or more generally, the equation of autocatalytic chemical reaction (3.4) (assuming that $a_0 + b_0 = 1$):

$$d\Delta u + u^p(1-u) = 0, \quad x \in \Omega, \quad u = 0, \quad x \in \partial\Omega, \quad p > 1. \tag{3.17}$$

The bifurcation diagram of (3.17) is similar to Figure 3.7-c, and a proof can be found in Ouyang and Shi (1998, 1999) or Zhao, Shi and Wang (2007). Precise global bifurcation diagrams can also been given for the reaction-diffusion systems of autocatalytic chemical reaction (3.3) and epidemic model (3.6), and we will discuss them in the next section along with the associated dynamics.

The threshold value λ_* is important biologically as λ_* could give early warning of extinction for the species. Usually it is difficult to give a precise estimate of λ_* and it seems that there is no existing result on that problem. Here we only give an estimate of λ_* for the equation (3.16) with $N = 1$ and $M \in (0, 1/2)$. Hence we consider

$$d\Delta u + u(1-u)(u-M) = 0, \quad x \in \Omega, \quad u(x) = 0, \quad x \in \partial\Omega. \tag{3.18}$$

Here we have $f(u) = u(1-u)(u-M)$. From an idea in Shi and Shivaji (2006), $\lambda_* > \lambda_1/f_*$, where $f_* = \max_{u\in[0,1]} f(u)/u$, or the maximal growth rate per capita. An upper bound of λ_* can be obtained if (3.18) has a nontrivial solution for that d. We define an associated energy functional

$$I(u) = \frac{d}{2}\int_\Omega |\nabla u|^2 dx - \int_\Omega F(u)dx, \tag{3.19}$$

where $F(u) = \int_0^u f(t)dt = -\frac{1}{4}u^4 + \frac{1+M}{3}u^3 - \frac{M}{2}u^2$. It is well-known that a solution u of (3.18) is a critical point of the functional $I(u)$ in a certain function space (see Rabinowitz (1986) or Struwe (2000) for more details.) In particular, if $\inf I(u) < 0$, then (3.18) has a nontrivial positive solution. For small d, it is apparent that $\inf I(u) < 0$ if $M \in (0, 1/2)$. Hence for largest $d = \tilde{d}$ so that $\inf I(u) < 0$, we must have $\lambda_* < \tilde{d}^{-1}$. For the case $\Omega = (0, L)$, we can obtain that

$$\frac{2\pi^2}{L^2(1+M)} < \lambda_* < \frac{48}{L^2(3-M)}. \tag{3.20}$$

Here the upper bound is obtained by using a test function $u(x) = x/l$ for $x \in [0, l]$, $u(x) = 1$ for $x \in [l, L/2]$ and $u(x) = u(L-x)$ for $x \in [L/2, L]$, then optimizing among all possible value of l. The estimate (3.20) is indeed quite sharp. For example, for $L = 1$ and $M = 0.2$, the estimate (3.20) becomes $16.45 < \lambda_* < 17.14$. A numerical calculation using `Maple` and the algorithm in Lee et.al. (2006) shows that $\lambda_* \approx 16.61$. The threshold value for other problems can be estimated similarly, and in general the determination of the threshold value remains an interesting open question.

Next we turn to bifurcation diagrams with hysteresis. The hysteresis diagrams in Section 1 (Fig. 3.2 and 3.4) are generated with parameter r, which is the herbivore density in (3.8) or the rate of internal nutrient recycling in (3.9). In this subsection, we consider the corresponding reaction-diffusion models. First the steady state reaction-diffusion grazing model

$$d\Delta V + V(1-V) - \frac{rV^p}{h^p + V^p} = 0, \quad x \in \Omega, \quad V = 0, \quad x \in \partial\Omega, \tag{3.21}$$

was considered in Ludwig, Aronson and Weinberger (1979). For the case $n = 1$, by using the quadrature method, they show that the rough bifurcation diagram goes from a monotone curve with a unique large steady state, to an S-shaped curve, to a disconnected S-shaped curve, and finally a monotone curve with a unique small steady state, when r increases from near 0 to a large value (see Fig. 3.8 or the ones in Ludwig et.al. (1979)). Note that the bifurcation diagrams in Ludwig et.al. (1979) are not exact, and it is only shown that the equation has at least three positive solutions but not exactly three. An exact multiplicity result like the one in Ouyang and Shi

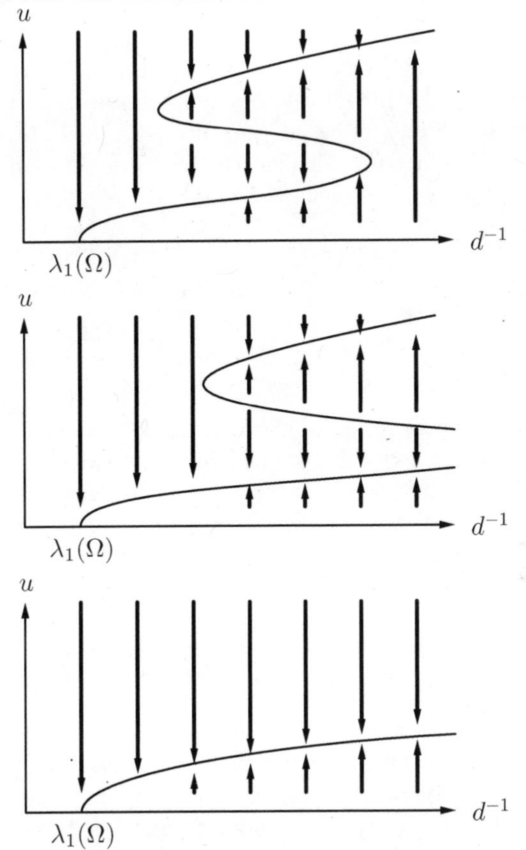

Figure 3.8 Bifurcation diagrams for (3.21): (a) weak hysteresis, r small but close to the first break point in ODE hysteresis loop, corresponding to f in Fig 3.3-a (upper); (b) strong hysteresis, corresponding to f in Fig 3.3-b (middle); (c) "collapsed", r larger than the second break point, corresponding to f in Fig 3.3-c (lower).

(1998, 1999) is not known even when $n = 1$. But it is known that in Fig. 3.8-b, the upper bound of the lower branch is the first zero of $f(u)$, and the lower bound of the upper branch is the smallest zero of $F(u) = \int_0^u f(t)dt = 0$ such that $f(u) > 0$; in Fig. 3.8-a, the lower turning point $\lambda^* \to \infty$ if the positive local minimum value of $f(u)$ tends to zero.

The transition of rough bifurcation diagrams suggests a bistable structure exists for intermediate range of r (see Fig. 3.2) when the nonlinearity is of strong hysteresis type, but a bistable structure could also exist when r is smaller when the nonlinearity is of weak hysteresis type (see Fig. 3.8-a). Indeed the S-shaped bifurcation diagram implies a hysteresis loop even though the weak hysteresis nonlinearity is positive until the zero at the "carrying capacity". Hence this is a hysteresis induced by the diffusion. Back to the context of shrinking habitat size, this suggests that for a seemingly safe ecosystem with the grazing is not too big so that the ODE model predicts a large stable equilibrium, the addition of diffusion can endanger the ecosystem if the habitat keeps shrinking, and a sudden drop to the small steady state is possible if the habitat size passes a critical value. Note that we do not exclude the possibility of catastrophic shift due to the increase of the grazing effect r, but the results in reaction-diffusion model offer another possible cause for such a sudden collapse, namely the decreasing natural vegetative habitat.

For the model (3.9) of lake turbidity, a reaction-diffusion model can also be proposed:

$$\begin{cases} u_t = d\Delta u + a - bu + \dfrac{ru^p}{h^p + u^p}, & t > 0, \quad x \in \Omega, \\ u(x,t) = 0, & x \in \partial\Omega, \\ u(x,t) = u_0(x), & t > 0, \quad x \in \Omega. \end{cases} \tag{3.22}$$

A similar argument can be made to offer another possible cause of the turbidity in shallow lakes, *i.e.* the shrinking that has occurred for many freshwater lakes because of the expanding of agriculture or industry. Here the bifurcation diagram of the steady state equation is not readily available in the existing literature, but similar problems with S-shaped bifurcation diagrams can be found in (Brown et.al. (1981), Du and Lou (2001), Korman and Li (1999), and Wang (1994)), to name a few. Indeed the nonlinearity $f(u)$ in (3.22) is qualitatively similar to the one in (3.21) (comparing Fig. 3.3 and Fig. 3.5), hence their bifurcation diagrams are similar.

In our discussion to this point, we have used a homogeneous Dirichlet boundary condition ($u = 0$ on the boundary). While diffusion plays an instrumental role in inducing bistability, the Dirichlet boundary condition also plays an important role. In some rough sense, a Dirichlet boundary condition is much more "spatially heterogeneous" than a Neumann boundary condition (or no flux, reflection boundary condition), and is more rigid than Neumann boundary condition. Here we also comment briefly on reaction-diffusion models with Neumann boundary condition:

$$\begin{cases} \dfrac{\partial u}{\partial t} = d\Delta u + f(u), & t > 0, \quad x \in \Omega, \\ \dfrac{\partial u}{\partial n} = 0, & t > 0, \quad x \in \partial\Omega, \\ u(0,x) = u_0(x) \geq 0, & x \in \Omega. \end{cases} \tag{3.23}$$

A classical result of Matano (1979), Casten and Holland (1978) is that (3.23) has no stable nonconstant equilibrium solution provided that the domain Ω is convex. A direct consequence is that the reaction-diffusion equation (3.23) has same number of stable equilibrium solutions as the ODE $u' = f(u)$, hence diffusion does not induce "more"stability. However the geometry of the domain Ω is also an important factor in the stability problem. Matano (1979) shows that if $f(u)$ is of bistable type, say $f(u) = u(1 - u^2)$, then (3.23) has a stable nonconstant equilibrium solution if Ω is dumbbell-shaped, see also Alikakos, Fusco and Kowalczyk (1996) for more intricate results in that direction. Indeed it was recently shown that the geometry of the domain is even important for the magnitude of the first non-zero eigenvalue of Laplacian operator under Neumann boundary condition, see Ni and Wang (2007). The work of Matano (1979) has been extended to two species competition models (Matano and

Mimura (1983)) for nonconvex domains and to cooperative models (Kishimoto and Weinberger (1985)) for convex domains. More results on Neumann boundary value problems can be found in Ni (1989, 1998).

To summarize, we have examined the reaction-diffusion ecological models of bistability or hysteresis in this section. When the diffusion coefficient d is small, or equivalently the habitat is large, we show the existence of multiple spatial heterogeneous steady states, so that the system possesses alternative stable spatial equilibrium solutions. Moreover, even when the non-spatial model is not bistable, the reaction-diffusion model may be bistable as we show in the weak Allee effect or weak hysteresis case. Hence diffusion enhances the stability of certain states in such systems.

The bifurcation diagrams can also be explained with habitat size as the bifurcation parameter. Indeed habitat fragmentation has been identified as one of the possible causes of the regime shift in the ecosystems [122]. The results here provide theoretical evidence to support that claim via the reaction-diffusion model approach. Other mathematical approaches concerning the implications of spatial heterogeneity in the catastrophic regime shifts have been taken. van Nes and Scheffer (2005) investigated lattice models with same nonlinearities in (3.21) and (3.22), but their numerical bifurcation diagrams have r or a as bifurcation parameters, just as in the ODE models (see Fig. 3.2 and Fig. 3.4). Bascompte and Solé (1996, 2006) consider spatially explicit metapopulation models to show the existence of extinction thresholds when a given fraction of habitat is destroyed.

Another question is as follows. When the existence of multiple steady states indicates bistability, what is the global dynamics of the system? We present some mathematical results in that direction in the following section.

3.3 Threshold manifold

For an ordinary differential equation such as (3.2) with strong Allee effect, $u = M$ is a threshold point so that the extinction and persistence depends on whether the initial value $u_0 < M$ or $> M$. Bistable dynamics in higher dimensional systems are characterized by a separatrix or threshold manifold. Sometimes such dynamics is also called saddle point behavior (Capasso and Maddalena (1982), Capasso and Wilson (1997)). This can be illustrated by considering the classical Lotka-Volterra competition model (in nondimensionalized form):

$$u' = u(1 - u - Av), v' = v(B - Cu - v), \tag{3.24}$$

where $A, B, C > 0$ satisfy $C > B > A^{-1} > 0$. The system is bistable since it possesses two locally stable equilibrium points $(1, 0)$ and $(0, B)$, and a separatrix—the stable manifold of the unstable coexistence equilibrium $(u_*, v_*) = ((AB-1)/(AC-1), (C-B)/(AC-1))$, which separates the basins of attraction of two stable equilibria, see Fig. 3.9. We also note that (3.24) possesses another invariant manifold connecting $(1, 0)$, $(0, B)$ and (u_*, v_*), called carrying simplex, see more remarks about it in later part of this section.

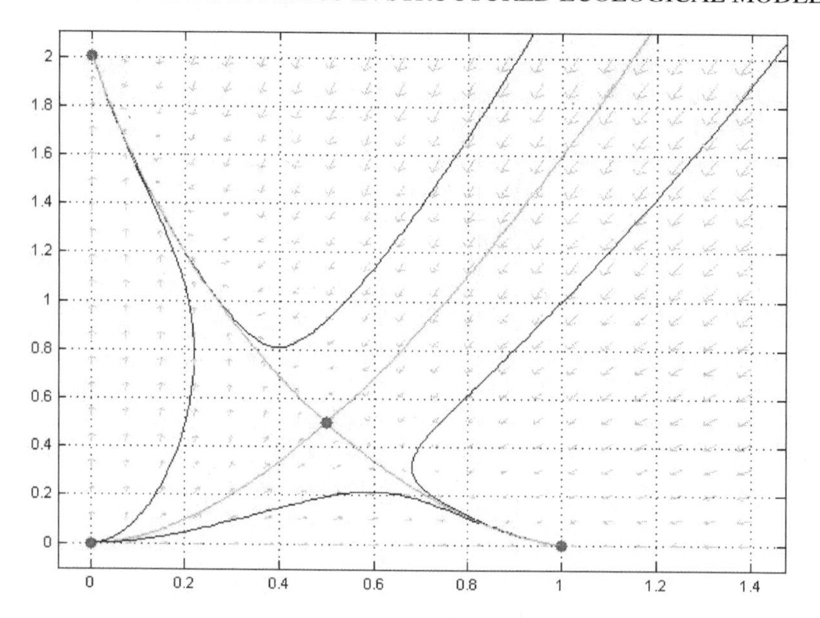

Figure 3.9 Phase portrait of the competition model (3.24). The stable manifold of (u_*, v_*) (connecting orbit from the origin) is the threshold manifold which separates the basins of attraction of two stable equilibria; and the unstable manifold of (u_*, v_*) (connecting orbits from stable equilibria) is the carrying simplex.

An abstract mathematical result about the threshold manifold has been recently given by Jiang, Liang and Zhao (2004). They prove that in a strongly order preserving or strongly monotone semiflow in a Banach space, if there are exactly two locally stable steady states, and any other possible steady state is unstable, then the set which separates the basins of attraction of two stable steady states is a codimension-one manifold (see more precise statement in Jiang et.al. (2004)). A scalar reaction-diffusion equation such as (3.10) and (3.11) generates a strongly monotone semi-flow in some function space. Thus this result is immediately applicable to the scalar reaction-diffusion equation. Hence the existence of a codimension-one manifold for the Nagumo equation or all examples discussed in Section 2 with exactly two stable steady state solutions follows from Jiang et.al. (2004). The existence of the threshold manifolds relies on earlier results of Takáč (1991, 1992). We also mention that the earliest example of threshold manifold was given by McKean and Moll (1986), and Moll and Rosencrans (1990) where the Nagumo equation

$$u_t = du_{xx} + u(a-u)(u-b), \quad x \in (0, L), \quad u(0) = u(L) = 0, \tag{3.25}$$

with $0 < b < a$, was considered. They also examined the case when the cubic function is replaced by a piecewise linear function, suggested by McKean (1970) as an alternative to the FitzHugh-Nagumo model. We remark that the existence of exactly two stable steady state solutions for (3.10) and (3.11) heavily depends on the geometry of the domain Ω. Most exact multiplicity results in Section 2 hold for the ball domains but not general bounded domain Ω, as shown by Dancer (1988) in the example of dumbbell shaped domains. A similar remark can be applied to Neumann boundary value problem (3.23). For the convex domains Ω, the bistable reaction-diffusion equation (3.23) with $f(u) = u(1-u^2)$ (Allen-Cahn equation from material science) has exactly two stable steady state solutions $u = \pm 1$ from the results of Casten and Hollnad (1978) and Matano (1979). Hence the existence of a threshold manifold follows from Jiang et.al. (2004). But for dumbbell shaped domain, it could have more stable steady state solutions from the result of Matano (1979).

The two locally stable equilibrium points in Jiang-Liang-Zhao's theorem can also be replaced by one locally stable steady state and "infinity" which is locally stable. An abstract formulation of this kind has been obtained in Lazzo and Schmidt (2005), but concrete examples have been given much earlier. For a matrix population model, Schreiber (2004) proved the existence of a threshold manifold that separates the initial values leading to extinction or unbounded growth. A more famous example in partial differential equations is the Fujita equation (Fujita (1966)):

$$u_t = d\Delta u + u^p, \quad x \in \mathbf{R}^n, \quad p > 1. \tag{3.26}$$

Fujita (1966) observed that for $p > (n+2)/(n-2)$ and $n \geq 3$, then the solution to (3.26) with certain initial values blows up in finite time, while some other solutions tend to zero as $t \to \infty$. Since the solution of the ordinary differential equation $u' = u^p$ with $p > 1$ always blows up, then the bistability in the Fujita equation is a combined effect of diffusion (stabilization) and growth (blow up). Aronson and Weinberger (1978) obtained some criteria on the extinction and blow-up of similar type equations, and they called the sensitivity of initial value between the extinction and blow-up the "hair-trigger effect". Mizoguchi (2002) proved the existence of the unique threshold between extinction and complete blow-up for radially symmetric compactly-supported initial values, although the existence of a threshold manifold cannot directly follow from Lazzo and Schmidt (2005) due to the lack of compactness when the domain is the whole space. Similar results have also been proved for bounded domain, see for example Ni, Sacks and Tavantzis (1984).

An intriguing question is whether such a precise bistable structure is still valid for systems of equations. When the system is still a monotone dynamical system, apparently this is true. For example, it holds for the reaction-diffusion counterpart of (3.24): the diffusive competition system with two competitors and no-flux boundary condition:

$$\begin{cases} u_t = d_u \Delta u + u(1 - u - Av), & t > 0, \ x \in \Omega, \\ v_t = d_v \Delta v + v(B - Cu - v), & t > 0, \ x \in \Omega, \\ \dfrac{\partial u}{\partial n} = \dfrac{\partial v}{\partial n} = 0, & t > 0, \ x \in \partial\Omega, \\ u(0,x) = u_0(x) \geq 0, \ \ v(0,x) = v_0(x) \geq 0, & x \in \Omega. \end{cases} \tag{3.27}$$

Here $d_u \geq 0$ and $d_v \geq 0$. The steady states of (3.24) are still (constant) equilibrium solutions of (3.27). Moreover it is known that any stable steady state of (3.27) is constant if Ω is convex from Kishimoto and Weinberger (1985). Thus a threshold manifold of codimension-one exists when Ω is convex following Jiang et.al. (2004) although the dynamics on the threshold is not clear. In a more general setting, Smith and Thieme (2001) studied abstract two species (u, v) competition systems with the origin being a repeller. Assuming that the unique nontrivial boundary steady state on each axis is stable and there is a unique positive steady state, they showed that there is an invariant threshold manifold through the positive steady state separating the attracting domains for both axis steady states. See Jiang and Liang (2006) and Castillo-Chavez, Huang and Li (1999) for more about threshold manifold of bistability in competition models. It should be noted that the results of Jiang et.al. (2004) are not valid for general competition systems with more than two competitors.

By way of contrast, for non-monotone dynamical systems, in general there is no such structure even with only two stable steady states. Some systems may however inherit threshold structure from their limiting systems or subsystems. Consider the reaction and diffusion of the two reactants A and B in an isothermal autocatalytic chemical reaction. We have the system

$$\begin{cases} a_t = D_A \Delta a - ab^p, \quad b_t = D_B \Delta b + ab^p, & t > 0, \ x \in \Omega, \\ a(x,t) = a_0 > 0, \quad b(x,t) = 0, & t > 0, \ x \in \partial\Omega, \\ a(x,0) = A_0(x) \geq 0, \quad b(x,0) = B_0(x) \geq 0, & x \in \Omega. \end{cases} \tag{3.28}$$

where a and b are the concentrations of the reactant A and the autocatalyst B, $p > 1$, D_A and D_B are the diffusion coefficients of A and B respectively, and Ω is a bounded reaction zone in $\mathbf{R}^n$ (Gray and Scott (1990)). It is known that when reactor Ω is a ball in $\mathbf{R}^n$, (3.28) has either only the trivial steady state $(a_0, 0)$, or exactly three non-negative steady state solutions with two of them stable. Under the additional assumption of equal diffusion coefficients ($D_A = D_B$), Jiang and Shi (2008) shown that in the latter case, the global stable manifold for the intermediate steady state (a_2, b_2) is a codimension-one manifold which separates the basin of attraction of the two stable steady states, and moreover every solution converges to one of three steady state solutions. Here we use the fact that the asymptotic limit of (3.28) is an autonomous scalar reaction-diffusion equation, which is a monotone dynamical system, see Chen and Poláčik (1995), Mischaikow, Smith and Thieme (1995). Although rather special, this is a rare example where the complete dynamics is known for a non-monotone dynamical system in infinite dimensional space. A different bistability result for (3.28) in $\mathbf{R}^n$ is also obtained in Shi and Wang (2006) which uses some ideas from Aronson and Weinberger (1978).

Capasso and Wilson (1997) analyzed the spread of infectious diseases with a reaction-diffusion system:

$$\begin{cases} u_{1t} = d\Delta u_1 - a_{11}u_1 + a_{12}u_2, & t > 0, \ x \in \Omega, \\ u_{2t} = -a_{22}u_2 + g(u_1), & t > 0, \ x \in \Omega, \\ u_1(x,t) = u_2(x,t) = 0, & t > 0, \ x \in \partial\Omega, \\ u_1(x,0) = U_1(x) \geq 0, \quad u_2(x,0) = U_2(x) \geq 0, & x \in \Omega. \end{cases} \tag{3.29}$$

This system models random dispersal of a pollutant while ignoring the small mobility of the infective human population. Here $u_1(x,t)$ denotes the spatial density of the pollutant, and $u_2(x,t)$ denotes the density of the infective human population. With $g(u)$ being the monotone sigmoid function discussed in Section 1, the steady state equation can be reduced to

$$d\Delta u_1 - a_{11}u_1 + \frac{a_{12}}{a_{22}}g(u_1) = 0, \quad x \in \Omega, \quad u_1 = 0, \quad x \in \partial\Omega. \tag{3.30}$$

The nonlinearity here $f(u_1) = -a_{11}u_1 + \frac{a_{12}}{a_{22}}g(u_1)$ is of strong Allee effect using the term introduced in the last subsection. Hence under some reasonable conditions and Ω being a ball, the bifurcation diagram of (3.30) is the one in Fig.3.7-c. This is shown in Capasso and Wilson (1997) for the case of $n = 1$, and the general case when $n \geq 2$ can be deduced from the results in Ouyang and Shi (1998). Since (3.29) is a monotone dynamical system, then again (3.29) admits a codimension-one manifold which separates the basin of attraction of the two stable steady states (Jiang et.al. (2004)), which confirms the conjecture in Capasso and Wilson (1997). But it is still not known that whether every solution on the threshold manifold converges to the intermediate steady state solution.

Even less is known about the dynamical behavior of FitzHugh-Nagumo system:

$$\begin{cases} \epsilon v_t = d_v\Delta v + v(v-a)(1-v) - w, & t > 0, \ x \in \Omega, \\ w_t = d_w\Delta w + cv - bw, & t > 0, \ x \in \Omega, \\ v(x,t) = w(x,t) = 0, & t > 0, \ x \in \partial\Omega, \\ v(x,0) = V(x) \geq 0, \ \ w(x,0) = W(x) \geq 0, & x \in \Omega. \end{cases} \tag{3.31}$$

Here $d_v > 0$ and $d_w \geq 0$. When $c = 0$, it follows that $w \to 0$, and the dynamics of (3.31) is reduced to that of Nagumo equation (3.25) (in higher dimensional domain). Since (3.25) has the saddle point behavior, then (3.31) still possesses this saddle point behavior for $0 < c \ll 1$ by structural stability theory. For more general parameter ranges, the existence of multiple positive steady state solutions of (3.31) is known, see for example Matsuzawa (2005) for a nice summary. Notice that (3.31) is not a monotone dynamical system, so even the information of stable steady state solutions cannot imply the saddle point behavior.

Threshold manifolds are a class of invariant manifolds in applied dynamical systems, and they are sensitively unstable in the dynamic sense as a small perturbation will shift it to the basin of attraction of a stable equilibrium. If one reverses the time t to $-t$ to a system with threshold manifold, then the manifold becomes an attracting manifold, or vice versa. For example, in the logistic model (3.1), if time is reversed, then it has the exactly same dynamical behavior as Fujita equation or the abstract formulation in Lazzo and Schmidt (2005): both the origin and the infinity are stable and the carrying capacity N becomes a threshold point. Similarly, if one reverses the time in the classical Lotka-Volterra competition system (3.24) without diffusion, then the origin and the infinity become stable, and there is a threshold manifold containing the boundary steady state $(1,0)$, $(0,B)$ and coexistence steady state on which "hair-trigger effect" occurs, which is deduced from Hirsch (1988) or an analysis for phase pictures. Of course it is not realistic to reverse the time in logistic model or Lotka-Volterra competition system. Nevertheless, in logistic model (3.1) or Lotka-Volterra system (3.24), both the origin and the infinity are repellers, and there is a threshold manifold separating the repelling domains for the origin and the infinity. Such a threshold manifold plays the role of carrying capacity in the logistic model, so it is often called *Carrying Simplex*.

The first example of a carrying simplex was given by Hirsch (1988) in his seminal paper. For a dissipative and strongly competitive Kolmogorov system:

$$\frac{dx_i}{dt} = x_i F_i(x_1, x_2, \cdots, x_n), \quad x_i \geq 0, \quad i = 1, 2, \cdots, n, \tag{3.32}$$

Hirsch (1988) proved that if the origin is a repeller, then there exists a carrying simplex which attracts all nontrivial orbits for (3.32) and it is homeomorphic to probability simplex by radial projection. Note that dissipation implies that the infinity is also a repeller.

Smith (1986) investigated C^2 diffeomorphisms T on the nonnegative orthant K which possesses the properties (see the hypotheses in Smith (1986)) of the Poincaré map induced by C^2 strong competition system

$$\frac{dx_i}{dt} = x_i F_i(t; x_1, x_2, \cdots, x_n), \quad x_i \geq 0, \quad i = 1, 2, \cdots, n, \tag{3.33}$$

where F_i is 2π-periodic in t, $F_i(t;0) > 0$, and (3.33) has a globally attracting 2π-periodic solution on each positive coordinate axis. This implies that the origin is a repeller for T and it has a global attractor Γ. He proved that the boundaries of the repulsion domain of the origin and the global attractor relative to the nonnegative orthant are a compact unordered invariant set homeomorphic to the probability simplex by radical projection. He conjectured both boundaries coincide, serving as a unique carrying simplex. Introducing a mild additional restriction on T, which is generically satisfied by the Poincaré map of the competitive Kolmogorov system (3.33), Wang and Jiang (2002) proved this conjecture and that the unstable manifold of $m-$periodic point of T is contained in this carrying simplex. Diekmann, Wang and Yan (2008) have showed the same result holds by dropping one of the hypotheses in Smith's original conjecture so that the result is easier to use in the setting of competitive mappings. Hirsch (2008) introduces a new condition—strict sublinearity in a neighborhood of the global attractor, to give a new existence criterion for the unique carrying simplex. The uniqueness of the carrying simplex is important in classifying the dynamics of lower dimensional competitive systems, for example the 3-dimensional Lokta-Volterra competition system (Zeeman (1993)). The classification of many three dimensional competitive mappings (see Davydova, Diekmann and van Gils (2005a, 2005b), Hirsch (2008) and references therein) are still open, and the uniqueness of the carrying simplex is one of the reasons.

Note that if one reverses the time t to $-t$ in the n-dimensional competition system (3.32), then the system becomes a monotone system with both the origin and the infinity stable (under the assumption that the origin and the infinity are repellers). However this new system is not strongly monotone as required in Jiang et.al. (2004) and Lazzo and Schmidt (2005). Thus the existence of the carrying simplex cannot follow from Jiang et.al. (2004) and Lazzo and Schmidt (2005) except in the case of

$n = 2$. Indeed this is one of the main difficulties in Hirsch (1988), Wang and Jiang (2002), and Diekmann, Wang and Yan (2008).

We conclude our discussion of threshold manifolds with a model of biochemical feedback control circuits. More details on the modeling can be found in, for example, Murray (2003) or Smith (1995). A segment of DNA is assumed to be translated to mRNA which in turn is translated to produce an enzyme and it in turn is translated to another enzyme and so on until an end product molecule is produced. This end product acts on a nearby segment of DNA to produce a feedback loop, controlling the translation of DNA to mRNA. Let x_1 be the cellular concentration of mRNA, let x_2 be the concentration of the first enzyme, and so on, finally let x_n be the concentration of their substrate. Then this biochemical control circuit is described by the system of equations

$$x_1{}' = g(x_n) - \alpha_1 x_1, \quad x_i{}' = x_{i-1} - \alpha_i x_i, \quad 2 \le i \le n, \tag{3.34}$$

where $\alpha_i > 0$ and the feedback function $g(u)$ is a bounded continuously differentiable function satisfying

$$0 < g(u) < M, \; g'(u) > 0, \; u > 0. \tag{3.35}$$

Hence it models a positive feedback. For the Griffith model (Griffith (1968)) we have

$$g(x_n) = \frac{x_n^p}{1 + x_n^p} \tag{3.36}$$

where p is a positive integer (the Hill coefficient). For the Tyson-Othmer model (Tyson and Othmer (1978)) we have

$$g(x_n) = \frac{1 + x_n^p}{K + x_n^p} \tag{3.37}$$

where p is a positive integer and $K > 1$. The solution flow for (3.34) is strongly monotone (see Smith (1995) for detail). The steady states for (3.34) are in one-to-one correspondence with solutions of

$$g(u) = \alpha u \tag{3.38}$$

where $\alpha = \prod \alpha_i$. Suppose that the line $v = \alpha u$ intersects the curve $v = g(u)$ $(u \ge 0)$ transversally. Then every non-negative steady state for (3.34) is hyperbolic, which implies that the number of steady states for (3.34) is odd for either the Griffith or Tyson-Othmer model. For most of biological parameters in the Griffith or Tyson-Othmer model, there are exactly three steady states (Selgrade (1979, 1980, 1982) and Jiang (1992, 1994)). In this case, the least steady state and the greatest steady state are asymptotically stable and intermediate one is a saddle point through which there is an invariant threshold manifold whose norm is positive. In the multistable case, there are $\left[\frac{n-1}{2}\right]$ invariant threshold manifolds which separate the attracting domains for stable steady states (see Jiang et.al. (2004)). From a general result of Mallet-Paret and Smith (1990), we know that on each invariant threshold manifold every orbit either converges to the saddle point or is asymptotic to a nontrivial unstable periodic orbit. For $n \le 3$, all orbits tend to the corresponding saddle point on threshold manifolds, which was proved by using topological arguments in Selgrade (1979,1980), the Dulac criterion for 3-dimensional cooperative system in Hirsch (1989) and a Lyapunov function in Jiang (1992); for $n \geq 5$, in the bistable case for the Griffith or Tyson-Othmer model, there may exist Hopf bifurcation on the unique threshold manifold (see Selgrade (1982)). But for $n = 4$, whether there is a nontrivial periodic orbit or not on threshold manifold is an open problem. In Jiang (1994), it was proved that for 4-dimensional Griffith or Tyson-Othmer model all orbits are convergent to a steady state via Lyapunov method for parameters with biological significance.

Hetzer and Shen (2005) added a third equation to the classical Lotka-Volterra equations for two competing species, which describes explicitly the evolution of toxin, called an inhibitor. The equations in rescaled form are

$$\begin{cases} \dot{u} = u(1 - u - d_1 v - d_2 w), \\ \dot{v} = \rho v(1 - fu - v), \\ \dot{w} = v - (g_1 u + g_2) w, \end{cases} \tag{3.39}$$

where $d_1, d_2, \rho, f, g_1, g_2 > 0$. Note that $O(0,0,0)$, $E_x(1,0,0)$, and $E_y(0,1,g_2^{-1})$ are non-negative steady states of (3.39). Observing that O is a saddle, not a repeller, Hetzer and Shen (2005) studied the long-time behavior for (3.39) and the existence of threshold manifold in the bistable case, where they called a "thin separatrix" following Hsu, Smith and Waltman (1996), Smith and Thieme (2001). Jiang and Tang (2008) gave a complete classification for dynamical behavior for (3.39) and proved that the bistability occurs if and only if

$$a^* > 0,\ b^* < 0,\ c^* > 0,\ \Delta^* = (b^*)^2 - 4a^*c^* > 0, 2a^* + b^* > 0,\ a^* + b^* + c^* > 0, \tag{3.40}$$

where a^*, b^*, c^* are given by

$$a^* = g_1(1 - d_1 f),\ \ c^* = g_2(d_1 + \frac{d_2}{g_2} - 1), \tag{3.41}$$

and

$$a^* + b^* + c^* = (1 - f)(d_1 g_1 + d_1 g_2 + d_2). \tag{3.42}$$

In this case the system (3.39) has exactly two hyperbolic positive steady states, one of which is stable, denoted by E^*, while the other is a saddle point, denoted by E_*. (3.39) has exactly two stable steady states E_y and E^*. The stable manifold for the saddle point E_*, which is a 2-dimensional smooth surface, separates the basins of attraction for E_y and E^*. Hence this smooth surface is a threshold manifold.

The production of the various proteins in the biochemical control circuit model (3.34) is, of course, not instantaneous and it is reasonable to introduce time delays into these terms. If one does so, (3.34) becomes a delay differential equation:

$$x_1{}' = g(x_n(t - r_n)) - \alpha_1 x_1,\ \ x_i{}' = x_{i-1}(t - r_{j-1}) - \alpha_i x_i,\ \ 2 \leq i \leq n, \tag{3.43}$$

with all delays r_i positive. It is easy to see that all steady states for (3.43) are the same as (3.34) and if a steady state for (3.34) is linearly stable (unstable) then it is also linearly stable (unstable) for (3.43) (Smith (1995) p.111). Thus in the bistable case for

(3.43), there is a codimension-one threshold manifold through a saddle point separating the attracting domains for the two steady states. The only difference is that such a threshold manifold in the space of continuous functions is infinite dimensional and less information is known for the dynamics on the threshold manifold. The results are similar for the multistable case (see Jiang et.al. (2004)). Of course another way to have an infinite dimensional threshold manifold is to add diffusion to bistable (multistable) monotone ODEs or FDEs with no-flux boundary condition on a smooth and convex domain, so that codimension-one threshold manifolds still exist (see Jiang et.al. (2004)).

3.4 Concluding Remarks

Sharp regime shifts occur in some large-scale ecosystems such as lakes, coral reefs, grazed grasslands and forests. Mathematical models have been set up to explain the sudden changes and hysteresis cycles in these systems. In this article, we review some of these models with a focus on the impact of spatial dispersal and habitat fragmentation. The rich dynamics of these problems share some common mathematical features such as multiple steady states, threshold manifold (separatrix), and non-monotone bifurcation diagrams. Mathematical tools from partial differential equations, bifurcation theory, and monotone dynamical systems have been applied and further developed in studying these important problems rooted from various applied areas.

Establishing the basic structure of multiple steady states and threshold manifold is the first step in a complete understanding of the bistable dynamics, regime shifts and ecosystems resilience. The dynamics on the separatrix could be very complicated, and there is also evidence that bistability in a reaction-diffusion predator-prey system could imply existence of more complex patterns (see Morozov, Petrovskii and Li (2004,2006), Petrovskii, Morozov and Li (2005)). Another important question is how to make early warning of the regime shifts. The bifurcation diagrams suggest that the regime shifts occur at saddle-node bifurcation points, at which the largest eigenvalue (principal eigenvalue) of the linearized system is zero. Near bifurcation points, the principal eigenvalue is small. It has been recognized that the principal eigenvalue at a steady state is related to the return time, which is another definition of resilience of the system (see Pimm (1991)). The return time is how fast a variable that has been displaced from equilibrium returns to it. For the dynamical models described here, such return time to the equilibrium is characterized by $\exp(\lambda_1 t)$, where λ_1 is the principal eigenvalue at the equilibrium. Hence early warning for regime shifts in large scale could be triggered by a change in return time, provided that information on the return time is obtained from small scale experiments.

3.5 Acknowledgements

J.S. would like to thank Steve Cantrell, Chris Cosner and Shigui Ruan for the invitation to give a lecture at the Workshop on Spatial Ecology: The Interplay between

Theory and Data, University of Miami, Jan. 2005, and this article is partially based on that lecture. Part of this work was done when the authors visited National Tsinghua University in Dec. 2007, and they would like to thank Sze-Bi Hsu and Shin-Hwa Wang for their hospitality. The authors also thank the anonymous referee for many helpful comments and suggestions which improved the earlier version of the manuscript, and they also thank the editors for careful reading of the manuscript. J.J. is supported by Chinese NNSF grants 10671143 and 10531030, and J.S. is supported by United States NSF grants DMS-0314736, DMS-0703532, EF-0436318, Chinese NNSF grant 10671049, and Longjiang scholar grant.

3.6 References

N.D. Alikakos, G. Fusco, and M. Kowalczyk (1996), Finite-dimensional dynamics and interfaces intersecting the boundary: Equilibria and quasi-invariant manifold, *Indiana Univ. Math. J.* **45**(4): 1119–1155.

W. C. Allee (1938), *The Social Life of Animals*, W.W Norton, New York.

H. Amann (1976), Fixed point equations and nonlinear eigenvalue problems in ordered Banach space, *SIAM Review* **18**: 620–709.

D. G. Aronson and H. F. Weinberger (1978), Multidimensional nonlinear diffusion arising in population genetics, *Adv. in Math.* **30**(1): 33–76.

J. Bascompte and R. V. Solé (1996), Habitat fragmentation and extinction thresholds in spatially explicit models, *J. Anim. Ecol.* **65**(4): 465–473.

B. E. Beisner, D. T. Haydon and K. Cuddington (2003), Alternative stable states in ecology, *Frontiers in Ecology and the Environment* **1**(7): 376–382.

E. Bradford and J. P. Philip (1970a), Stability of steady distributions of asocial populations dispersing in one dimension, *J. Theor. Biol.* **29**(1): 13–26.

E. Bradford and J. P. Philip (1970b), Note on asocial populations dispersing in two dimensions, *J. Theor. Biol.* **29**(1): 27–33.

K. J. Brown, M.M.A. Ibrahim, and R. Shivaji (1981), S-shaped bifurcation curves, *Nonlinear Anal.* **5** (5): 475–486.

R. S. Cantrell and C. Cosner (1989), Diffusive logistic equations with indefinite weights: Population models in disrupted environments, *Proc. Roy. Soc. Edinburgh Sect. A* **112**(3/4): 293–318.

R. S. Cantrell and C. Cosner (2003a), Conditional persistence in logistic models via nonlinear diffusion, *Proc. Roy. Soc. Edinburgh Sect. A* **132**(2): 267–281.

R. S. Cantrell and C. Cosner (2003b), *Spatial Ecology via Reaction-Diffusion Equation*, Wiley Series in Mathematical and Computational Biology, John Wiley & Sons Ltd.

S. R. Carpenter, D. Ludwig, and W. A. Brock (1999), Management of eutrophication for lakes subject to potentially irreversible change, *Ecol. Appl.* **9**: 751–771.

C. Castillo-Chavez, W. Huang, and J. Li (1999), Competitive exclusion and coexistence of multiple strains in an SIS STD model, *SIAM J. Appl. Math.* **59**: 1790–1811.

V. Capasso and L. Maddalena (1981/82), Convergence to equilibrium states for a reaction-diffusion system modelling the spatial spread of a class of bacterial and viral diseases, *J. Math. Biol.* **13**(2): 173–184.

V. Capasso and L. Maddalena (1982), Saddle point behavior for a reaction-diffusion system: application to a class of epidemic models, *Math. Comput. Simulation* **24**(6): 540–547.

V. Capasso and R. E. Wilson (1997), Analysis of a reaction-diffusion system modeling man-environment-man epidemics, *SIAM J. Appl. Math.* **57**(2): 327–346.

R.G. Casten and C. J. Holland (1978), Instability results for reaction diffusion equations with Neumann boundary conditions, *J. Differential Equations* **27**(2): 266–273.

D. S. Cohen and T. W. Laetsch (1970), Nonlinear boundary value problems suggested by chemical reactor theory, *J. Differential Equations* **7**: 217–226.

X.-Y. Chen and P. Poláčik (1995), Gradient-like structure and Morse decompositions for time-periodic one-dimensional parabolic equations, *J. Dynam. Differential Equations* **7**: 73–107.

P. T. Church and J. G. Timourian (1997), Global structure for nonlinear operators in differential and integral equations. I. Folds; II. Cusps, in *"Topological Nonlinear Analysis"* (Frascati, 1995), Progr. Nonlinear Differential Equations Appl. **27**, Birkhäüser, Boston, MA, pp. 109–160; pp. 161–245.

C. W. Clark (1991), *Mathematical Bioeconomics: The Optimal Management of Renewable Resources*, John Wiley & Sons, Inc. New York.

E. D. Conway (1983), Diffusion and predator-prey interaction: Steady states with flux at the boundaries, *Contemporary Mathematics* **17**: 217–234.

E. D. Conway (1984), Diffusion and predator-prey interaction: Pattern in closed systems, in *"Partial Differential Equations and Dynamical Systems"*, Res. Notes in Math. **101**, Pitman, Boston-London, pp. 85–133.

E.N. Dancer (1988), The effect of domain shape on the number of positive solutions of certain nonlinear equations, *J. Differential Equations* **74**(1): 120–156.

B. Dennis (1989), Allee effects: population growth, critical density, and the chance of extinction, *Natur. Resource Modeling* **3**(4): 481–538.

N. V. Davydova, O. Diekmann, and S. van Gils (2005), On circulant populations. I. The algebra of semelparity, *Linear Alg. Appl.* **398**: 185–243.

O. Diekmann, N. Davydova, and S. van Gils (2005), On a boom and bust year class cycle, *J. Difference Equ. Appl.* **11**(4/5): 327–335.

O. Diekmann, Y. Wang, and P. Yan (2008), Carrying simplices in discrete competitive systems and age-structured semelparous populations, *Discrete Contin. Dyn. Syst. A* **20**(1): 37–52.

A. M. de Roos, E. McCawley, and W. G. Wilson (1998), Pattern formation and the spatial scale of interaction between predators and their prey, *Theo. Popu. Biol* **53**: 108–130.

Y. Du and Y. Lou (2001), Proof of a conjecture for the perturbed Gelfand equation from combustion theory, *J. Differential Equations* **173**(2): 213–230.

R. A. Fisher (1937), The wave of advance of advantageous genes, *Ann. Eugenics* **7**: 353–369.

R. FitzHugh (1961), Impulses and physiological states in theoretical models of nerve membrane, *Biophys. J.* **1**: 445–466.

C. Folke, S. Carpenter, B. Walker, M. Scheffer, T. Elmqvist, L. Gunderson, and C. S. Holling (2004), Regime shifts, resilience and biodiversity in ecosystem management, *Ann. Rev. Ecol. Evol. Syst.* **35**: 557–581.

H. Fujita (1966), On the blowing up of solutions of the Cauchy problem for $u_t = \Delta u + u^{1+\alpha}$, *J. Fac. Sci. Univ. Tokyo Sect. I* **13**: 109–124.

J. S. Griffith (1968), Mathematics of cellular control processes, II: Positive feedback to one gene, *J. Theor. Biol.* **20**: 209–216.

P. Gray and S. K. Scott (1990), *Chemical Oscillations and Instabilies: Nonlinear Chemical Kinectics*, Clarendon Press, Oxford.

M. J. Groom (1998), Allee effects limit population viability of an annual plant, *Amer. Naturalist* **151**: 487–496.

D. Henry (1981), *Geometric Theory of Semilinear Parabolic Equations*, Lecture Notes in Mathematics **840**. Springer-Verlag, Berlin-New York.

G. Hetzer and W. Shen (2005), Two species competition with an inhibitor involved, *Discrete Contin. Dyn. Syst. A* **12**: 39–57.

M. W. Hirsch (1988), Systems of differential equations that are competitive or cooperative. III: Competing species, *Nonlinearity* **1**: 51–71.

M. W. Hirsch (1989), Systems of differential equations that are competitive or cooperative. V: Convergence in 3-dimensional systems, *J. Differential Equations* **80**: 94–106.

M. W. Hirsch (2008), On existence and uniqueness of the carrying simplex for competitive dynamical systems, *J. Biol. Dyna.* **2**(2): 169–179.

C. S. Holling (1959), The components of predation as revealed by a study of small mammal predation of the European Pine Sawfly, *Canadian Entomologist* **91**: 293–320.

C. S. Holling (1973), Resilience and stability of ecological systems, *Ann. Rev. Ecol. Syst.* **4**: 1–23.

F. A. Hopf and F. W. Hopf (1985), The role of the Allee effect in species packing, *Theo. Popu. Biol.* **27**(1): 27–50.

S.-B. Hsu, H. L. Smith, and P. Waltman (1996), Competitive exclusion and coexistence for competitive systems on ordered Banach spaces, *Trans. Amer. Math. Soc.* **348**: 4083–4094.

J. Jiang (1992), A Liapunov function for 3-dimensional positive feedback systems, *Proc. Amer. Math. Soc.* **114**: 1009–1013.

J. Jiang (1994), A Liapunov function for four-dimensional positive feedback systems, *Quart. Appl. Math.* **LII**: 601–614.

J. Jiang and X. Liang (2006), Competitive systems with migration and Poincare-Bendixson theorem for a 4-dimensional case, *Quart. Appl. Math.* **64**(3): 483–498.

J. Jiang, X. Liang, and X.-Q. Zhao (2004), Saddle-point behavior for monotone semiflows and reaction-diffusion models, *J. Differential Equations* **203**(2): 313–330.

J. Jiang and J. Shi (2008), Dynamics of a reaction-diffusion system of autocatalytic chemical reaction, *Discrete Contin. Dyn. Syst. A* **21**(1): 245–258.

J. Jiang and F. Tang (2008), The complete classification on a model of two species competition with an inhibitor, *Discrete Contin. Dyn. Syst. A* **20**(3): 650–672.

C. A. Klausmeier (1999), Regular and irregular patterns in semiarid vegetation, *Science* **284**: 1826–1828.

K. Kishimoto and H. F. Weinberger (1985), The spatial homogeneity of stable equilibria of some reaction-diffusion systems on convex domains, *J. Differential Equations* **58**(1): 15–21.

A. Kolmogoroff, I. Petrovsky, and N. Piscounoff (1937), Study of the diffusion equation with growth of the quantity of matter and its application to a biological problem (in French), *Moscow Univ. Bull. Math.* **1**: 1–25.

P. Korman and Y. Li (1999), On the exactness of an S-shaped bifurcation curve, *Proc. Amer. Math. Soc.* **127**(4): 1011–1020.

P. Korman and J. Shi (2001), New exact multiplicity results with an application to a population model, *Proc. Roy. Soc. Edinburgh Sect. A* **131**(5): 1167–1182.

M. Lazzo and P. G. Schmidt (2005), Monotone local semiflows with saddle-point dynamics and applications to semilinear diffusion equations, *Discrete Contin. Dyn. Syst.* (suppl.), pp. 566–575.

Y. H. Lee, L. Sherbakov, J. G. Taber,and J. Shi (2006), Bifurcation diagrams of population models with nonlinear diffusion, *J. Comp. Appl. Math.* **194**(2): 357–367.

M. A. Lewis and P. Kareiva (1993), Allee dynamics and the spread of invading organisms, *Theo. Popu. Biol.* **43**: 141–158.

P.-L. Lions (1982), On the existence of positive solutions of semilinear elliptic equations, *SIAM Review* **24**(4): 441–467.

P. Liu, J. Shi, and Y. Wang (2007), Imperfect bifurcation with weak transversality, *J. Func. Anal.* **251**(2): 573–600.

A. J. Lotka (1920), Analytical note on certain rhythmic relations in organic systems, *Proc. Natl. Acad. Sci. USA.* **6**(7): 410-415.

D. Ludwig, D. G. Aronson, and H. F. Weinberger (1978), Spatial patterning of the spruce budworm, *J. Math. Biol.* **8**(3): 217–258.

D. Ludwig, D. Jones, and C. S. Holling (1978), Qualitative analysis of insect outbreak systems: The spruce budworm and the forest, *J. Ani. Ecol.* **47**: 315-332.

J. Mallet-Paret and H. L. Smith (1990), The Poincaré-Bendixson theorem for monotone cyclic feedback systems, *J. Dynamics Diff. Eqns.* **2**: 367–421.

H. Matano (1979), Asymptotic behavior and stability of solutions of semilinear diffusion equations, *Publ. Res. Inst. Math. Sci.* **15**(2): 401–454.

H. Matano and M. Mimura (1983), Pattern formation in competition-diffusion systems in nonconvex domains, *Publ. Res. Inst. Math. Sci.* **19**(3): 1049–1079.

H. Matsuzawa (2005), Asymptotic profiles of variational solutions for a FitzHugh Nagumo-type elliptic system, *Nonlinear Analysis* **63**(5-7): e2545–e2551.

R. M. May (1977), Thresholds and breakpoints in ecosystems with a multiplicity of stable states, *Nature* **269**: 471–477.

H. P. McKean (1970), Nagumo's equation, *Adv. in Math.* **4**: 209–223.

H. P. McKean and V. Moll (1986), Stabilization to the standing wave in a simple caricature of the nerve equation, *Comm. Pure Appl. Math.* **39**(4): 485–529.

K. Mischaikow, H. L. Smith, and H. R. Thieme (1995), Asymptotically autonomous semiflows: Chain recurrence and Lyapunov functions, *Trans. Amer. Math. Soc.* **347**: 1669–1685.

N. Mizoguchi (2002), On the behavior of solutions for a semilinear parabolic equation with supercritical nonlinearity, *Math. Z.* **239**(2): 215–229.

V. Moll and S. I. Rosencrans (1990), Calculation of the threshold surface for nerve equations, *SIAM J. Appl. Math.* **50**(5): 1419–1441.

A. Morozov, S. Petrovskii, and B.-L. Li (2004), Bifurcations and chaos in a predator-prey system with the Allee effect, *Proc. R. Soc. Lond. B* **271**: 1407–1414,.

A. Morozov, S. Petrovskii, and B.-L. Li (2006), Spatiotemporal complexity of patchy invasion in a predator-prey system with the Allee effect, *J. Theoret. Biol.* **238**(1): 18–35.

J. D. Murray (2003), *Mathematical Biology. I. An Introduction.* Interdisciplinary Applied Mathematics **17**; *II. Spatial Models and Biomedical Applications,* Interdisciplinary Applied Mathematics **18**, Springer-Verlag, New York.

J. Nagumo, S. Arimoto, and S. Yoshizawa (1962), An active pulse transmission line simulating nerve axon, *Proc IRE.* **50**: 2061-2070.

W.-M. Ni (1989), *Lectures on Semilinear Elliptic Equations*, National Tsing Hua University, Taiwan.

W.-M. Ni (1998), Diffusion, cross-diffusion, and their spike-layer steady states, *Notices Amer. Math. Soc.* **45**(1): 9–18.

W.-M. Ni, P. E. Sacks, and J. Tavantzis (1984), On the asymptotic behavior of solutions of certain quasilinear parabolic equations, *J. Differential Equations* **54**(1): 97–120.

W.-M. Ni and X. Wang (2007), On the first positive Neumann eigenvalue, *Discrete Contin. Dyn. Syst.* **17**(1): 1–19.

I. Noy-Meir (1975), Stability of grazing systems: An application of predator-prey graphs, *J. Ecol.* **63**(2): 459–481.

A. Okubo and S. A. Levin (2001), *Diffusion and Ecological Problems: Modern Perspectives,* Second edition, Interdisciplinary Applied Mathematics **14**, Springer-Verlag, New York.

T. Ouyang and J. Shi (1998), Exact multiplicity of positive solutions for a class of semilinear problem, *J. Differential Equations* **146**(1): 121–156.

T. Ouyang and J. Shi (1999), Exact multiplicity of positive solutions for a class of semilinear problem: II. *J. Differential Equations* **158**(1): 94–151.

S. Petrovskii, A. Morozov, and B.-L. Li (2005), Regimes of biological invasion in a predator-prey system with the Allee effect, *Bull. Math. Biol.* **67**(3): 637–661.

D.P.C. Peters, R. A. Pielke et. al. (2004), Cross-scale interactions, nonlinearities, and forecasting catastrophic events. *Proc. Nati. Acad. Soc. USA* **101**(42): 15130–15135.

J. R. Philip (1957), Sociality and sparse populations, *Ecology* **38**: 107–111.

S. L. Pimm (1991), *The Balance of Nature? Ecological Issues in the Conservation of Species and Communities,* University of Chicago Press, Chicago.

P. Polácik (2002), Parabolic equations: Asymptotic behavior and dynamics on invariant manifolds, *Handbook of Dynamical Systems* Vol. 2, North-Holland, Amsterdam, pp. 835–883.

P. H. Rabinowitz (1973/74), Variational methods for nonlinear elliptic eigenvalue problems, *Indiana Univ. Math. J.* **23**: 729–754.

M. Rietkerk, S. C. Dekker, P. C. de Ruiter, and J. van de Koppel (2004), Self-organized patchiness and catastrophic shifts in ecosystems, *Science* **305**(5692): 1926–1929.

M. L. Rosenzweig (1971), Paradox of enrichment: Destabilization of exploitation ecosystems in ecological time, *Science* **171**(3969): 385–387.

M. L. Rosenzweig and R. MacArthur (1963), Graphical representation and stability conditions of predator-prey interactions, *Amer. Natur.* **97**: 209–223.

M. Scheffer and S. R. Carpenter (2003), Catastrophic regime shifts in ecosystems: Linking theory to observation, *Trends Ecol. Evol.* **18**: 648–656.

M. Scheffer, M., S. Carpenter, J. A. Foley, C. Folke, and B. Walkerk (2001), Catastrophic shifts in ecosystems, *Nature* **413**: 591–596.

S. J. Schreiber (2004), On Allee effects in structured populations, *Proc. Amer. Math. Soc.* **132**(10): 3047–3053.

J. F. Selgrade (1979), Mathematical analysis of a cellular control process with positive feedback, *SIAM J. Appl. Math.* **36**: 219–229.

J. F. Selgrade (1980), Asymptotical behavior of solutions to single loop positive feedback systems, *J. Differential Equations* **38**: 80–103.

J. F. Selgrade (1982), A Hopf bifurcation in single-loop positive feedback systems, *Quart. Appl. Math.* **40**: 347–351.

J. G. Skellam (1951), Random dispersal in theoritical populations, *Biometrika* **38**: 196–218.

J. Shi (1999), Persistence and bifurcation of begenerate solutions, *J. Func. Anal.* **169**(2): 494–531.

J. Shi (2009), *Solution Set of Semilinear Elliptic Equations: Global Bifurcation and Exact Multiplicity,* World Scientific Publishing Co.

J. Shi and R. Shivaji (2006), Persistence in reaction diffusion models with weak Allee effect, *J. Math. Biol.* **52**(6): 807–829.

J. Shi and X. Wang (2006), Hair-triggered instability of radial steady states, spread and extinction in semilinear heat equations, *J. Differential Equations* **231**(1): 235–251.

H. L. Smith (1986), Periodic competitive differential equations and the discrete dynamics of competitive maps, *J. Differential Equations* **64**: 165–194.

H. L. Smith (1995), *Monotone Dynamical Systems: An Introduction to the Theory of Competitive and Cooperative Systems,* Mathematical Surveys and Monographs **41**, American Mathematical Society, Providence, RI.

H. L. Smith and H. R. Thieme (2001), Stable coexistence and bi-stability for competitive systems on ordered Banach space, *J. Differential Equations* **176**: 195–222.

J. Smoller and A. Wasserman (1981), Global bifurcation of steady-state solutions, *J. Differential Equations* **39**(2): 269–290.

R. V. Solé and J. Bascompte (2006), *Self-Organization in Complex Ecosystems,* Monographs in Population Biology **42**, Princeton University Press, Princeton.

M. Struwe (2000), *Variational Methods. Applications to Nonlinear Partial Differential Equations and Hamiltonian Systems,* Third edition, Ergebnisse der Mathematik und ihrer Grenzgebiete. 3. Folge., **34**, Springer-Verlag, Berlin.

P. Takáč (1991), Domains of attraction of generic ω-limit sets for strongly monotone semiflows, *Z. Anal. Anwendungen* **10**(3): 275–317.

P. Takáč (1992), Domains of attraction of generic ω-limit sets for strongly monotone discrete-time semigroups, *J. Reine Angew. Math.* **423**: 101–173.

J. J. Tyson and H. G. Othmer (1978), The dynamics of feedback control circuits in biochemical pathways, *Prog. Theor. Biol.* **5**: 1–62.

P. F. Verhulst (1838), Notice sur la loi que la population pursuit dans son accroissement, *Correspondance Mathèmatique et Physique* **10**: 113–121.

E. H. van Nes and M. Scheffer (2005), Implications of spatial heterogeneity for catastrophic regime shifts in ecosystems, *Ecology* **86**(7): 1797–1807.

R. R. Veit and M. A. Lewis (1996), Dispersal, population growth, and the Allee effect: Dynamics of the house finch invasion of eastern North America, *Amer. Naturalist* **148**: 255–274.

B. Walker, C. S. Holling, S. R. Carpenter, and A. P. Kinzig (2004), Resilience, adaptability, and transformability in social-ecological systems, *Ecology and Society* **9**(2): art.5.

B. Walker and J. A. Meyers (2004), Thresholds in ecological and social ecological systems: A developing database, *Ecology and Society* **9**(2): art.3.

S. H. Wang (1994), On S-shaped bifurcation curves. *Nonlinear Anal.* **22**(12): 1475–1485.

M.-H. Wang and M. Kot (2001), Speeds of invasion in a model with strong or weak Allee effects, *Math. Biosci.* **171**(1): 83–97.

M.-H. Wang, M. Kot, M., and M. G. Neubert (2002), Integrodifference equations, Allee effects, and invasions, *J. Math. Biol.* **44**(2): 150–168.

Y. Wang and J. Jiang (2002), Uniqueness and attractivity of the carrying simplex for discrete-time competitive dynamical systems, *J. Differential Equations* **186**(2): 611–632.

W. G. Wilson and R. M. Nisbet (1997), Cooperation and competition along smooth environment gradients, *Ecology* **78**: 2004–2017.

S. Yoshizawa (1970), Population growth process described by a semilinear parabolic equation, *Math. Biosci.* **7**: 291–303.

M. L. Zeeman (1993), Hopf bifurcation in competitive three-dimensional Lotka-Volterra systems, *Dynam. Stability Systems* **8**(3): 189–217.

Y. Zhao, J. Shi, and Y. Wang (2007), Exact multiplicity and bifurcation of a chemical reaction model, *J. Math. Anal. Appl.* **331**(1): 263–278.

CHAPTER 4

Modeling animal movement with diffusion

Otso Ovaskainen
University of Helsinki

Elizabeth E. Crone
University of Montana

Abstract. Diffusion models have long been used in theoretical ecology, but they have often been considered too simplistic to be applied to real data. In this chapter, we discuss how diffusion-advection-reaction models can be used to analyze animal movement. We consider a family of models that apply to heterogeneous landscapes by assuming that the model parameters (diffusion, advection, and reaction) depend on the landscape features at the present location of the animal. The landscape features may include both discrete (e.g., a classification to habitat types) and continuous (e.g., elevation) variation. We pay special attention to linear landscape features, discussing how behavioral responses to one-dimensional movement corridors and barriers and edge-mediated behavior can be built into the diffusion framework. We illustrate the application of the modeling framework by using diffusion to mimic wolf movements in a mountainous landscape. We conclude that recent developments in mathematical and computational methods make it possible to bridge the gap between theory and data by using diffusion models to facilitate the analysis of movement data acquired from heterogeneous landscapes, and discuss future research priorities in this area.

4.1 Introduction

Animal movement in heterogeneous environments involves the interplay of landscape features, such as resource patches, barriers, and different land cover types, with behavioral responses of animals to those features and their interactions with other animals. To account for this complexity, ecologists typically describe movements using rule-based simulation models such as variants of correlated random walks. Random walk models are a natural way to describe animal movement in that they can easily be modified to incorporate increasingly complex behaviors and landscapes (e.g., Revilla et al., 2004; Vuilleumier and Perrin, 2006; Matanoski and Hood, 2006; Peer et al.,

2006), and they provide rules that can be used to simulate long-term consequences of these behaviors under current conditions, or to examine the potential responses of animals to novel environments (e.g., Tischendorf et al., 2003). However, at the same time, the ability of these models to incorporate system-specific behavior makes them potentially idiosyncratic and data-intensive descriptors of particular populations, that can be difficult to generalize, or even compare, using a common currency (Grimm et al., 1999).

General theoretical models of animal movement often approximate random walks with advection-diffusion models (Turchin, 1998; Okubo and Levin, 2001; Cantrell and Cosner, 2003). Diffusion is a mathematical approximation of a random walk, so biologically they can be considered the same model. However, in many cases, using diffusion instead of random walks facilitates the parameterization and analysis of movement models. Diffusion approximations are simpler to interpret than random walk models because they aggregate the many parameters of a simulation model (e.g., distributions of turning angles, step lengths, and movements speeds) into relevant summary statistics (e.g., diffusion coefficient), and hence provide a common currency that facilitates comparison among studies (Turchin, 1998). Using diffusion instead of random walks also makes models more transparent, as diffusion can be written down as an exact equation, whereas simulation models are often described in terms of a complex computer algorithm. However, diffusion models have typically been used only for relatively simple scenarios, such as population dynamics and species interactions in homogeneous environments or movement in landscapes composed of two habitat types. Therefore, diffusion models are often viewed as too simple to describe animal responses to heterogeneous environments.

Random walks and their corresponding diffusion models have been related to animal movement in three ways. The first and most common is to work in the context of random walk by quantifying distributions of move lengths, turning angles, directed orientation, and so forth, possibly as a function of habitat type (Revilla et al., 2004; Stevens et al., 2004; Haynes and Cronin, 2006). If one wishes to do so, the resulting random walk model can then be approximated by a diffusion model (e.g., Patlak, 1953; Turchin, 1998). In general, the limitation of this approach has been estimating the necessary parameters for animal movement directly, and perhaps secondarily by the availability of mathematical formulae that translate these parameters into meaningful statistics (such as diffusion rates through heterogeneous environments) (c.f. Okubo and Kareiva, 2001). Second, continuous-time diffusion models can be approximated by stochastic discrete-time difference equations (e.g., Preisler et al., 2004; Fieberg, 2007). This approach is in many ways identical to simulation of random walks, but with the advantage of explicitly specifying the model using a diffusion equation from the very beginning. Third, a diffusion model can be analyzed by solving directly how the probability density of an individual's location evolves in time. This approach involves the same kinds of models as the second approach, but the models are used differently. Solutions over finite time periods lead to a probability density of an individual's location at some future time, conditioned on its current location (e.g., Ovaskainen, 2008; Ovaskainen et al., 2008a; Horne et al., 2007). The stationary state (or the quasi-stationary state) describes an individual's location in the absence of information about the individual's past history, or the distibution of representative locations (Fieberg, 2007; Ovaskainen, 2008). In this chapter, we focus on this third approach because working directly with probability densities facilitates the use of likelihood-based approaches in relating models to data.

In this context, diffusion is not as simplistic a model as is often thought, but it can be used to model many of the complex behaviors that have been previously built into random walk models. We discuss a framework from which it is natural and mathematically feasible to extend the diffusion approximation to complex behaviors and heterogeneous environments (c.f. Ovaskainen and Cornell, 2003; Ovaskainen, 2008). We start by describing a general model of animal behavior in heterogeneous environments, and then explore how habitat features such as patch edges and corridors interact with behavior to affect animal movement. To demonstrate the flexibility of the modeling framework, we finish with a specific example, in which we use diffusion to mimic wolf movements in a mountainous landscape (see, e.g., Whittington et al., 2005; Hebblewhite et al., 2005).

4.2 Advection-diffusion in heterogeneous environments

In general, models of movement are specified using two components: random undirected movement (diffusion), and directed movement towards particular objects or locations (advection, also known as biased movement). If two-dimensional movement is affected by diffusion and advection, the probability density of an individual's location at time t, $v = v(\mathbf{x}, t)$, changes as follows (e.g., Turchin, 1998; Ovaskainen and Cornell, 2003; Ovaskainen, 2008):

$$\frac{\partial v(\mathbf{x},t)}{\partial t} = \sum_{i,j=1}^{2} \frac{\partial^2 [a_{ij}(\mathbf{x},t)v(\mathbf{x},t)]}{\partial x_i \partial x_j} - \sum_{i=1}^{2} \frac{\partial [b_i(\mathbf{x},t)v(\mathbf{x},t)]}{\partial x_i} - c(\mathbf{x},t)v(\mathbf{x},t). \quad (4.1)$$

Here $\mathbf{x} = \{x_1, x_2\}$ refers to the spatial location, $\mathbf{A}(\mathbf{x},t) = \{a_{ij}(\mathbf{x},t)\}_{i,j=1}^{2}$ is the matrix of diffusion coefficients that may vary in space ($\mathbf{x}$) and time (t) (the random component of movement); $\mathbf{b}$ is a vector of advection coefficients, again with the potential to vary in space ($\mathbf{x}$) and time (t) (the deterministic component of movement); and c is mortality at location $\mathbf{x}$ and time t. Because v is the probability density for an individual's location at time t, $P_X(t) = \int_X v(\mathbf{x},t)d\mathbf{x}$ is the probability that the individual is within a region X at time t. If we include the possibility of mortality ($c > 0$) or an absorbing boundary condition, $P_{\mathcal{R}^2}(t) \leq 1$ gives the probability that the individual is still alive at time t, and $v(\mathbf{x},t)/P_{\mathcal{R}^2}(t)$ is the probability density conditional on the individual being alive. In this chapter, we focus on the biological interpretation rather than on mathematical issues such as existence and uniqueness of solutions. However, we note that for the diffusion model to be mathematically well-posed, the coefficients $\mathbf{A}$ and $\mathbf{b}$ need to have a given degree of smoothness through space.

If the individual is initially at time t_0 at a location $\mathbf{x}_0$, the initial condition is given by $v(\mathbf{x}, t_0) = \delta(\mathbf{x} - \mathbf{x}_0)$. Here δ is the Dirac delta distribution, i.e., a narrow peak concentrated at location $\mathbf{x}_0$ describing that the individual is known to be at this location with certainty. We note that the probability density v has also a population level interpretation. Assuming that N individuals have been released to a point $\mathbf{x}_0$ at time t_0, and that the individuals move independently of each other, $Nv(\mathbf{x}, t)$ gives the expected density of individuals at time t, and hence $NP_X(t)$ the expected number of individuals in a region X. However, if the individuals do not move independently, a population level model would need to describe the joint distribution of all individuals, and would hence become much more complex than the model for a single individual.

Using the population level interpretation, the probability density at a given location changes over time if the number of individuals arriving to the location does not equal the number of individuals leaving from that location. The net flow of movement is called the flux (Turchin, 1998). In the context of the general model 4.1, the flux $\mathcal{F}$ at a location $\mathbf{x}$ (Ovaskainen and Cornell, 2003) is given by

$$\mathcal{F}(\mathbf{x}, t) = \sum_{i,j=1}^{2} \partial_j [a_{ij}(\mathbf{x}, t) v(\mathbf{x}, t)] \mathbf{n}_i - \sum_{i=1}^{2} b_i(\mathbf{x}, t) v(\mathbf{x}, t) \mathbf{n}_i, \tag{4.2}$$

where $\mathbf{n}_i$ is the unit vector in direction i, so that in two dimensions $\mathbf{n}_1 = \{1, 0\}$ and $\mathbf{n}_2 = \{0, 1\}$. Consider a curve S with normal vector $\mathbf{z}$. Then $\int_S \mathcal{F} \cdot \mathbf{z} dS$ gives the net rate at which individuals cross the curve in the direction of $\mathbf{z}$.

The diffusivity matrix needs to be positive definite, meaning that both of its eigenvalues are greater than zero. Such a matrix can be visualized as an ellipse, the lengths of the semi-axes being given by the eigenvalues, and the orientation by the eigenvectors. Intuitively, the next location of the animal is randomized from such an ellipse centered at the animal's current location, more precisely the ellipse representing a quantile of an underlying multinormal distribution. Allowing the diffusion coefficient ($\mathbf{A}$) to vary in space allows spread rates to differ as a function of landscape features such as discrete habitat types (e.g., Ovaskainen, 2004; Ovaskainen et al., 2008a), or continuous characteristics (Turchin, 1998) such as ruggedness (Fortin et al., 2005; Whittington et al., 2005), density of food resources (Kareiva and Odell, 1987; Fortin, 2003; Forester et al., 2007), or density of predators (Fortin et al., 2005; Forester et al., 2007).

Allowing the advection parameter ($\mathbf{b}$) to vary in space allows animals to move in a particular geographic direction, or to be attracted toward a particular point in space (e.g., Moorcroft et al., 1999; Moorcroft et al., 2006; Moorcroft and Lewis, 2006). The mortality rate c may vary in space if the mortality risk is different in different habitat types. Allowing these parameters to vary in time accounts for switching among behavioral states such as sleeping and waking hours, or foraging and traveling modes (e.g., Morales et al., 2004). Such composite random walks can also be used to approximate Lévy walks (Benhamou, 2007).

If diffusion, advection, and mortality are constant in space and time, the advection-diffusion model reduces to

$$\frac{\partial v(\mathbf{x},t)}{\partial t} = \sum_{i,j=1}^{2} a_{ij} \frac{\partial^2 v(\mathbf{x},t)}{\partial x_i \partial x_j} - \sum_{i=1}^{2} b_i \frac{\partial v(\mathbf{x},t)}{\partial x_i} - cv(\mathbf{x},t). \tag{4.3}$$

The constant advection term $\mathbf{b}$ represents a deterministic force pushing the individual to a given direction. In this case, the diffusion equation solves in the closed form, giving

$$v(\mathbf{x},t) = e^{-c(t-t_0)} N(\mathbf{x}|\mathbf{x}_0 + \mathbf{b}(t-t_0), 2\mathbf{A}(t-t_0)), \tag{4.4}$$

where $N(\mathbf{x}|\mu,\Sigma)$ denotes the probability density of the multivariate normal distribution with mean μ and variance-covariance matrix Σ. The time evolution of this probability density is illustrated in Fig. 4.1. As time goes by, the most likely position moves due to the advection term, but uncertainty about the individual's location increases due to cumulative effects of the diffusion term.

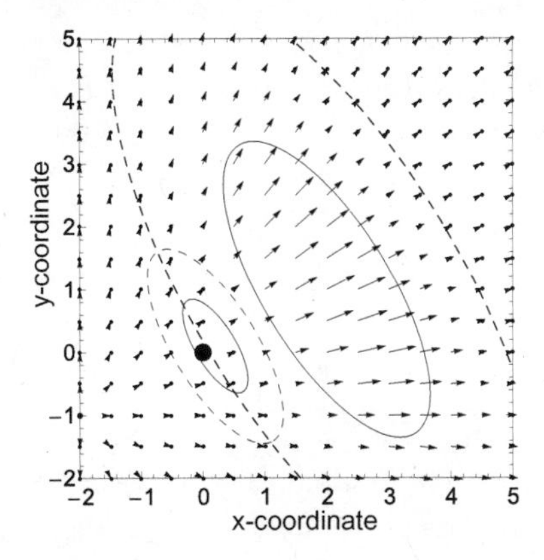

Figure 4.1 Time evolution of the probability density for an individual's location, assuming that the individual is originally (at time $t = 0$) in the origin (indicated by the black dot). The circles show the 50% (solid) and 95% (dashed) ellipsoid quantiles of the probability density v at times $t = 0.1$ (smaller ellipsoids) and $t = 1$ (larger ellipsoids). The vector field shows the flux at time $t = 1$. Parameters $\mathbf{A} = (1, -1; -1, 2)$, $\mathbf{b} = (2; 1)$, $c = 0$.

In the isotropic case $\mathbf{A} = A\mathbf{I}$, where A is a scalar and $\mathbf{I}$ is the identity matrix corresponding to a circle. In this case the diffusion part of the model reduces to the familiar form

$$\sum_{i,j=1}^{2} a_{ij} \frac{\partial^2 v(\mathbf{x},t)}{\partial x_i \partial x_j} = A\left(\frac{\partial^2 v(\mathbf{x},t)}{\partial x_1^2} + \frac{\partial^2 v(\mathbf{x},t)}{\partial x_2^2}\right) = A\Delta v(\mathbf{x},t), \tag{4.5}$$

where Δ is the Laplacian operator.

The general model (Eq. (4.1)) allows the diffusion and advection terms to vary in space and time, and it can hence be used to describe arbitrarily complex behavioral responses to heterogeneous landscapes. However, to connect models to data in a fruitful way, one needs to simplify the general modeling framework by making biologically plausible assumptions. In the rest of this section, we introduce some possible simplifications, focused particularly on describing behavioral responses of animals to heterogeneous environments.

4.2.1 Edge behavior and habitat selection

In real landscapes, habitat boundaries are often to some degree abrupt. Many species of insects specialize on particular host plants or habitat types, and are likely to turn back to their preferred habitat at patch boundaries, or return to habitat patches shortly after leaving them (Crone and Schultz, 2008). Such behavior could be modeled with Eq. (4.1) by defining a boundary zone, in which the advection term $\mathbf{b}$ points towards the preferred habitat with some smooth functional variation in space (e.g., Fig. 4.2). However, incorporating this kind of behavior into Eq. (4.1) in a complex landscape becomes numerically challenging. In addition, although it is clear that many species display edge mediated behavior (e.g., Kindvall, 1999; Ries and Debinski, 2001; Schultz and Crone, 2001; Conradt and Roper, 2006; Schtickzelle et al., 2007; Crone and Schultz, 2008), the exact functional form of advection towards preferred habitat types is rarely known (e.g., Crone and Schultz, 2008).

If the boundary in which edge-mediated behavior takes place is narrow compared to the dimensions of the landscape, it is convenient to simplify edge behavior by taking the limit in which the width of the boundary zone decreases to zero. To obtain a nontrivial limit, one needs to assume that, at the same time as the boundary gets more and more narrow, the per unit area strength of the bias increases, so that its integral over the boundary zone remains constant (Fig. 4.2). At the limit, this assumption leads to a discontinuity in the probability density across the edge (the black line in Fig. 4.2),

$$\lim_{H_1 \ni \mathbf{x}' \to \mathbf{x}} \frac{v(\mathbf{x}', t)}{k_1} = \lim_{H_2 \ni \mathbf{x}' \to \mathbf{x}} \frac{v(\mathbf{x}', t)}{k_2}, \tag{4.6}$$

where k_1 and k_2 are the habitat preferences for habitat types H_1 and H_2, and $\mathbf{x}$ is a point at the boundary (Ovaskainen and Cornell, 2003; Ovaskainen, 2004; Ovaskainen, 2008). The second matching condition is given by the fact that the flux $\mathcal{F}$ needs to be continuous across the boundary,

$$\lim_{H_1 \ni \mathbf{x}' \to \mathbf{x}} \mathcal{F}(\mathbf{x}', t) \cdot \mathbf{n}_{H_1}(\mathbf{x}) + \lim_{H_2 \ni \mathbf{x}' \to \mathbf{x}} \mathcal{F}(\mathbf{x}', t) \cdot \mathbf{n}_{H_2}(\mathbf{x}) = 0, \tag{4.7}$$

where $\mathbf{n}_H(\mathbf{x})$ denotes the corridor's normal vector at location $\mathbf{x}$ towards the habitat type H. We note that if the flux would not be continuous, the individuals would accumulate at the boundary.

In a landscape consisting of a mosaic of different habitat types, the diffusion equation (Eq. (4.1)) holds in the interior of each habitat type, and the two matching conditions (discontinuous probability density and continuous flux) describe the net result of edge-mediated behavior.

Figure 4.2 and the discussion above relate to one particular behavioral mechanism

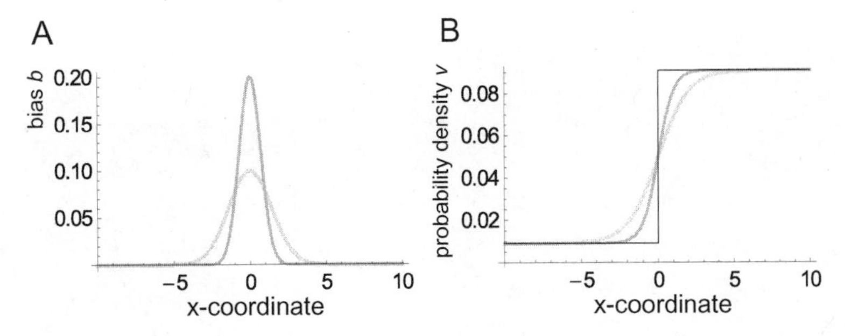

Figure 4.2 One-dimensional illustration of edge-mediated behavior. The positive x-axis ($x > 0$) represents preferred habitat, the negative x-axis ($x < 0$) avoided habitat. Panel A shows the advection term b describing the tendency of the individuals to move towards the preferred habitat when close to the boundary. Panel B shows the resulting profile of the probability density v. In both panels, the lighter lines correspond to a wide boundary zone, the darker lines to a narrow boundary zone. In both cases the net effect of edge-mediated behavior (spatial integral of the bias b) is the same. The thin black line in panel B shows the limiting probability density obtained when the width of the boundary zone tends to zero.

at a boundary between two habitat types. Because the net result is simply a proportional difference in animal densities between the two habitat types, we believe that a number of other particular mechanisms could lead to exactly the same matching condition. If researchers are primarily interested in larger scale phenomena such as rates of movement through heterogeneous environments, or the proportion of time spent in particular habitat patches, it is not necessary to find out the exact behavioral mechanism that generates the density difference between different habitat types.

4.2.2 *Responses to linear landscape features*

Many important landscape features, such as roads, trails, fencerows, and manmade wildlife movement corridors, are similar to patch boundaries in that they are narrow compared to their length. We next consider such a narrow linear element L surrounded by habitat types H_1 and H_2 (Fig. 4.3A). As with patch boundaries, it is possible in principle to use Eq. (4.1) to model responses to these elements by specifying movement rules in two-dimensional space. For example, one may assume that the animal shows edge-mediated behavior at both edges of a corridor, being more likely to turn back to the corridor than to leave it. However, as with our derivation related to edge-mediated behavior, it may be simpler to approximate the net effect of a corridor with a one-dimensional landscape element. In this case, the animals would switch between 1-dimensional and 2-dimensional modes of movement as they move into and out of corridors or other linear landscape features. Such a simplification can be derived by taking the limit in which the width of the linear element $\Delta \to 0$.

In what follows, we will assume that the diffusion, advection, and mortality coefficients within the linear element remain fixed, and focus on the effect of the relative

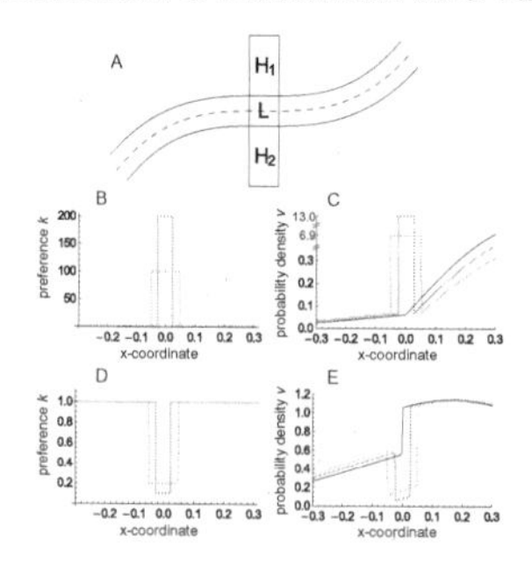

Figure 4.3 (A) A linear element L surrounded by habitat types H_1 and H_2. The panels (B-E) illustrate the matching conditions for a corridor (B,C) and for a barrier (D,E). The 1-dimensional domain can be considered as a line perpendicular to the corridor in a 2-dimensional domain, the linear element being presented by a single point at $x = 0$. Panels (B) and (D) depict the habitat preference k, and panels (C) and (E) the resulting profile of the probability density v at time $t = 1$, assuming that the individual is at time $t = 0$ at location $x = 0.2$. The lighter lines correspond to a wide element ($\Delta = 1/10$), the darker lines to a narrow element ($\Delta = 1/20$). The black lines in panels (C,E) correspond to the limiting case $\Delta \to 0$. Note the discontinuities in the y-axis in the panel (C). In panel (C), the limiting probability that the individual is in the corridor (exactly at the origin) is 0.62. Preference outside the element ($x < 0$ or $x > 0$) was set to $k_H = 1$, preference for the corridor to $z_L = 10$ and for the barrier to $z_L = 1/2$. Other parameters $A = 0.1$, $b = 0$, $c = 0$. Numerical solution in panels (C,E) was obtained by the finite element method.

density k_L within the corridor (resulting, e.g., from edge-mediated behavior at the edges between the corridor and the surrounding habitats). If the preference k_L for the linear element remains fixed, the net effect of the linear element would disappear at the limit $\Delta \to 0$. In the case of edge-mediated behavior, we assumed that the spatial integral of the advection term remained constant when the width of the boundary zone decreased to zero. Similarly, in the case of the narrow linear element, we scale preference with width, $k_L = z_L \Delta^{\alpha}$, where the parameter z_L is independent of the width of the linear element. The properties of the linear element are thus described by the parameter pair (α, z_L). If $\alpha > 0$ the preference for the linear element decreases as its width gets smaller, whereas for $\alpha < 0$ the preference increases as the width gets smaller. It turns out that a nontrivial limit appears only if $\alpha = \pm 1$. Hence, we define preferred elements ($\alpha = -1$) as structural corridors, and avoided elements ($\alpha = 1$) as structural barriers. If $\alpha < -1$, the individual prefers the corridor so much that at the limit it never leaves the corridor after first encountering it (hence the corridor becomes an absorbing boundary condition), whereas for $\alpha > 1$ the individual would never enter the linear element (hence the barrier would become a reflecting boundary condition). If $-1 < \alpha < 1$, the effect of the linear element disappears at the limit $\Delta \to 0$.

Structural corridors ($\alpha = -1$)

Let the rectangle in Fig. 4.3A be centered at a location $\mathbf{x} \in L$. Assume that the rectangle is small enough so that the probability density v and the flux $\mathcal{F}$ can be considered constants within the regions H_1 and H_2 representing the habitats surrounding the corridor from the two sides. Assume then that the rectangle remains fixed, but the width Δ of the linear element goes to zero. Then the probability that the individual is within the corridor part of the rectangle scales as $k_L\Delta = z_L\Delta^{\alpha+1}$. Hence a nontrivial limit is obtained if $\alpha = -1$, in which case we denote by $p(\mathbf{x}, t)$ the 1-dimensional probability density of the individual being at location $\mathbf{x}$ at time t. The probability $P_{L_X}(t)$ that the individual is within a finite part L_X of the corridor is given by $P_{L_X}(t) = \int_{L_X} p(\mathbf{x}, t)dS(\mathbf{x})$, where dS refers to 1-dimensional integration along the corridor.

The flux describes the rate at which individuals drift towards the corridor, hence the difference in the flux at the two sides gives the rate at which the individuals accumulate to or dissolve from the corridor. The probability of the individual being at location $\mathbf{x}$ within the corridor hence evolves as

$$\begin{aligned} \frac{dp(\mathbf{x},t)}{dt} &= \lim_{H_1 \ni \mathbf{x}' \to \mathbf{x}} \mathcal{F}(\mathbf{x}',t) \cdot \mathbf{n}_{H_1}(\mathbf{x}) + \lim_{H_2 \ni \mathbf{x}' \to \mathbf{x}} \mathcal{F}(\mathbf{x}',t) \cdot \mathbf{n}_{H_2}(\mathbf{x}) \\ &+ \text{terms representing diffusion, advection and mortality inside the corridor,} \end{aligned} \quad (4.8)$$

where $\mathbf{n}_H(\mathbf{x})$ denotes the corridor's normal vector at location $\mathbf{x}$ towards the habitat type H. The second matching condition is given by

$$\frac{p(\mathbf{x},t)}{z_L} = \lim_{H_1 \ni \mathbf{x}' \to \mathbf{x}} \frac{v(\mathbf{x}',t)}{k_1} = \lim_{H_2 \ni \mathbf{x}' \to \mathbf{x}} \frac{v(\mathbf{x}',t)}{k_2}, \quad (4.9)$$

where k_1 and k_2 are the habitat preferences for habitat types H_1 and H_2. This matching condition is identical to the discontinuity of the probability density in case of edge-mediated behavior, applied over the edges from H_1 to L, and from L to H_2. The two matching conditions are illustrated in Fig. 4.3B-C.

To make the connection between two-dimensional (k_L) and one-dimensional (z_L) preference values, assume that the preference z_L for a corridor of finite width Δ has been measured using the 1-dimensional approximation. In this case, the corresponding 2-dimensional preference k_L can be calculated simply by the equation $k_L = z_L/\Delta$.

Structural barriers ($\alpha = 1$)

If $\alpha = 1$, so that $k_L = z_L\Delta$, the individual avoids entering the linear element more and more the more narrow it becomes (Fig. 4.3D). At the limit, we obtain the matching conditions

$$\lim_{H_1 \ni \mathbf{x}' \to \mathbf{x}} \mathcal{F}(\mathbf{x}',t) \cdot \mathbf{n}_{H_1}(\mathbf{x}) + \lim_{H_2 \ni \mathbf{x}' \to \mathbf{x}} \mathcal{F}(\mathbf{x}',t) \cdot \mathbf{n}_{H_2}(\mathbf{x}) = 0, \quad (4.10)$$

and

$$\lim_{H_1 \ni \mathbf{x}' \to \mathbf{x}} \frac{v_1(\mathbf{x}, t)}{k_1} = \lim_{H_2 \ni \mathbf{x}' \to \mathbf{x}} \frac{v_2(\mathbf{x}, t)}{k_2} + \lambda(\mathbf{x}) \mathcal{F}(\mathbf{x}(H_1), t) \mathbf{n}_{H_1}, \tag{4.11}$$

where $\lambda(\mathbf{x}) = 1/(z_L A(\mathbf{x}))$. The first matching condition follows from the observation that the probability of the individual being in the barrier is zero, hence the flux must be continuous across the barrier. In the second matching condition the parameter λ measures the time delay that the barrier causes to the stabilization of the probability density at the two sides of the boundary. To see this, we note that $\lambda(\mathbf{x}) = 0$ would correspond to the usual effect of the edge-mediated behavior between the habitat types H_1 and H_2 without the intervening barrier. The delaying effect is determined both by the preference for the corridor (z_L) and by the diffusion rate within the corridor ($A(\mathbf{x})$), which we have for simplicity assumed to be isotropic. The condition (4.11) can be derived informally from a simple graphical considerations in a 1-dimensional domain. Consider the shape of the probability density in Fig. 4.3E, and let the negative x-axis represent habitat H_1 and the positive x-axis habitat H_2. The probability density inside the corridor changes from $(k_L/k_1)v_1$ to $(k_L/k_2)v_2$, where $v_i = \lim_{H_i \ni x' \to 0} v(x', t)$ denotes the probability density at the two sides of the barrier. Hence the spatial derivative of the probability density v within the corridor is

$$\frac{\partial v(x,t)}{\partial x} = \frac{(k_L/k_2)v_2 - (k_L/k_1)v_1}{\Delta} = (z_L/k_2)v_2 - (z_L/k_1)v_1. \tag{4.12}$$

Assuming that there is no advection term pushing the individual across the barrier, the flux (Eq. (4.2)) is given by $\mathcal{F}(x,t) = A(x)\partial v(x,t)/\partial x$, leading to the second matching condition.

4.3 Application: Wolf movement in a mountainous landscape

To illustrate how diffusion models can be made complex enough to describe animal movement in real landscapes, we next relate the diffusion and advection terms to wolfpack movements in a mountainous landscape near Banff, AB, Canada. In this region, mountaintops typically consist of rock and ice, whereas the valleys are forested and also often contain roads or trails (Franklin et al., 2001). Typical responses of wolves to landscape features in the southern Canadian Rocky Mountains include preference for valley bottoms over mountain tops and tendency to avoid steep slopes (Whittington et al., 2005; Hebblewhite et al., 2005). In addition, wolves appear to respond to roads and trails, although this response ranges from preference to avoidance, possibly depending on the spatial scale of development (Whittington et al., 2005). Based on these observations, we constructed advection-diffusion models to include possible behavioral responses of wolves to landscape structure using maps derived from Franklin et al. (2001). We solved the models numerically using the finite element method (Ovaskainen, 2004; Ovaskainen, 2008), modified to account for 1-dimensional barriers and corridors. To begin, we created a diffusion model corresponding to a pure random walk, i.e., isotropic diffusion with no advection or habitat preference,

$$\begin{aligned} k(\mathbf{x},t) &= 1 \\ \mathbf{A}(\mathbf{x},t) &= 1000\mathbf{I} \\ \mathbf{b}(\mathbf{x},t) &= (0,0). \end{aligned}$$

As our example is not rooted to real data, the units are somewhat arbitrary, but for the sake of illustration the spatial unit can be considered to be a meter and the temporal unit an hour. Then, we modified the model to represent possible responses to slope, elevation, and roads that might lead to a higher proportion of locations in valleys than mountains:

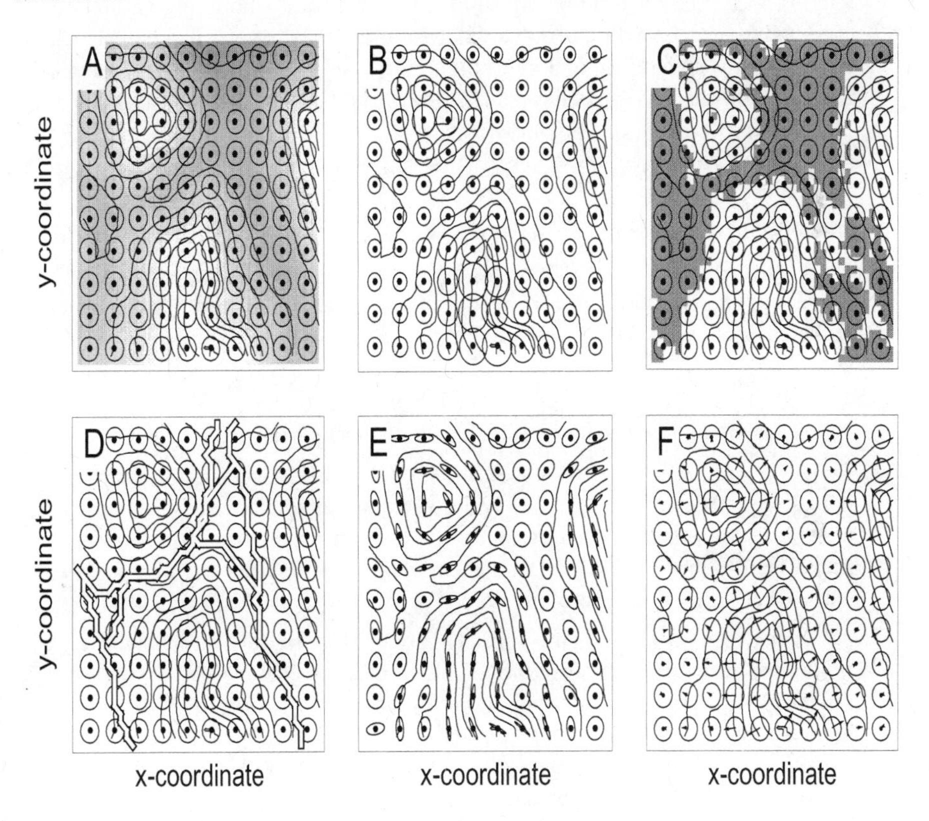

Figure 4.4 Graphical illustrations of six models describing how wolves might respond to elevation (see text). In all panels, the grey color shows the preference k (the lighter, the more preferred), the ellipses correspond to the diffusivity matrix $\mathbf{A}$, and the arrows show the advection term $\mathbf{b}$. The lower left corner is located at (x=5,640, y=57,060), and the size of the area is 4*4 km. In this part of the landscape, the altitude varies between 2081 and 2912 meters, and the slope varies between 0 and 55 degrees.

1. *Preference for low altitudes over high altitudes.* The most straightforward mecha-

nism of habitat preference is to make the habitat selection parameter k a function of altitude. As the preferences are restricted to positive values, and as only their relative values count, we model k in Fig. 4.4A as

$$\log k(\mathbf{x}, t) = c_1 a(\mathbf{x}), \tag{4.13}$$

where $a(\mathbf{x})$ is the altitude (in meters) at location $\mathbf{x}$, and the parameter c_1 is set to $c_1 = -0.003$.

2. *Faster diffusion at high altitudes.* Alternatively, animals might spend less time at high elevations because they move more quickly through high elevation sites. This mechanism is analogous to a random walk in which animals have shorter move lengths and/or larger turning angles at lower elevations. Foraging animals typically move more slowly through areas with more prey, so this mechanism might result if wolves were responding to higher prey availability at low elevations. To illustrate this mechanism, we replaced the constant diffusion rate with an increasing function of elevation (Fig 4.4B):

$$\mathbf{A}(\mathbf{x}, t) = e^{c_0 + c_1 a(\mathbf{x})}\mathbf{I}, \tag{4.14}$$

where we have set the parameters to $c_0 = 0.8$ and $c_1 = 0.0025$.

3. *Preference for forests.* As a third example, wolves might not respond to altitude per se, but might prefer forests over rock and ice, and hence show habitat selection at the edge between these two habitat types. Fig. 4.4C illustrates this assumption with

$$k(\mathbf{x}, t) = \begin{cases} 1 & \text{for } \mathbf{x} \text{ in forest}, \\ 1/10 & \text{for } \mathbf{x} \text{ in rock and ice.} \end{cases} \tag{4.15}$$

4. *Preference for roads.* In this case, the animal would not respond to altitude or habitat type, but just prefer roads and trails, for example because they create accessible terrain. Because roads and trails tend to occur in valley bottoms, this might lead to more locations in adjacent valley bottoms. To test this possibility, we set the 1-dimensional preference z_L for roads to $z_L = 1000$ (Fig. 4.4D).

5. *Following contour lines.* Some animals may tend to follow contour lines, e.g., to minimize the energetic costs of movement (c.f. Fortin et al., 2005). To incorporate this mechanism into our diffusion model, we assumed anisotropic diffusion, with higher preference for following contour lines at higher slopes (Fig. 4.4E). In the anisotropic case, it is convenient to describe the matrix $\mathbf{A}$ by its eigenvalues and eigenvectors. We set the eigenvalues to

$$\begin{aligned} \lambda_1(\mathbf{x}) &= 1000 \\ \lambda_2(\mathbf{x}) &= \lambda_1 e^{-c_1 \sin(s(\mathbf{x}))}, \end{aligned}$$

and the eigenvectors to

$$\begin{aligned} w_1(\mathbf{x}) &= (-\cos(\alpha(\mathbf{x})), \sin(\alpha(\mathbf{x}))) \\ w_2(\mathbf{x}) &= (\sin(\alpha(\mathbf{x})), \cos(\alpha(\mathbf{x}))), \end{aligned}$$

where $s(\mathbf{x})$ is the angle of inclination (the slope is given as $\tan s$), and $\alpha(\mathbf{x})$ is the aspect, measured as the angle from due north in the clockwise direction, and we have set the parameter to $c_1 = -6.5$.

6. *Advection downhill.* As a final example, it is possible that animals spend more time at low elevations because they prefer to move downhill when possible. This behavior could be modeled by adding advection towards downhill directions. We illustrate this behavior by assuming in Fig. 4.4F

$$b(\mathbf{x}, t) = c_1 \sin(s(\mathbf{x})) \cdot (\sin a(\mathbf{x}), \cos a(\mathbf{x})), \tag{4.16}$$

where we have set the parameter to $c_1 = -6$.

Figure 4.5 illustrates the movement patterns that would follow from these six alternative models. In general, three out of the six behaviors lead to higher probabilities of animal locations at low altitudes: preference for low altitudes, faster diffusion at high altitudes, and advection downhill. In this landscape, preference for forests also leads to higher probabilities at low altitudes, because forests are mainly found in the valley bottoms. Following contour lines did not cause animals to spend time in valleys, at least not as implemented in this example. Because we assumed that the strength of the tendency to follow the contour line increases with increasing slope, individuals tended to spend most time following contour lines in steep areas and were less likely to return to valley bottoms. Preference for roads increased the possibility of long-distance movements along the roads. The probability density is very high exactly at the locations of the roads, but locally not affected by the proximity of the roads, because the possibility that individuals leave the road is exactly compensated by the attraction back to the roads. In order for preference for roads to lead to increased locations in valley bottoms also outside the roads, it might be necessary to add advection from a distance towards roads, e.g., due to spatial memory and knowledge that roads represent accessible terrain, or attraction towards prey items that congregate on roads.

To demonstrate that it is possible to construct and solve diffusion models with combinations of behavioral responses, we constructed a model with three of the above behaviors: preference for forests, preference for roads, and tendency to travel downhill. In addition, we added home range behavior, modeled as advection towards the release point, with the strength of the advection term increasing with increasing distance from the release point (Okubo and Levin, 2001; Moorcroft and Lewis, 2006). We assumed the functional form

$$f(\mathbf{x}, \mathbf{x_0}, h_1, h_2) = h_1 \frac{(\mathbf{x} - \mathbf{x_0})}{|\mathbf{x} - \mathbf{x_0}|}(1 - e^{-|\mathbf{x}-\mathbf{x_0}|/h_2}), \tag{4.17}$$

where $\mathbf{x}$ is the current location, and $\mathbf{x_0}$ is the home-range center. The parameter h_1 is the maximum speed towards the home-range, obtained when far away from the home-range center, and h_2 measures how quickly the bias increases with increasing distance.

These modifications lead to $k = 1$ for forests, $k = 1/10$ for rock and ice, $z = 1000$ for roads, $\mathbf{A} = 1000\mathbf{I}$, and

$$b(\mathbf{x}, t) = c_1 \sin(s(\mathbf{x})) \cdot (\sin a(\mathbf{x}), \cos a(\mathbf{x})) + f(\mathbf{x}, \mathbf{x_0}, h_1, h_2), \tag{4.18}$$

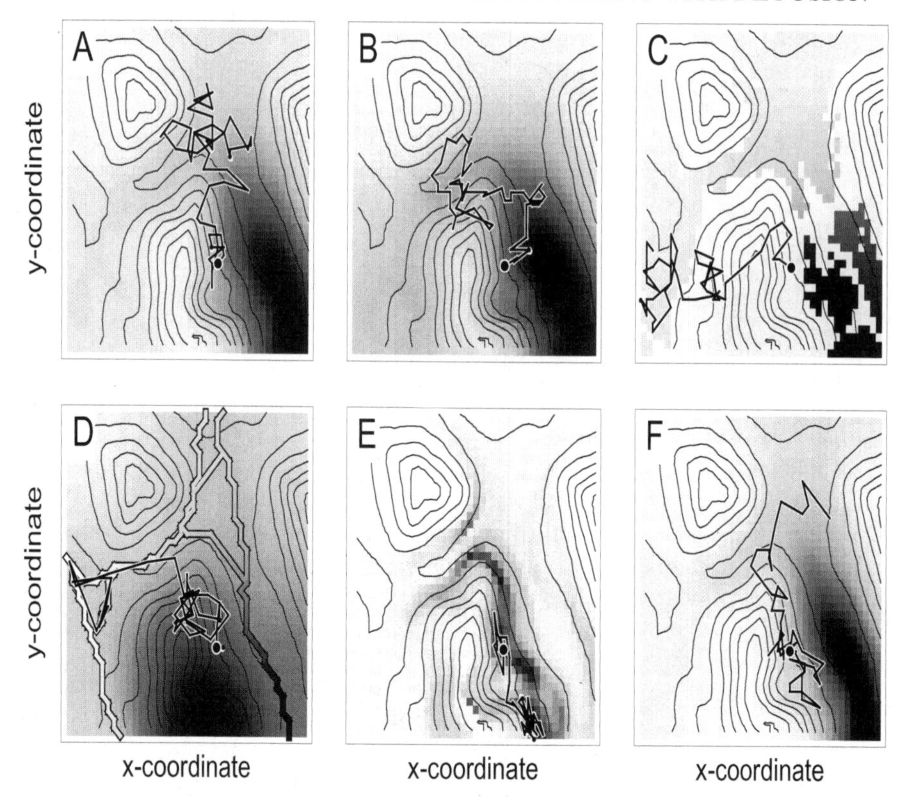

Figure 4.5 Movement behaviors corresponding to the six models described in Fig. 4.4. The black lines show a simulation track from $t = 0$ to $t = 50$, and the shading illustrates the probability density at time $t = 50$, with the initial location shown by the dot.

The prediction of this model is illustrated in Fig. 4.6A.

Finally, we added autocorrelated movement directions to the above model. In the context of correlated random walks, correlation refers to a tendency to continue in the current movement direction, i.e., turning angles that are not uniform. At longer time scales, the correlation fades out, and correlated random walks can be approximated by diffusion. All other things being equal, a correlated random walk simply results in a higher diffusion coefficient than an uncorrelated random walk (Patlak, 1953; Turchin, 1998). However, the basic diffusion model (4.1) fails to approximate the structure of a correlated random walk at small time scales. This failure is a fundamental shortcoming of the diffusion model, if we want to analyze or predict fine scale responses to environmental features using detailed movement data. However, it is possible to bring a correlation structure to fine scale predictions by modifying the advection term in the diffusion model to include a tendency to continue moving towards a previous direction. For example, we may add to the advection $\mathbf{b}$ the term $f(\mathbf{x}_t, \mathbf{x}_{t-\Delta t}, q_1, q_2)$, where the functional form of f is as in Eq. (4.17).

Simulating diffusion with such an autocorrelation structure is as easy as simulating a corresponding random walk (Fig. 4.6B). Due to the correlated nature of the movements, the area covered by the movement track is larger than with the same model parameters but without autocorrelation. However, this implementation of diffusion with autocorrelation is not a Markov process, so the mathematical and numerical tools that make diffusion models more tractable than the underlying random walk models do not apply.

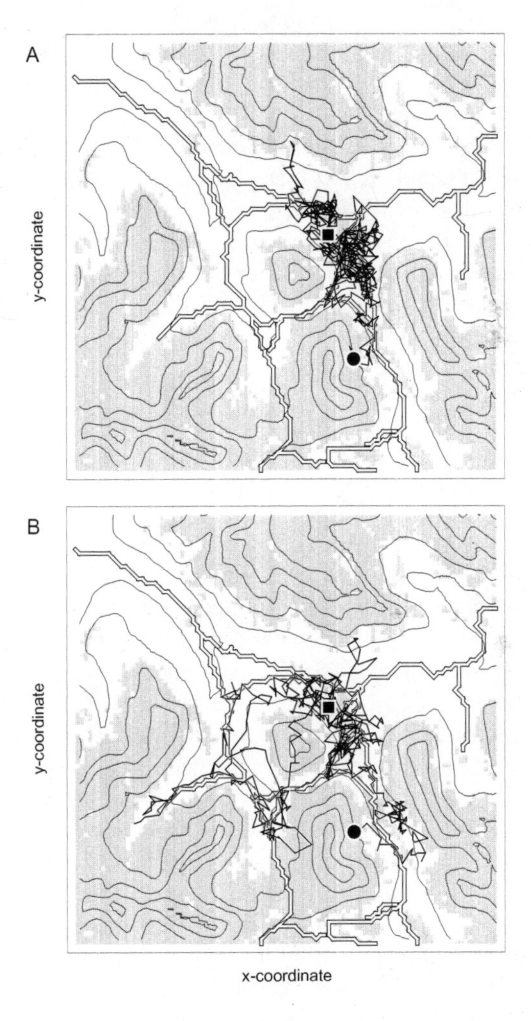

Figure 4.6 Simulation of the diffusion model with preference for forests, preference for roads, tendency to travel downhill, and home range behavior. Panel A shows a simulation run with a model without temporal autocorrelation, panel B for a model with temporal autocorrelation. The initial location is shown by the dot, the center of the home-range by a square. The parameters of the model (see text) were set to $c_1 = -6$, $-h_1 = 4$, $h_2 = 2$, $q_1 = 3$, $q_2 = 0.5$, $\Delta t = 5$. The size of the area is 10*10 km.

4.4 Applications of diffusion models

As shown in the examples above, diffusion models can not only be analyzed by simulating them, but also by calculating probability densities of an animal's location over time, as a function of its location at a particular starting time. This feature makes the diffusion framework particularly powerful for both predicting and analyzing animal movement. In this section, we discuss in more detail why this is the case.

4.4.1 Predictions from diffusion models

One difficulty of synthesizing the literature and data about animal movement models is that different people report many different statistics about animal movement, making it difficult to synthesize or compare numerous case studies. Diffusion models lend themselves naturally to ecologically meaningful parameters that summarize potentially complex movement behavior in heterogeneous environments. The most important of these is the habitat-specific diffusion coefficient, which integrates move lengths, frequencies, and turning angles. A higher diffusion rate within a particular habitat type means faster movement through it, and hence as a side product, less time spent in that habitat type. Similarly, the "habitat preference" parameter, k, combines the effect of biased behavior at habitat boundaries, and the fact that habitat-specific diffusion rates can lead to habitat-specific differences in animal densities even without any behavioral bias at the boundaries.

The probability density v is the fundamental solution to the diffusion model, and in a sense sufficient to describe all aspects of the model's behavior. However, it is often convenient to summarize model predictions in terms of parameters that integrate the probability density over time or space (Turchin, 1998; Ovaskainen and Cornell, 2003; Ovaskainen, 2008; Gardiner, 2002). These include, for example, the occupancy time, which describes how long the individual is expected to stay within a given region (Ovaskainen, 2004). Hitting probabilities describe the probability that an individual ever visits a given region, and first passage times the time it takes to do so (Frair et al., 2005). Conditional diffusion processes can be used to reconstruct the likely movement path of an individual between two observations (Horne et al., 2007). As many of these derived quantities are independent of time, they are actually often easier to calculate than the time evolution of the probability density v. While analytical solutions are available only in the very simplest cases, numerical solutions using the finite-element method or similar techniques can be calculated in very complex domains (Preisler et al., 2004; Ovaskainen, 2008).

Residence times and hitting probabilities predicted by diffusion models can be interpreted as measures of functional connectivity (Ovaskainen et al., 2008a). It is more interesting to compare these measures across studies than, say, move lengths and turning angles, because they are independent of the time-scale used for their measurement, and relate to behaviors at larger scales relevant to management issues such as conservation planning. Further, these kinds of integrated measures of movement are ecologically significant since they can be directly used as components in population dynamic models (Ovaskainen and Hanski, 2004).

4.4.2 Data analysis with diffusion models

Though diffusion models can readily be defined and numerically solved for realistically complex scenarios, relating them to data from animal populations is still somewhat in its infancy. Currently, most of the statistical methods parameterizing movement models are developed in the context of random walks rather than diffusion. As movement data seldom come without error, much of the recent developments have been focused on state space models, which explicitly model the observation process on top of the movement process (Jonsen et al., 2005; Patterson et al., 2008). State-space models are most naturally parameterized using maximum likelihood or Bayesian methods. State-space models can be as readily used for random walks (Patterson et al., 2008) as for diffusion models (Horne et al., 2007; Ovaskainen, 2004; Moorcroft and Lewis, 2006; Ovaskainen et al., 2008a). In fact, diffusion models can be viewed as a more natural match to these statistical techniques because they, unlike random walk models, explicitly generate probability densities of locations, as a function of the landscape between animal locations as well as the observed locations (see, e.g., Horne et al., 2007; Ovaskainen et al., 2008a; Ovaskainen et al., 2008b). In other words, a key difference between parameter estimation from random walk vs diffusion models is that diffusion integrates over all possible movement tracks, hence integrates the unobserved locations (which can be thought of as a nuisance parameter) out of the likelihood expression (Patterson et al., 2008). This integration makes diffusion computationally efficient, and solves the problem of how to deal naturally with habitat characteristics between starting and final locations in random walk models (c.f. Whittington et al., 2005; Fortin et al., 2005).

4.5 Conclusions

In this chapter, we have presented a general diffusion model for animal movement, and demonstrated that it can be made complex enough to include many aspects of animal behavior and, especially, responses to environmental heterogeneity. As noted by Kareiva and Odell (1987) and Moorcroft and Lewis (2006), an advantage of working with mechanistic models in general is that it may be possible to extrapolate beyond observations, e.g., predicting responses to habitat alteration or asking how parameters might evolve in current or altered landscapes. In the context of the diffusion model, this was demonstrated by Ovaskainen et al. (2008a), who showed that a model parameterized with data from a reference landscape successfully predicted clouded apollo butterfly movements in a structurally dissimilar landscape. If behaviors are built phenomenologically into simulation models or purely statistical models, it is less reliable to predict how these behaviors would change if the landscape changes.

Given the advantages of diffusion models, why aren't they used more widely to describe animal movement? We speculate that one reason is that theoretical ecologists working with these models have tended to focus on general, analytically tractable examples as case studies, rather than working with the complexities of the interactions of behavior and environmental features. At the same time, animal ecologists collecting movement data have tended to be most interested in the complexities of environmental responses, so have used statistical, rather than mechanistic, models to relate animal movement or locations to the environment. Also, many ecologists may not be comfortable working with differential equations, or not familiar with numerical methods needed for solving them. Finding solutions to diffusion models with complex behaviors and environments is computationally intensive, which, in the short term, might limit application to data analysis. Finally, it is clear that all kinds of movement behaviors cannot be fruitfully described by diffusion. The main assumption of the basic diffusion model (4.1) is that of a pure Markov process. If the animal's past movement history is needed to predict its future behavior, diffusion may not be likely to be the most successful modeling approach. Examples of such behavior include autocorrelation at fine spatial scales, and learning from experience moving in a particular landscape.

So where to go from here? We encourage readers of this book to consider diffusion as a simple though biologically plausible framework for describing and especially analyzing animal movement. The most important area for future research is relating models to data. This flows naturally from using diffusion models to calculate probability densities of animal's location, but is not yet widely recognized. In tandem with developing such techniques, we can take advantage of the growing number of animal movement studies and advances in remote tracking technologies, such as radio and satellite tracking. Important conservation questions to address in this area include the effects of habitat loss and fragmentation on animal movement, the role of matrix habitat in determining among-site dispersal, and the effects of landscape alteration on encounters between animals, such as predators and prey, or human-wildlife conflicts. Basic ecological questions include the consequences of different foraging strategies, the evolution of dispersal under different landscape structures, and the interaction of movement behavior and landscape structure in shaping species interactions. We will only learn from case studies how well different mechanisms of movement can be distinguished for real populations in real landscapes.

4.6 Acknowledgments

We thank Mark Hebblewhite and Juho Pennanen for stimulating discussions, and an anonymous referee for helpful comments. OO was supported by the Academy of Finland (grants 271173 and 213457). EC was supported by a Fulbright fellowship for research at the University of Helsinki.

4.7 References

S. Benhamou (2007), How many animals really do the Levy walk? *Ecology* **88**: 1962–1969.

S. Cantrell and C. Cosner (2003), *Spatial Ecology via Reaction-Diffusion Equations*, Wiley, John & Sons, Inc.

L. Conradt and. T. J. Roper (2006), Nonrandom movement behavior at habitat boundaries in two butterfly species: Implications for dispersal, *Ecology* **87**: 125–132.

E. E. Crone and C. B. Schultz (2008), Old models explain new observations of butterfly movement at patch edges, *Ecology* (in press).

J. Fieberg (2007), Kernel density estimators of home range: Smoothing and the autocorrelation red herring, *Ecology* **88**: 1059–1066.

J. D. Forester, A. R. Ives, M. G. Turner, D. P. Anderson, D. Fortin, H. L. Beyer, D. W. Smith, and M. S. Boyce (2007), State-space models link elk movement patterns to landscape characteristics in Yellowstone National Park, *Ecological Monographs* **77**: 285–299.

D. Fortin (2003), Searching behavior and use of sampling information by free-ranging bison (bos bison), *Behavioral Ecology and Sociobiology* **54**: 194–203.

D. Fortin, H. L. Beyer, M. S. Boyce, D. W. Smith, T. Duchesne, and J. S. Mao (2005), Wolves influence elk movements: Behavior shapes a trophic cascade in Yellowstone National Park, *Ecology* **86**: 1320–1330.

J. L. Frair, E. H. Merrill, D. R. Visscher, D. Fortin, H. L. Beyer, and J. M. Morales (2005), Scales of movement by elk (cervus elaphus) in response to heterogeneity in forage resources and predation risk, *Landscape Ecology* **20**: 273–287.

S. E. Franklin, G. B. Stenhouse, M. J. Hansen, C. C. Popplewell, J. A. Dechka, and D. R. Peddle (2001), An integrated decision tree approach (idta) to mapping landcover using sattelite remote sensing in support of grizzly bear habitat analysis in the Alberta Yellowhead ecosystem, *Canadian Journal of Remote Sensing* **27**: 579–592.

C. W. Gardiner (2002), *Handbook of Stochastic Methods: For Physics, Chemistry and the Natural Sciences*, 2nd Edition, Springer-Verlag, Berlin Heidelberg.

V. Grimm, T. Wyszomirski, D. Aikman, and J. Uchmanski (1999), Individual-based modelling and ecological theory: synthesis of a workshop, *Ecological Modelling* **115**: 275–282.

K. J. Haynes and J. T. Cronin (2006), Interpatch movement and edge effects: the role of behavioral responses to the landscape matrix, *Oikos* **113**: 43–54.

M. Hebblewhite, E. H. Merrill, and T. L. McDonald (2005), Spatial decomposition of predation risk using resource selection functions: an example in a wolf-elk predator-prey system, *Oikos* **111**: 101–111.

J. S. Horne, E. O. Garton, S. M. Krone, and J. S. Lewis (2007), Analyzing animal movements using brownian bridges, *Ecology* **88**: 2354–2363.

I. D. Jonsen, J. M. Flenming, and R. A. Myers (2005), Robust state-space modeling of animal movement data, *Ecology* **86**: 2874–2880.

P. Kareiva and G. Odell (1987), Swarms of predators exhibit preytaxis if individual predators use area-restricted search, *American Naturalist* **130**: 233–270.

O. Kindvall (1999), Dispersal in a metapopulation of the bush cricket, metrioptera bicolor (orthoptera : Tettigoniidae), *Journal of Animal Ecology* **68**: 172–185.

J. C. Matanoski and R. R. Hood (2006), An individual-based numerical model of medusa swimming behavior, *Marine Biology* **149**: 595–608.

P. R. Moorcroft and M. A. Lewis (2006), *Mechanistic Home Range Analysis*, Princeton University Press, Princeton.

P. R. Moorcroft, M. A. Lewis, and R. L. Crabtree (1999), Home range analysis using a mechanistic home range model, *Ecology* **80**: 1656–1665.

P. R. Moorcroft, M. A. Lewis, and R. L. Crabtree (2006), Mechanistic home range models capture spatial patterns and dynamics of coyote territories in Yellowstone, *Proceedings of the Royal Society B-Biological Sciences* **273**: 1651–1659.

J. M. Morales, D. T. Haydon, J. Frair, K. E. Holsiner, and J. M. Fryxell (2004), Extracting more out of relocation data: Building movement models as mixtures of random walks, *Ecology* **85**: 2436–2445.

A. Okubo and P. Kareiva (2001), Some examples of animal diffusion, in *"Diffusion and Ecological Problems : Modern Perspectives,"* ed. by A. Okubo and S. A. Levin, Springer-Verlag, New York, pp. 170–196.

A. Okubo and S. A. Levin (2001), *Diffusion and Ecological Problems: Modern Perspectives*, Springer-Verlag, New York.

O. Ovaskainen (2004), Habitat-specific movement parameters estimated using mark-recapture data and a diffusion model, *Ecology* **85**: 242–257.

O. Ovaskainen (2008), Analytical and numerical tools for diffusion based movement models, *Theoretical Population Biology* **73**: 198-211.

O. Ovaskainen and S. J. Cornell (2003), Biased movement at a boundary and conditional occupancy times for diffusion processes, *Journal of Applied Probability* **40**: 557–580.

O. Ovaskainen and I. Hanski (2004), Metapopulation dynamics in highly fragmented landscapes, in *"Ecology, Genetics, and Evolution in Metapopulations,"* ed. by I. Hanski and O. Gaggiotti, Academic Press, New York, pp. 73–103.

O. Ovaskainen, M. Luoto, H. Rekola, E. Meyke, and M. Kuussaari (2008a), An empirical test of a diffusion model: predicting clouded apollo movements in a novel environment, *American Naturalist* **171**: 610-619.

O. Ovaskainen, H. Rekola, E. Meyke, and E. Arjas (2008b), Bayesian methods for analyzing movements in heterogeneous landscapes from mark-recapture data, *Ecology* **89**: 542-554.

C. S. Patlak (1953), Random walk with persistence and external bias, *Bulletin of Mathematical Biophysics* **15**: 311–338.

T. Patterson, L. Thomas, C. Wilcox, O. Ovaskainen, and J. Matthiopoulos (2008), State-space models of individual animal movement, *Trends in Ecology & Evolution* (in press).

G. Pe'er, S. K. Heinz, and K. Frank (2006), Connectivity in heterogeneous landscapes: Analyzing the effect of topography, *Landscape Ecology* **21**: 47–61.

H. K. Preisler, A. A. Ager, B. K. Johnson, and J. G. Kie (2004), Modeling animal movements using stochastic differential equations, *Environmetrics* **15**: 643–657.

E. Revilla, T. Wiegand, F. Palomares, P. Ferreras, and M. Delibes (2004), Effects of matrix heterogeneity on animal dispersal: From individual behavior to metapopulation-level parameters, *American Naturalist* **164**: E130–E153.

L. Ries and D. M. Debinski (2001), Butterfly responses to habitat edges in the highly fragmented prairies of central Iowa, *Journal of Animal Ecology* **70**: 840–852.

N. Schtickzelle, A. Joiris, H. Van Dyck, and M. Baguette (2007), Quantitative analysis of changes in movement behaviour within and outside habitat in a specialist butterfly, *BMC Evolutionary Biology* **7**:4 doi:10.1186/1471-2148-7-4.

C. B. Schultz and E. E. Crone (2001), Edge-mediated dispersal behavior in a prairie butterfly, *Ecology* **82**: 1879–1892.

V. M. Stevens, E. Polus, R. A. Wesselingh, N. Schtickzelle, and M. Baguette (2004), Quantifying functional connectivity: experimental evidence for patch-specific resistance in the natterjack toad (bufo calamita), *Landscape Ecology* **19**: 829–842.

L. Tischendorf, D. J. Bender, and L. Fahrig (2003), Evaluation of patch isolation metrics in mosaic landscapes for specialist vs. generalist dispersers, *Landscape Ecology* **18**: 41–50.

P. Turchin (1998),Quantitative analysis of movement: measuring and modeling population redistribution in animals and plants, Sinauer Associates, Sunderland, Massachusetts, USA.

S. Vuilleumier and N. Perrin (2006), Effects of cognitive abilities on metapopulation connectivity, *Oikos* **113**: 139–147.

J. Whittington, C. C. St Clair, and G. Mercer (2005), Spatial responses of wolves to roads and trails in mountain valleys, *Ecological Applications* **15**: 543–553.

CHAPTER 5

Riverine landscapes: Ecology for an alternative geometry

William F. Fagan
University of Maryland

Evan H. Campbell Grant
University of Maryland

Heather J. Lynch
University of Maryland

Peter J. Unmack
Brigham Young University

Abstract. Ecologists interested in spatial processes are increasingly turning to models and sampling efforts that are spatially explicit. By definition, such explicitness necessitates a conceptualization of the underlying geometry of the landscapes in which important ecological processes (e.g., habitat loss, fragmentation, transport) are seen to operate. Perhaps because humans are fundamentally a terrestrial species, the default perspective in much of ecology—and in theoretical ecology in particular—is of two-dimensional terrestrial landscapes in which habitat patches of various types are interspersed within a habitat matrix. However, riverine landscapes (including riparian systems as well as creeks and rivers themselves) exhibit fundamentally different geometric properties than do 2-D terrestrial landscapes. These geometric properties likely have important consequences for population, community, and ecosystem ecology, but they have been relatively little explored. Drawing upon several examples, we lay out a rationale for increased research on the linkages between the branching geometry of riverine landscapes and ecological dynamics, focusing on the fundamental issue of the 'branchiness' of riverine networks. Given the rich biodiversity of riverine landscapes and the pervasive threats that these key systems face, extensive opportunities exist for theoretical and empirical research in this alternative geometry.

5.1 Dendritic networks as a problem in spatial ecology

Whether one considers the evolution of new species, the dynamics of invading species, or the maintenance of biodiversity, spatial processes play central roles in ecology. A

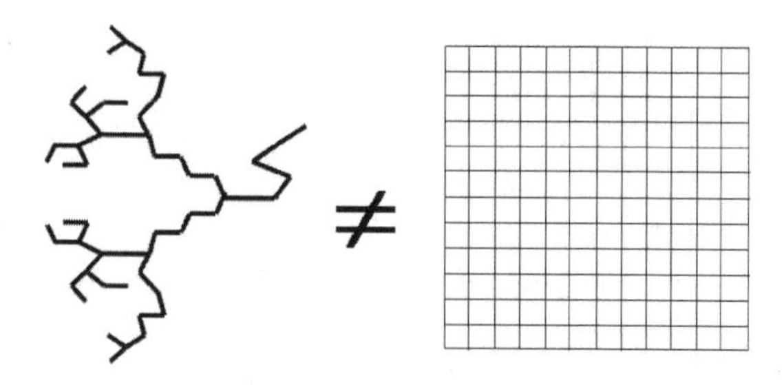

Figure 5.1 A dendritic landscape (left) differs fundamentally from the standard two dimensional landscape often featured in spatially explicit investigations in ecology.

key aspect of such spatial processes is the degree to which subunits of a system or network are connected to one another, because connectivity is often a linchpin for population persistence, patterns of biodiversity, and ecosystem function (Calabrese and Fagan 2004).

Thus far, however, ecological studies of landscape connectivity have dealt almost exclusively with 'planar' geometries, wherein habitat units or patches (such as forests, fields, and cities) extend in two dimensions and can completely 'fill up' a landscape. In contrast, other natural landscape geometries have received far less attention. For example, dendritic networks, such as river systems, which consist of effectively linear (rather than 2D) habitat units sequentially arranged, have inherently different geometries than planar landscapes (Fagan 2002, Grant et al. 2007) (Fig. 5.1). River systems (and their associated riparian zones: Gregory et al. 1991, Malanson 1993, Naiman and Décamp 1997) are perhaps the most obvious dendritic networks in nature, but caves, plant structures, and animal migratory pathways exhibit similar topologies. Geometry is a critical feature of these dendritic networks because it is intimately tied to network dynamics. For example, branching, hierarchical networks may slow down movement, altering opportunities for interactions between individuals or network components (Cuddington and Yodzis 2002, Campos et al. 2006).

Despite their potential importance, the unique contributions of dendritic geometry to the dynamics and emergent functions of networks have only rarely been studied by theoreticians (Johnson et al. 1995, Charles et al. 1998a,b, Fagan 2002, Anderson et al. 2005, Muneepeerakul et al. 2007). Empirical studies addressing dendritic geometry are also scarce, but the extant few highlight the potential importance of hierarchical geometry for species persistence and patterns of biodiversity (Bornette et al. 1998, Crabbe and Fausch 2000, Cottenie and de Meester 2003, Muneepeerakul et al. 2008). For example, in the Amazon, river confluences exhibit dramatically higher diversities of predatory 'electric' fishes than do other reaches (Fernandes et al. 2004), suggesting a key link between connectivity and community structure. Likewise, in Sonoran stream networks, fish species with highly fragmented distributions have exhibited markedly increased rates of local extinction compared to species whose historical distributions were more connected (Fagan et al. 2002, 2005a,b).

5.2 Unique features of dendritic landscapes and their consequences for ecological theory

Given the profound lack of research on the ecology of dendritic geometries, even fundamental issues remain unresolved. For example, within a branching network, what are the relative contributions of linear components and branching frequency to individual movements, population persistence, and species diversity? When species move through a dendritic landscape, do transient changes in density influence population persistence, competition, predation, and the transmission of pathogens, and what is the contribution of network geometry to those changes? What are the functional differences between rooted networks (e.g., rivers or ant trails) and other dendritic geometries where hierarchical branching occurs on both ends (e.g., avian flyways)? These and many other interesting problems remain to be explored.

Clearly, issues of dendritic geometry are related in certain ways to the increasingly popular 'network theory' approaches that have been used in studies of telecommunication (Albert et al. 2000), epidemiology (Grenfell and Bolker 1998), foodwebs (Brose et al. 2004, Garlaschelli et al. 2003), and elsewhere. However, several key differences set the problem of dendritic geometries apart from network theory more generally (Grant et al. 2007). The most important of these is that dendritic ecological networks exist as physical entities, whereas in network theory the 'branches' or links of a network represent rates or magnitudes of connections among entities (e.g., patches). Dendritic networks require separate investigations because organisms actually live and interact in those alternative geometries (Grant et al. 2007). Consequently, the important issues in the ecological dynamics of dendritic networks are not easily addressable via graph theory or similar approaches that are so popular with network theorists (Vincent and Myerscough 2004, Grant et al. 2007). Instead, to explore these issues theoretically, one must often build model landscapes of patches arranged in a variety of branching, hierarchical fashions and then simulate populations or communities of species that interact on those landscapes (e.g., Fagan 2002, Muneepeerakul et al. 2007).

Despite commonalities with other network-related topics, dendritic ecological networks present several unique complications that, together, constitute a novel research frontier in spatial ecology. Four of these sources of complexity are:

1. Intrinsic effects of configuration
   Even without any additional complications (i.e., even if the three sources of complexity listed below are not present), the hierarchical, branching arrangement of local communities per se can affect ecological patterns and dynamics in dendritic networks. In dendritic geometries, confluences and spatial sequencing are important considerations because they can act as impediments to spatial averaging as the scale changes (Guo et al. 2003, Kuby et al. 2005). For example, temporal variation in a natural process (e.g., water retention) may have starkly different consequences depending on whether the process occurs upstream or downstream within a river network (Guo et al. 2003).
2. Directional biases
   River networks feature directionally biased flows (e.g., river flows) that introduce systematic anisometries and noncommutativities into problems of dispersal in branching networks (e.g., the 'distance' or 'ease of travel' from patch A to B is not necessarily the same as from patch B to A). Directionality has received some theoretical attention via advection-diffusion models focusing on questions of population persistence, critical patch sizes, and the 'drift paradox' (Anderson et al. 2005, Lutscher et al. 2005, Pachepsky et al. 2005), but these studies considered linear habitats, not branching geometries.
3. Out-of-network connections
   Some processes in dendritic networks are out-of-network by nature. For example, forest fires and other disturbances, which need not follow the geometry of river networks, represent situations in which out-of-network processes are mismatched against the geometry of organisms' in-network dispersal (Fagan 2002). Likewise, human trucking of salmon and overland 'walking' by invasive snakehead fish are good examples of situations where out-of-network movement is critical to network-level dynamics. Some species (e.g., fish) are restricted to travel along the network, while others (e.g., stream insects) may occasionally make overland movements.
4. Transient connectivity
   The connectivity among patches in a river network may be transient (that is, time-dependent) rather than static. For example, river networks featuring regional droughts (Arizona) or episodic flooding (Amazonia) exhibit reduced or enhanced connectivity, periodically altering opportunities for dispersal and redistribution of resources.

To investigate how hierarchical, dendritic geometries influence ecological dynamics and patterns of biodiversity will require the development of a series of models of varying complexity, detail, and focus. Explicit dendritic landscapes should be at the core of these models, providing a common framework that transcends differences in model structure and purpose. For example, to explore the interface between dendritic geometry and network dynamics, one could vary the geometric properties of those landscapes (e.g., branching frequency, rooted versus nonrooted topology, hierarchical form of spatial heterogeneity) and impose one or more of the four complications above to examine their joint impacts on ecological patterns and dynamics. Of the four sources of complexity, the last item, transient connectivity, is arguably the most novel and most likely to yield results that generalize in important ways to network problems far beyond theoretical ecology. In such models, a difference or differential equation (such as those routinely used to study local population dynamics and species interactions) could operate within each compartment, and the compartments would then be linked to other compartments within the hierarchy. Given their complexity, such models will typically be solved via extensive numerical simulations, but in some cases variable aggregation methods may be useful (Charles et al. 1998a,b). Across model runs, outputs could be interpreted in terms of scaling laws for such metrics as population persistence times, average abundance or occupancy, or rates of spatial spread (Muneepeerakul et al. 2008, unpublished ms.). This approach is commonly used in ecohydrology (Rodriguez-Iturbe and Rinaldo 1997).

5.3 The 'branchiness' of a river network influences colonization opportunities and extinction risk

To illustrate the importance of dendritic geometry for ecological systems, we focus in this chapter on one important geometric factor, namely the 'branchiness' of a river network. Branchiness refers to the arrangement of a number of patches in a hierarchical network. With an increasing mean number of connections for each branch, the network is seen as being more complex (i.e., branchiness high). Extending some ideas about riverine metapopulation dynamics that were initially laid out in Fagan (2002), we first use a simulation model to explore how branchiness of a network alters opportunities for recolonization and consequently extinction risk. We then draw upon a database of fish distributions to illustrate the effects of network branchiness in a real system where fragmentation is already known to be an important driver of extinction risk.

5.4 Modeling the effects of network branchiness for metapopulation dynamics

We investigated the relationship among network branchiness, movement probabilities, and extinction risk of a metapopulation within networks of 15 stream reaches ('habitat patches'). In our model, all patches are of equal habitat quality, and we assume a uniform distribution of a population in the network. After investigating extinction risk in general, we look closer at a particular metapopulation scenario using parameter values guided by a mark-recapture study of a stream salamander species.

We created two 15-patch dendritic networks with different topologies: (1) a fractal network ('Full') with bifurcations at each branch node, and (2) a network with reduced complexity ('Pruned'), where only one bifurcation is present at each depth in the network (Fig. 5.2). The first configuration corresponds to the model in Fagan

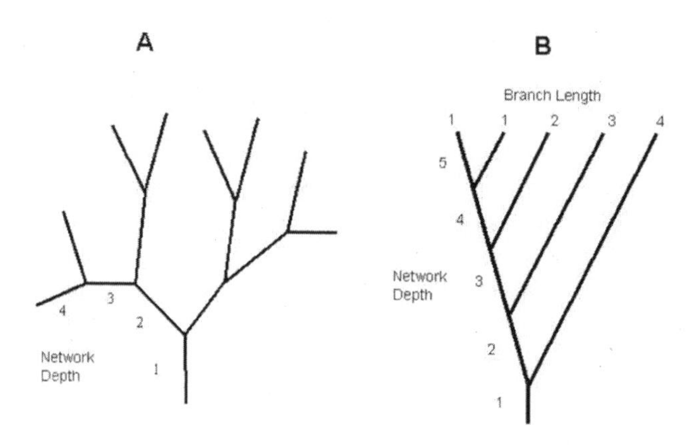

Figure 5.2 Fifteen patch network configurations (A = Full, B = Pruned) considered in investigations of extinction risk in dendritic metapopulations. The network depth is used to index position in the network.

(2002), whereas the second configuration results in a network with branches of differing lengths from the mainstem (Fig. 5.2). Realistic networks in nature may fall between the dendritic network topologies considered here. Location in the network is indexed by specifying a network 'depth,' or location along the mainstem of the network, and the horizontal position in the network (Fig. 5.2). In our ordering schema, starting from the downstream terminus of the network, a patch in position (3,2) is located 3 steps along the mainstem, and 2 branches from the leftmost patch (at the first bifurcation point, keep left, at the second, keep right).

For each model run, we initialized full occupancy of all patches in the network and fixed the time-specific extinction probability in each patch for each model run. At each time step, we allowed colonization of extinct patches via three movement routes: (1) upstream, (2) downstream, and (3) overland (out-of-network) colonization from one of the two closest neighboring patches within the same depth. We investigated three probabilities for extinction probability (0.1, 0.01, 0.001), and four movement probabilities (0, 0.1, 0.01, 0.001). The model was run for a maximum of 10,000 time steps (or until full extinction of the network) for all parameter combinations, and each combination was replicated 100 times. We present here results for the case with upstream = downstream movement probabilities.

For the full dendritic network, the presence of out-of-network connectivity has a large effect on the time to extinction. This effect was most prominent with high levels of within-network movement, suggesting that out-of-network movement is not the sole driver of extinction risk (Fig. 5.3, left panels). When per-patch extinction risk was low (0.001), the network persisted for a wide range of both within- and out-of-network dispersal probabilities (Fig. 5.3, left panels). At intermediate levels of extinction probability (0.01), the network had a reduced time to extinction when there was at least a small amount of out-of-network movement compared to the scenario without out-of-network movement. The metapopulation persisted when both within network dispersal was high (0.1), and out-of-network dispersal was moderate to high (0.01–0.1). Extinction risk in the Pruned network (Fig. 5.3, right panels) was similar to that in the Full network but featured a damped pattern that was especially evident at intermediate levels of extinction probability (Fig. 5.3, middle panels, extinction = 0.01). With high extinction probability (0.1), the Pruned network goes extinct rather quickly (Fig. 5.3 bottom right panel; note different axis scale), regardless of the level of out-of-network dispersal.

Finally, we compared 15-patch networks in two configurations (Full vs. Pruned networks) guided by empirical within-network movement data on a species of stream salamander, *Gyrinophilus porphyriticus* (Lowe 2003). Little is known about rates of out-of-network movements in stream amphibians, though populations of some species are more closely associated with stream networks with confluent first order branches (Lowe and Bolger 2002, Rissler et al. 2004, Grant et al. in press), suggesting that this type of movement may be naturally low in some species. Stream networks in altered landscapes typically lose complexity via loss of small headwater streams (Dunne and Leopold 1978). In species that are adapted to live in streams, the loss of network complexity may result in an increased extinction risk, especially when within-network movements are the predominant mode of dispersal. Some species may be capable of making out-of-network movements, which may be important for stabilizing populations (a type of weak link, Csermely 2004). Further, in undisturbed populations, stream salamanders likely have low rates of extinction (Hairston and Riley 1993), though with increasing landscape disturbance, rates of extinction are likely higher (Price et al. 2006). Using our model, we found that at low rates of extinction (0.01), both network complexities have similar extinction risk when there is a small amount of out-of-network dispersal (Fig. 5.4, top). However, at higher extinction rates (0.1), the Full dendritic network has a greater potential for population persistence, when out-of-network movements are proportional to or greater than other modes of dispersal (Fig. 5.4, bottom).

From our simulation results, it is apparent that the spatial layout of a stream network helps determine the risk of metapopulation extinction. Consequently, understanding how network complexity interacts with population extinction risk may be important for managing stream network habitats. More complex patterns of extinction (e.g., correlated disturbances, Fagan 2002, Lowe 2002) and biases in animal movements or habitat preferences in the network (e.g., preference for higher order branch locations) may alter the results from our simple model discussed here. However, we expect

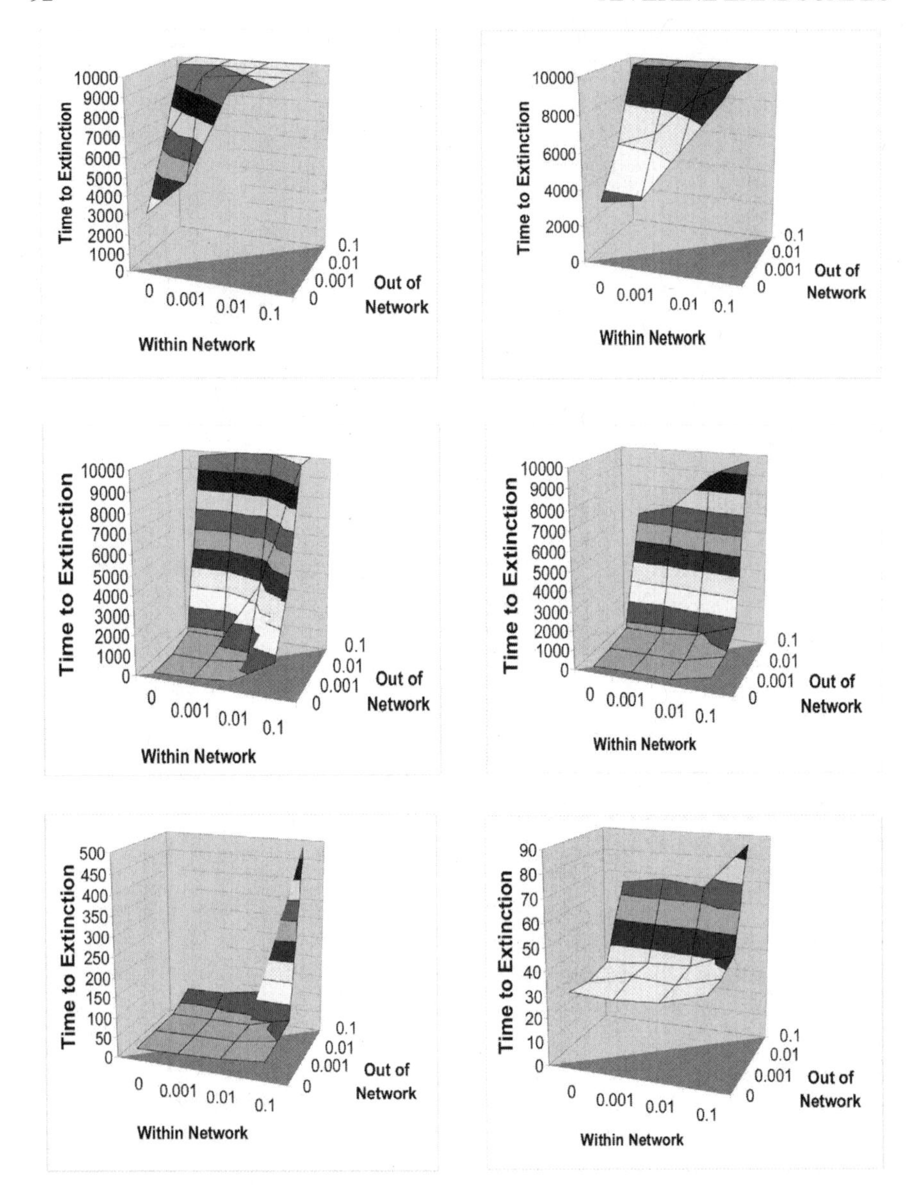

Figure 5.3 Effects of river network 'branchiness' on extinction risk in 15-patch dendritic metapopulations. Panels on the left are from a Full dendritic network, and on the right are from the Pruned network. Three extinction probabilities were modeled (0.001, top row; 0.01, middle; 0.1 bottom row), under combinations of within- and out-of-network dispersal probabilities (0, 0.001, 0.01, 0.1). (See color insert following page 202.)

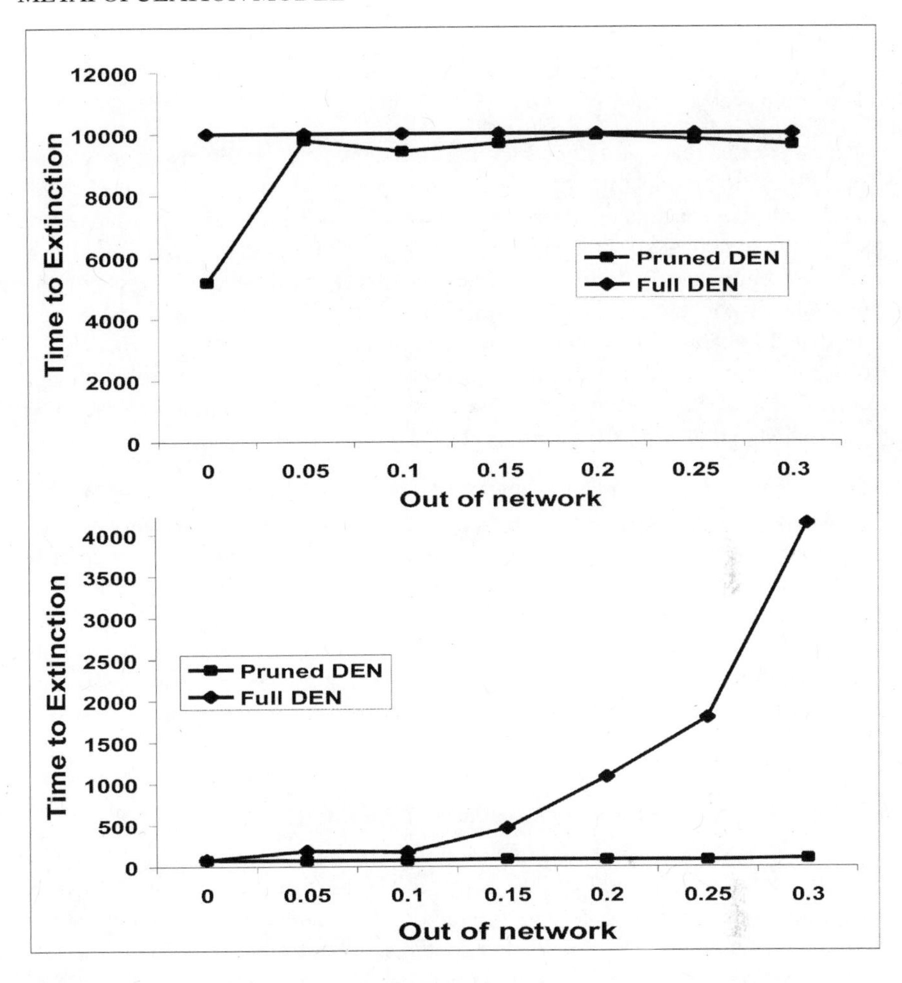

Figure 5.4 Effects of river network 'branchiness' on extinction risk for the spring salamander *Gyrinophilus porphyriticus* under different levels of out-of-network dispersal (upstream dispersal probability = 0.15, downstream dispersal probability = 0.05). Top panel, extinction probability = 0.01. Bottom panel, extinction probability = 0.1.

that more realistic models will strengthen the relationship among movement probabilities and network complexity, especially considering variation in out-of-network colonization probabilities.

For most species that live in dendritic networks, empirical estimates of movement probabilities are a critical information need for managing populations in these habitats, though these estimates are largely unavailable at large scales. As our results suggest, the specific combination of movement probabilities is important for assessing metapopulation extinction risk. While out-of-network connectivity generally increases the time to metapopulation extinction, the effect of increasing this movement is mediated by the within-network movement probabilities. Few long-term data exist to test our model in existing dendritic network systems at large scales, though recently established monitoring programs that recognize the potential importance of the spatial layout of dendritic networks should prove useful.

Finally, we note that the modeling approach employed here may be useful for planning repatriation, translocation or stocking programs in dendritic stream networks. Viewing out-of-network dispersal as a translocation or stocking event, alternative scenarios could be explored in advance of implementing a management action. Extensions to our model could specify stocking or translocation frequency via modification of the out-of-network colonization probability, consider the impact of stocking location within the network hierarchy, and allow for a greater range of colonization distances (e.g., allowing for long distance dispersal events in the network).

5.5 Network branchiness and extinction risk for desert fishes

A key prediction emerging from the above modeling scenarios is that 'branchier' networks should facilitate recolonization among subpopulations and thereby buffer the system as a whole from regional extinctions. To test this prediction in the real world, we investigated the link between network branchiness and local extirpation risk in an assemblage of fish species native to the Sonoran Desert ecoregion.

Occurrence records for this group of species are summarized in the Sonoran Fishes (or 'SONFISHES') database, initially developed by the late ichthyologist W. L. Minckley. This GIS database provides extensive distributional data for native freshwater fishes in the southwestern USA and northwestern Mexico. Within the Sonoran ecoregion, the Lower Colorado River basin, and within that, the Gila River, feature the most detailed biogeographical coverage and the greatest density of collecting records. Parts of this landscape are highly fragmented due to a lack of perennial water resulting from the interplay among precipitation, discharge, and substrate, and more recently as a result of diversion and drawdown by human activities (Brown et al. 1981). Moreover, even when contiguous stretches of surface water exist, the widespread introduction of multiple, nonnative, invasive fish species induce a type of biological fragmentation due to larvivory, in which nonnative species prey on juvenile native fish and greatly limit their recruitment (Unmack and Fagan 2004).

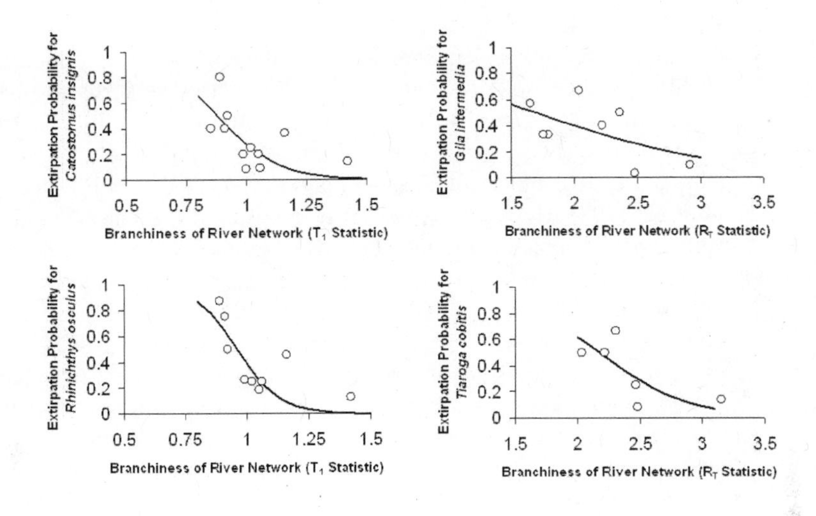

Figure 5.5 Measures of river network branchiness as predictors of extinction risk in fish species of the Gila River in the Sonoran Desert ecoregion.

The SONFISHES database encompasses ~160 years (from 1843 to ~2005) and contains incidence, identity, and collection data for the complete holdings of the major museum collections from this region, numerous smaller collections of southwestern fishes, records from the Non-Game Branch of the Arizona Game and Fish Department, and peer-reviewed and 'gray' literature sources. Due to the intensity and time span of sampling, SONFISHES summarizes virtually all that is known about past and present distributions of fishes in the region and represents an unusually comprehensive resource for examining changes in species' spatial distributions over time.

Previous analyses of the SONFISHES database have demonstrated that, among 25 fish species native to the Lower Colorado River basin, the degree to which a species' distribution was fragmented historically is a strong predictor of the frequency of local extirpations that the species has since experienced (Fagan et al. 2002a, 2005a,b). Although a species' historical frequency of occurrence (i.e., number of localities at which it was found) is also correlated with the risk of extirpation on local scales (e.g., 5 km or 25 km reach lengths), historical fragmentation of occurrences is a far stronger predictor of variation in extinction risk among species (Fagan et al. 2002), and this dependence manifests on small through large spatial scales (5 to 2500 km reach lengths; Fagan et al. 2005a). Thus the physical arrangement of species' populations and not just the number of those populations has been an important determinant of extinction risk in the Sonoran ecoregrion.

Here we seek to expand on this understanding by quantifying the relationship between the branching geometry of river networks in particular watersheds and the observed frequency of local extinctions in those watersheds. To quantify network branchiness, we will adopt two measures of riverine geometry from the theoretical hydrology literature (Rodriguez-Iturbe and Rinaldo 1997, Dodds and Rothman 1999, Turcotte et al. 1998). However, before introducing the branchiness metrics themselves, we first define some important hydrological terms that provide context. Using the conventional methodology for characterizing watershed geomorphology (Strahler 1967), a stream's 'order' is an index that relates to both flow capacity and network position. Starting from tiny trickles far upstream (first order streams), stream order increases when two streams of equal order merge together. For a given stream of order n, a 'major side tributary' is a stream of order $n - 1$ that merges with the parent stream partway along its course (rather than at its upstream confluence). Likewise, a 'stream segment' is defined as a contiguous reach of stream with the same order (i.e., a stream segment is bounded by upstream and downstream confluences where order changes). In idealized watersheds, these concepts are related by Tokunaga's Law (Dodds and Rothman 1999) which states

$$T_n = T_1 R_T^{n-1} \tag{5.1}$$

where T_n is the expected number of tributaries of order n in a given watershed, T_1 is the average number of major side tributaries per stream segment, and R_T is a multiplicative factor describing the average rate at which numbers of side tributaries of successively lower orders accumulate in a watershed. Example calculations of these branchiness metrics appear in Dodds and Rothman (1999). Although Equation (5.1) is typically used in theoretical hydrology problems, the metrics T_1 and R_T, which quantify different, but complementary, aspects of stream network complexity, can also be calculated for real watersheds via tedious effort.

To characterize the branchiness of different river networks, we quantified T_1 and R_T for 13 watersheds within the Gila River drainage (central Arizona and western New Mexico, USA, plus small portions of northern Sonora, Mexico). We used watersheds defined at the HUC-8 scale (Hydrological Unit Code 8; Seaber et al. 1987), and given the monotony involved in calculating the branchiness metrics, chose a subset (52%) of the HUC-8 watersheds that spanned the range of watershed complexity evident in the Gila drainage. We focused our analyses on six species of small- and medium-sized fish: *Agosia chrysogaster*, *Catostomus insignis*, *Gila intermedia*, *Meda fulgida*, *Rhinichthys osculus*, and *Tiaroga cobitis*. These species were all widespread in the Gila River drainage historically, and, unlike other fish native of the region, were not restricted to particular elevational zones (e.g., *Onchorhynchus* spp.) or river flow volumes (e.g., *Xyrauchen texanus*, *Ptychocheilus lucius*). We then used logistic regression to quantify relationships between watershed branchiness and the observed frequency of extirpation at the local scale (=5 km of reach).

Gila River watersheds vary substantially in branchiness, whether that geometric complexity is measured in terms of the average number of major side tributaries (T_1) or the rate at which reaches accumulate lower order tributaries (R_T). On average, a given stream segment in Gila River watersheds has $T_1 = 1.00$ major tributaries (range: 0.85–1.42) and finer scale branching occurs at an average rate of $R_T = 2.21$ branches per segment (range: 1.44–3.15).

For four of our six focal species, the frequency of local extirpation was strongly and significantly dependent on one or both measures of network branchiness (Fig. 5.5). *Catostomus* and *Rhinichthys* both exhibited lower local extinction risk in those watersheds with relatively high T_1 scores, whereas observed extinction risk in Gila and Tiaroga were more strongly related to R_T. In contrast, extinction risk in neither *Agosia* nor *Meda* was significantly related to watershed branchiness, although extinction risk for *Agosia* trended downward with increasing branchiness for both R_T and T_1.

In a system like the Sonoran ecoregion, where connectivity may be determined largely by in-stream proximity of individual populations, it is intuitive that the extent of fragmentation in populations is a strong predictor of extinction risk, and this has been borne out by several analyses (Fagan et al. 2002, 2005a,b). Our analysis here suggests that watershed 'branchiness' may contribute to those previously observed relationships between fragmentation and extinction risk, with branchier watersheds being less prone to local population extinctions. Consequently, conservationists and resource managers may want to consider the branching geometry of riverine networks when seeking to identify watersheds that will yield a high probability of local population persistence for Sonoran fishes.

5.6 Conclusion

Although this chapter has focused on riverine geometry, river networks are only one example within a broader class of ecological networks involving dendritic geometries. For example, caves feature network-like geometry, but exist in three dimensions rather than just two (Curl 1986, Palmer 1991). Likewise, avian flyways, ungulate migratory pathways, and ant trails possess branching, hierarchical geometries but exist at a functional level (for migration or resource acquisition) rather than in a structural sense (Watmough and Edelstein-Keshet 1995, Speirs and Gurney 2001, Hindmarch and Kirby 2002, Jackson et al. 2004, Xia et al. 2004). The architecture of individual plants also involves dendritic geometries that may alter species interactions and drive emergent food web dynamics (Kareiva and Sahakian 1990, Cuddington and Yodzis 2002). Unfortunately, links between geometry and dynamics in these other dendritic systems have received even less theoretical attention than have river networks. Consequently, the ecology of alternative geometries will afford rich research opportunities for years to come.

5.7 Acknowledgments

Funding for this work came from the USGS-Northeast ARMI program and the James S. McDonnell Foundation.

5.8 References

R. Albert, H. Jeong, and A.L. Barabasi (2000), Error and attack tolerance of complex networks, *Nature* **406**: 378-382.

K. E. Anderson, R. M. Nisbet, S. Diehl, and S. D. Cooper (2005), Scaling population responses to spatial environmental variability in advection-dominated systems, *Ecology Letters* **8**: 933-943.

G. Bornette, C. Amoros, and N. Lamouroux (1998), Aquatic plant diversity in riverine wetlands: the role of connectivity, *Freshwater Biology* **39**: 267-283.

U. Brose, A. Ostling, K. Harrison, and N. D. Martinez (2004), Unified spatial scaling of species and their trophic interactions, *Nature* **428**: 167-171.

D. E. Brown, N. B. Carmony, and R. M. Turner (1981), Drainage map of Arizona showing perennial streams and some important wetlands, Phoenix: Arizona Game and Fish Department.

J. M. Calabrese and W. F. Fagan (2004), A comparison shoppers' guide to connectivity metrics: trading off between data requirements and information content, *Frontiers in Ecology and the Environment* **2**: 529-536.

D. Campos, J. Fort, and V. Méndez (2006), Transport on fractal river networks. Application to migration fronts, *Theoretical Population Biology* **69**: 88-93.

S. Charles, R. B. de la Parra, J. P. Mallet, H. Persat, and P. Auger (1998a), A density dependent model describing *Salmo trutta* population dynamics in an arborescent river network: Effects of dams and channeling, *Comptes Rendus de L'Academie des Sciences Serie III - Sciences De La Vie - Life Sciences* **321**: 979-990.

S. Charles, R.B. de la Parra, J.P. Mallet, H. Persat, and P. Auger (1998b), Population dynamics modelling in an hierarchical arborescent river network: An attempt with *Salmo trutta*, *Acta Biotheoretica* **46**: 223-234.

J. Chave, H. C. Muller-Landau, and S. A. Levin (2002), Comparing classical community models: Theoretical consequences for patterns of diversity, *American Naturalist* **159**: 1-23.

K. Cottenie and L. de Meester (2003), Connectivity and cladoceran species richness in a metacommunity of shallow lakes, *Freshwater Biology* **48**: 823-832.

P. Csermely (2004), Strong links are important, but weak links stabilize them, *Trends in Biochemical Sciences* **29**: 331-334.

K. Cuddington and P. Yodzis (2002), Predator-prey dynamics and movement in fractal environments, *American Naturalist* **160**: 119-134.

R. L. Curl (1986), Fractal dimensions and properties of caves, *Mathematical Geology* **18**: 765-783.

P. S. Dodds and D. H. Rothman (1999), Unified view of scaling laws for river networks, *Physical Review E* **59**: 4865-4877.

T. Dunne and L. B. Leopold (1978), *Water in Environmental Planning*, New York: W. H. Freeman and Co.

W. F. Fagan (2002), Fragmentation and extinction risk in dendritic metapopulations, *Ecology* **83**: 3243-3249.

W. F. Fagan, P. Unmack, C. Burgess, and W. L. Minckley (2002), Rarity, fragmentation, and extinction risk in desert fishes, *Ecology* **83**: 3250-3256.

W. F. Fagan, C. Aumann, C. M. Kennedy, and P. J. Unmack (2005a), Rarity, fragmentation and the scale-dependence of extinction-risk in desert fishes, *Ecology* **86**: 34-41.

W. F. Fagan, C. M. Kennedy, and P.J. Unmack (2005b), Quantifying rarity, losses, and risks for lower Colorado River Basin fishes: Implications for conservation listing, *Conservation Biology* **19**: 1872-1882.

C. C. Fernandes, J. Podos, and J.G. Lundberg (2004), Amazonian ecology: tributaries enhance the diversity of electric fishes, *Science* **305**: 1960-1962.

M. A. Fortuna, C. Gomez-Rodriguez, and J. Bascompte (2006), Spatial network structure and amphibian persistence in stochastic environments, *Proceedings of the Royal Society B - Biological Sciences* **273**: 1429-1434.

D. Garlaschelli, G. Caldarelli, and L. Pietronero (2003), Universal scaling relations in food webs, *Nature* **423**: 165-168.

E. H. C. Grant, L. E. Green, and W. H. Lowe (in press), Salamander occupancy in headwater stream networks, *Freshwater Biology*.

E. H. C. Grant, W. Lowe, and W.F. Fagan (2007), Living in the branches: population dynamics and ecological processes in dendritic networks, *Ecology Letters* **10**: 165-175.

S. V. Gregory, F.J. Swanson, W.A. McKee and K.W. Cummins (1991), An ecosystem perspective of riparian zones, *Bioscience* **41**(8): 540-551.

B. T. Grenfell and B. M Bolker (1998), Cities and villages: infection hierarchies in a measles metapopulation, *Ecology Letters* **1**: 63-70.

Z. W. Guo, Y. L. Gan, and Y. M. Li (2003), Spatial pattern of ecosystem function and ecosystem conservation, *Environmental Management* **32**: 682-692.

S. Hairston, G. Nelson, and R. H. Wiley (1993), No decline in salamander (Amphibia: Caudata) populations: A twenty-year study in the southern Appalachians, *Brimleyana* **18**: 59-64.

C. Hindmarch and J. Kirby (2002), Corridors for birds within a pan-European ecological network, Council of Europe, Strasbourg, France.

D. E. Jackson, M. Holcombe, and F.L.W. Ratnieks (2004), Trail geometry gives polarity to ant foraging networks, *Nature* **432**: 907-909.

A. R. Johnson, C.A. Hatfield, and B.T. Milne (1995), Simulated diffusion dynamics in river networks, *Ecological Modelling* **83**: 311-325.

P. Kareiva and R. Sahakian (1990), Tritrophic effects of a simple architectural mutation in pea-plants, *Nature* **345**: 433-434.

W. H. Lowe (2002), Landscape-scale spatial population dynamics in human-impacted stream systems, *Environmental Management* **30**: 225-233.

W. H. Lowe (2003), Linking dispersal to local population dynamics: A case study using a headwater salamander system, *Ecology* **84**: 2145-2154.

W. H. Lowe and D. T. Bolger (2002), Local and landscape-scale predictors of salamander abundance in New Hampshire headwater streams, *Conservation Biology* **16**: 183-193.

F. Lutscher, E. Pachepsky, and M. A. Lewis (2005), The effect of dispersal patterns on stream populations, *SIAM Journal on Applied Mathematics* **65**: 1305-1327.

M. J. Kuby, W.F. Fagan, C. ReVelle, and W. Graf (2005), A multiobjective optimization model for dam removal: An example trading off salmon passage with hydropower and water storage in the Willamette basin, *Advances in Water Resources* **28**: 845-855.

T. R. Labbe and K.D. Fausch (2000), Dynamics of intermittent stream habitats regulate persistence of a threatened fish at multiple scales, *Ecological Applications* **10**: 1774-1791.

G. P. Malanson (1993), *Riparian Landscapes*, Cambridge University Press, Cambridge, UK.

R. R. Miller, W.L. Minckley, and S.M. Norris (2005); *Freshwater Fishes of Mexico*. University of Chicago Press, Chicago.

R. Muneepeerakul, J.S. Weitz, S. A. Levin, A. Rinaldo, and I. Rodriguez-Iturbe (2007), A neutral metapopulation model of biodiversity in river networks, *Journal of Theoretical Biology* **245**: 351-363.

R. Muneepeerakul, E. Bertuzzo, H. Lynch, W. F. Fagan, A. Rinaldo, and I. Rodriguez-Iturbe (2008), Neutral metacommunity models predict fish diversity patterns in Mississippi-Missouri basin, *Nature* **453**: 220-223.

R. J. Naiman and H. Décamp (1997), The ecology of interfaces: riparian zones, *Annual Review of Ecology Systematics* **28**: 621-658.

E. Pachepsky, F. Lutscher, R. M. Nisbet, and M. A. Lewis (2005), Persistence, spread and the drift paradox, *Theoretical Population Biology* **67**: 61-73.

A. N. Palmer (1991), Origin and morphology of limestone caves, *Geological Society of America Bulletin* **103**: 1-21.

S. J. Price, M. E. Dorcas, A. D. Gallant, R. W. Klaver, and J. D. Willson (2006), Three decades of urbanization: Estimating the impact of land-cover change on stream salamander populations, *Biological Conservation* **133**: 436-441.

L. J. Rissler, H. M. Wilbur, and D. R. Taylor (2004), The influence of ecology and genetics on behavioral variation in salamander populations across the Eastern Continental Divide, *American Naturalist* **164**: 201-213.

I. Rodriguez-Iturbe and A. Rinaldo (1997), *Fractal River Basins: Chance and Self-organization*, Cambridge University Press, Cambridge, UK.

P.R. Seaber, F. P. Kapinos, and G. L. Knapp (1987), Hydrologic unit maps, US Geological Survey, Water Supply Paper 2294.

R. P. Smith, P.W. Rand, E.H. Lacombe, S.R. Morris, D.W. Holmes, and D.A. Caporale (1996), Role of bird migration in the long-distance dispersal of *Ixodes dammini*, the vector of Lyme disease, *Journal Of Infectious Diseases* **174**: 221-224.

D. C. Speirs and W. S. C. Gurney (2001), Population persistence in rivers and estuaries, *Ecology* **82**: 1219-1237.

A. N. Strahler (1967), Quantitative analysis of watershed geomorphology, *Transactions of the American Geophysical Union.* **38**: 913-920.

D. L. Turcotte, J. D. Pelletier, and W. I. Newman (1998), Networks with side branching in biology, *Journal of Theoretical Biology* **193**: 577-592.

P. J. Unmack and W. F. Fagan (2004), Convergence of differentially invaded systems toward invader-dominance: Time-lagged invasions as a predictor in desert fish communities, *Biological Invasions* **6**: 233-243.

D. Urban and T. Keitt (2001), Landscape connectivity: A graph-theoretic perspective, *Ecology* **82**: 1205-1218.

U.S. Environmental Protection Agency (USEPA) and U.S. Geological Survey(USGS) (2005), http://www.horizon-systems.com/nhdplus/index.php.

A. D. Vincent and M. R. Myerscough (2004), The effect of a non-uniform turning kernel on ant trail morphology, *Journal of Mathematical Biology* **49**: 391-432.

J. Watmough and L. Edelstein-Keshet (1995), Modelling the formation of trail networks by foraging ants, *Journal of Theoretical Biology* **176**: 357-371.

Y. C. Xia, O. N. Bjornstad, and B. T. Grenfell (2004), Measles metapopulation dynamics: A gravity model for epidemiological coupling and dynamics, *American Naturalist* **164**: 267-281.

CHAPTER 6

Biological modeling with quiescent phases

Karl P. Hadeler
University of Tübingen

Thomas Hillen
University of Alberta

Mark A. Lewis
University of Alberta

Abstract: Quiescence or dormancy plays an important role in biological systems, from spore formation in bacteria to predator-prey cycles. In a mathematical framework, quiescence is modeled by diffusive coupling of the active dynamics to quiescent phases. Although coupling a given vector field to the zero field may appear simple at first glance, quiescent phases have biologically relevant effects which can be cast into rigorous mathematical formulations: permanence of stationary points, stabilization against oscillations and Hopf bifurcations, decrease in amplitude of periodic orbits. These features are common to ordinary and partial differential equations and delay equations and persist to some extent even for density-dependent transition rates. Applications range from tumor growth to engineered bacteria.

6.1 Introduction

On all levels of biological organization we find quiescent phases although these may occur with different names. Genes may be suppressed, tumor cells quiescent, nerve cells at rest, animals hibernating or just inactive. Although these phenomena are quite diverse, there are some common general features. There is an active phase and a quiescent phase and there are transition laws which govern the exit to the quiescent phase and reentrance into the active phase.

In this chapter we investigate mathematical models for biological systems which have a sedentary, quiescent, removed or immobile phase. A quiescent phase typically describes immobile periods of mobile individuals, or refuges from predation, shelters and nests, as well as quiescent phases in a cell cycle, or bound state of diffusible proteins. For the purpose of the general analysis, we call all these phenomena *quiescent phases*. Later, in the application section, we come back to the more specific notions.

Modeling with quiescent phases can be summarized in a common mathematical framework. We will first introduce the general mathematical set up and then present a selection of applications, including ecological and epidemiological models, and cell and protein dynamics.

It is a general trend in all the results presented here, that a quiescent phase stabilizes the system; stable equilibria become more stable in the presence of a quiescent state, Hopf bifurcations become less likely, attractors become more stable, and traveling waves slow down.

In the following section, we introduce the class of models with quiescent phase and we summarize some basic mathematical properties.

6.2 Diffusive coupling and quiescence

Suppose n types of particles can exist in two different phases $v, w \in \mathbb{R}^n$ that are governed by two systems of ordinary differential equations

$$\begin{aligned} \dot{v} &= f(v) \\ \dot{w} &= g(w). \end{aligned} \tag{6.1}$$

Particles switch between phases according to Poisson processes with rates depending on the type of particle. Then we have a system in $\mathbb{R}^{2n}$,

$$\begin{aligned} \dot{v} &= f(v) - Pv + Qw \\ \dot{w} &= g(w) + Pv - Qw \end{aligned} \tag{6.2}$$

with diagonal matrices P, Q with positive entries. We say that the vector fields f, g are diffusively coupled. This type of coupling is very different from seasonal switching which leads to nonautonomous systems.

The vector of total particle densities $u = v + w$ and the vector of probability flows $z = Pv - Qw$ satisfy the equations

$$\begin{aligned} \dot{u} &= f(\tilde{Q}u + Sz) + g(\tilde{P}u - Sz) \\ S\dot{z} &= \tilde{P}f(\tilde{Q}u + Sz) - \tilde{Q}g(\tilde{P}u - Sz) - z \end{aligned} \tag{6.3}$$

with positive diagonal matrices

$$\tilde{P} = (P+Q)^{-1}P, \quad \tilde{Q} = (P+Q)^{-1}Q, \quad S = (P+Q)^{-1}.$$

If the particles switch frequently (rates going to infinity with fixed proportions) then we get the limiting system, again in $\mathbb{R}^n$,

$$\dot{u} = f(\tilde{Q}u) + g(\tilde{P}u). \tag{6.4}$$

The situation of quiescent phases occurs when g is the zero vector field. Then we compare the system

$$\dot{u} = f(u) \tag{6.5}$$

in $\mathbb{R}^n$ to the system

$$\begin{aligned} \dot{v} &= f(v) - Pv + Qw \\ \dot{w} &= Pv - Qw \end{aligned} \tag{6.6}$$

in $\mathbb{R}^{2n}$. One may think that adding a zero field does not change much. But from (6.6) we get the three following equations

$$\begin{aligned} \ddot{v} &= f'(v)\dot{v} - P\dot{v} + Q\dot{w} \\ Q\dot{w} &= QPv - Q^2 w \\ Q\dot{v} &= Qf(v) - QPv + Q^2 w. \end{aligned}$$

We add these equations, multiply by S, and get an equivalent second order equation in $\mathbb{R}^n$ for the active component v,

$$S\ddot{v} + (I - Sf'(v))\dot{v} = \tilde{Q}f(v). \tag{6.7}$$

This equation has the general form of a damped oscillator. Hence introducing a quiescent phase may lead to new phenomena. The following examples suggest that this is indeed the case.

From (6.4) we get the limiting equation $\dot{u} = f(\tilde{Q}u)$ for $u = v + w$ (the total population) and from (6.7) the limiting equation $\dot{v} = \tilde{Q}f(v)$ for v (the active population). These are equivalent by $v = \tilde{Q}u$.

Example 6.1 The equation for exponential growth, $\dot{u} = au$, with $a > 0$, leads to the system, with $p, q > 0$,

$$\begin{aligned} \dot{u} &= au - pu + qx \\ \dot{x} &= pu - qx. \end{aligned} \tag{6.8}$$

For the system (6.8) the exponent of growth is

$$\rho = \rho(a,p,q) = \frac{1}{2}\left[a - p - q + \sqrt{(a-p+q)^2 + 4pq}\right]. \tag{6.9}$$

It is easy to see that $0 < \rho < a$ and that ρ is a decreasing function of p and an increasing function of q. In the limiting cases we have

$$\rho(a,p,0) = \max(a-p,0), \quad \rho(a,0,q) = a.$$

The first formula shows that there may be population growth even if there is no return from the quiescent phase.

If we choose a negative, the result is reverted; we get $a < \rho < 0$.

Example 6.2 (Hadeler and Hillen, 2006) The logistic equation $\dot{u} = au(1 - u/K)$ is coupled to a quiescent phase and the limiting equation for the total population becomes $\dot{u} = a\tilde{q}u(1 - \tilde{q}u/K)$, where $\tilde{q} = q/(q+p)$.

Hence the growth rate is reduced by the factor $\tilde{q}$, and the carrying capacity K is enlarged to $K/\tilde{q}$. A quiescent phase slows down population growth and increases the capacity.

The equation with an Allee effect $\dot{u} = u(1-u)(u-\alpha)$, with $0 < \alpha < 1$, leads to the limiting equation $\dot{u} = \tilde{q}u(1-\tilde{q}u)(\tilde{q}u-\alpha)$. Here the threshold α is increased to $\alpha/\tilde{q}$.

Example 6.3 The harmonic oscillator (which can be seen as the linearization of a Volterra population system)

$$\begin{aligned} \dot{u} &= v \\ \dot{v} &= -u \end{aligned} \tag{6.10}$$

becomes

$$\begin{aligned} \dot{u} &= v - pu + qx \\ \dot{v} &= -u - pv + qy \\ \dot{x} &= pu - qx \\ \dot{y} &= pv - qy. \end{aligned} \tag{6.11}$$

The characteristic polynomial of the matrix is

$$\lambda^2(p+q+\lambda)^2 + (q+\lambda)^2$$

or

$$\lambda^4 + 2(p+q)\lambda^3 + (1+(p+q)^2)\lambda^2 + 2q\lambda + q^2$$

and hence the Routh-Hurwitz criterion tells that all roots have strictly negative real parts. The example shows that quiescence stabilizes the system.

In the following we show that the features of the examples are not accidental. In systems with quiescence (and equal rates) real eigenvalues move towards zero while purely imaginary eigenvalues move into the left half-plane (as has been observed first in Neubert *et al.* (2002)).

We mention in passing that quiescent phases need not be exponentially distributed. In fact, allowing other distributions and studying the stability properties of the resulting systems is a challenging problem (Hadeler and Lutscher, 2008). A case of particular interest is when exit to the quiescent phase is Poisson distributed with rate p and the length of the quiescent phase is exactly $\tau > 0$. Then the limiting equation is

$$\dot{v}(t) = f(v(t)) + p(v(t-\tau) - v(t)). \tag{6.12}$$

Again, the model is controlled by two parameters, p, τ, instead of p, q above.

6.3 Stationary states and stability

From a biological point of view we want to know how the dynamics of the system (6.5) is changed by introducing quiescent phases. This problem is also interesting from a mathematical point of view. Some aspects concerning global existence of solutions and of compact global attractors are presented in Hadeler and Hillen (2006).

General results on global attractors are surprisingly difficult. On the other hand we have some detailed results on stationary points and their stability and some preliminary results for periodic orbits.

At first glance introducing quiescent phases seems similar to introducing delays. For delay equations we know that combining a negative feedback with sufficiently large delays leads to oscillations and then periodic orbits. Quite on the contrary, quiescent phases stabilize against oscillations.

Suppose $\bar{u}$ is a stationary point of the system (6.5), i.e., $f(\bar{u}) = 0$. Then

$$(\bar{v}, \bar{w}) = (\bar{u},\, Q^{-1}P\bar{u}) \tag{6.13}$$

is a stationary point of (6.6). Let $A = f'(\bar{u})$ be the Jacobian matrix of (6.5) at the stationary point. Then the Jacobian matrix of (6.6) is given by

$$B = \begin{pmatrix} A - P & Q \\ P & -Q \end{pmatrix}. \tag{6.14}$$

The eigenvalue problem of the matrix B is equivalent to that of the matrix pencil

$$\lambda^2 I + \lambda(P + Q - A) - AQ. \tag{6.15}$$

Equal rates: In the case of equal rates we have $P = pI$, $Q = qI$, the matrices P, Q, A commute and we can apply the spectral mapping theorem to the pencil (6.15). To each eigenvalue μ of the matrix A there are two eigenvalues λ_1 and λ_2, ordered by $\Re\lambda_2 \le \Re\lambda_1$, which can be obtained from the equation

$$\lambda^2 + \lambda(p + q - \mu) - \mu q = 0. \tag{6.16}$$

This is a very simple quadratic equation. In principle the two solutions can be represented by an explicit formula. The problem is that μ is a complex number. The following can be shown. Always $\Re\lambda_2 < 0$. Hence λ_2 does not affect stability. Stability is governed by the eigenvalue λ_1.

Now there are three quite distinct cases: If $\mu = 0$ then $\lambda_1 = 0$. If μ is real then λ_1 is located between μ and 0. Hence, with respect to real eigenvalues, quiescence does not change stability. If μ is complex (with nonvanishing imaginary part) then, generally speaking, for eigenvalues with positive real parts the real parts are decreased by introducing quiescence and may eventually become negative. This effect is most prominent for eigenvalues with large imaginary parts, i.e., high frequency oscillations are damped. Detailed information is given by the following theorem.

Theorem 6.1 (Hadeler, 2008a) *Let $\mu = \alpha + i\beta$ be an eigenvalue of the linearization of (6.5) at a steady state $\bar{u}$. Then the linearization of (6.6) at $(\bar{u}, p\bar{u}/q)$ has two corresponding eigenvalues λ_1, λ_2 with $\Re\lambda_2 \le \Re\lambda_1$. The eigenvalues μ and λ_1, λ_2 are related as follows:*

(a) *Let $\mu = \alpha \in \mathbb{R}$. Then λ_1, λ_2 are real.*

(a.i) *If $\alpha < 0$ then $\lambda_2 < \alpha < \lambda_1 < 0$.*

(a.ii) *If $\alpha = 0$ then $\lambda_2 = -(p + q) < 0 = \lambda_1$.*

(a.iii) *If* $\alpha > 0$ *then* $\lambda_2 < 0 < \lambda_1 < \alpha$.

(b) *Let* $\mu = \alpha \pm i\beta$, $\beta > 0$. *Then* $\Re\lambda_2 < 0$.

(b.i) *If* $\alpha \leq 0$ *then* $\Re\lambda_1 < 0$.

(b.ii) *If* $\alpha > 0$ *then* $\Re\lambda_1 < \alpha$.

(b.iii) *If* $\alpha \leq 0$ *and*

$$\beta^2 + (p+q+\alpha)^2 + 4\alpha p > 0 \text{ and } \beta^2(q+\alpha) + \alpha(p+q+\alpha)^2 > 0,$$

then $\Re\lambda_1 < \alpha$.

(b.iv) *If* $\alpha > 0$ *and*

$$\beta^2 > 4\alpha q - (p+q-\alpha)^2 \text{ and } \beta(p-\alpha) > \alpha(p+q-\alpha)^2,$$

then $\Re\lambda_1 < 0$.

Unequal rates: If the matrices P and Q are not multiples of the identity and the various types of particles go quiescent and return with pairwise distinct rates, then the situation is quite different and the stability problem has about the same complexity as the Turing stability problem. Indeed, here as in the Turing problem we have a given stable matrix and a matrix pencil depending on positive diagonal matrices. So far only the case $n = 2$ of two types has been dealt with (Hadeler, 2008a). Recall that a 2×2 matrix $A = (a_{ij})$ is stable if $\operatorname{tr} A = a_{11} + a_{22} < 0$ and $\det A = a_{11}a_{22} - a_{12}a_{21} > 0$, and strongly stable (in the sense of Turing) if, in addition, $a_{11} \leq 0$, $a_{21} \leq 0$. (A is excitable if A is stable, but not strongly stable.) Suppose that A is stable. Then the matrix B is stable for all choices of P and Q if and only if A is strongly stable. Thus, if A is excitable in the sense of Turing, the system may become destabilized by introducing quiescent phases with suitably chosen distinct rates. The problem for $n > 2$ is open.

However, there are classes of problems for which additional mathematical tools are available (Hadeler and Thieme, 2008). For example, if the system (6.5) is cooperative then the system (6.6) is cooperative as well; or if the system (6.5) is competitive then the system for v and $-w$ is competitive as well.

6.4 Periodic orbits

Numerical simulation of standard biological systems like the MacArthur-Rosenzweig model (Holling type II predator response) as well as analytic results on highly symmetric systems show that limit cycles of the system (6.5) undergo some systematic changes if quiescent phases are introduced. From the local stability analysis at a stationary point it is evident that introducing a quiescent phase works against Hopf bifurcations. Suppose we have a system depending on some parameter α which undergoes a Hopf bifurcation. A stationary state is stable for $\alpha < 0$ and unstable for $\alpha > 0$ in such a way that a pair of eigenvalues crosses the imaginary axis at $\alpha = 0$. The stability Theorem 3.1 suggests that by introducing a quiescent phase the Hopf bifurcation is shifted to some parameter value $\alpha > 0$. This is what indeed happens in concrete examples.

Example 6.4 (Bilinsky and Hadeler, 2008) The MacArthur-Rosenzweig model with quiescence reads

$$\begin{aligned}
\dot{u} &= au(1-\frac{u}{K}) - b\frac{uv}{1+mu} - p_1u + q_1w \\
\dot{v} &= c\left(\frac{u}{1+mu} - \frac{B}{1+mB}\right)v - p_2v + q_2z \\
\dot{w} &= p_1u - q_1w \\
\dot{z} &= p_2v - q_2z.
\end{aligned} \tag{6.17}$$

It is known that the two-dimensional system without quiescence has either a stable coexistence point or a unique (stable) limit cycle. In the latter case the system with quiescence either has no limit cycle at all or again a limit cycle, this time in dimension four, whereby the "size" of the projection in the active phase gets smaller. If the coexistence state in the problem without quiescence is stable then it is strongly stable. For every choice of P and Q there is a gain in stability. Let δ and τ be the determinant and the trace at the coexistence state of the system without quiescence. In the τ, δ-plane the stability domain is given by $\delta > 0$, $\tau < 0$. For given P, Q the stability domain extends into the range $\tau > 0$. The boundary of the stability domain can be found explicitly as a curve $\tau = \phi(\delta)$ with $\phi(0) = 0$ and $\phi(\delta) > 0$ for $\delta > 0$ (Bilinsky and Hadeler, 2008).

In numerical experiments, the four-dimensional limit cycle of (6.17) can be visualized by presenting the total population sizes $u+w$ and $v+z$ for prey and predator. In this projection the effect of a quiescent phase is not easily recognized because the position of the (projection of) the stationary point is shifted. It is easier to project to the u, v-plane and also to the w, z-plane. Then one sees that the "size" of the projected limit cycle in the u, v-plane is smaller than the limit cycle in the system without quiescence and gets ever smaller if the rates are increased. Eventually the limit cycle may contract to the stationary point.

Here "size" is used as a phenomenological description. For the typical egg-shaped limit cycles of predator prey models area and circumference and diameter all shrink (see Figure 6.1) . It is interesting to observe that at the same time the projection onto the w, z-plane gets larger. For symmetric systems the shrinking of the periodic orbit can be rigorously proven, see Hadeler and Hillen (2006).

6.5 Rates depending on density

There are various biological models that can be interpreted in terms of quiescence and in which transition rates depend on the state. An example is Malik and Smith (2006, 2008), where chemostat models are extended by quiescent phases. A general

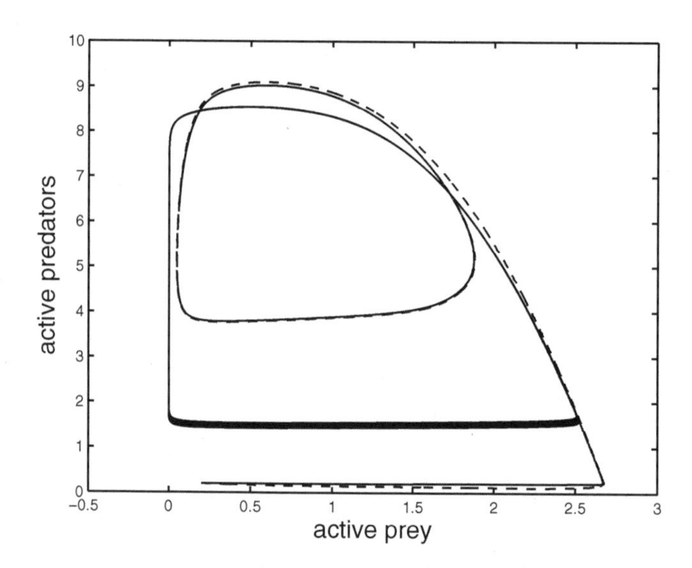

Figure 6.1 (Bilinsky and Hadeler, 2008) Phase plane for the MacArthur-Rosenzweig system (solid) and projection to the u, v-plane for the system with quiescence (dashed). Both systems have limit cycles. The projected limit cycle of the quiescent system is much smaller.

formulation of the problem is

$$\begin{aligned} \dot{v} &= f(v) - p(v,w)v + q(v,w)w \\ \dot{w} &= -q(v,w)w + p(v,w)v. \end{aligned} \tag{6.18}$$

Such a model, with $f(v) = \Delta v$, has been used in the discussion of swarming behavior (Edelstein-Keshet et al., 1998). For $f(v) = 0$ this system is equivalent to a scalar differential equation. Here we consider the case where particles in the active compartment avoid crowding, $p = p(v)$, with $p'(v) > 0$, and q constant,

$$\begin{aligned} \dot{v} &= f(v) - p(v)v + qw \\ \dot{w} &= p(v)v - qw. \end{aligned} \tag{6.19}$$

This system is equivalent to the second order equation for the active phase

$$\ddot{v} + [q + (p(v)v)' - f'(v)]\dot{v} = qf(v) \tag{6.20}$$

and then, for large transition rates, we get the limiting equation

$$\dot{v} = \frac{q}{q + (p(v)v)'} f(v) \tag{6.21}$$

which may be used (for instance in ecological applications) to estimate the total population from the observed active phase. The equations (6.5) and (6.21) have the same stationary points. The derivative at a stationary point of (6.21) becomes

$$q \frac{f'(\bar{v})}{q + p(\bar{v})}. \tag{6.22}$$

Hence the sign of the derivative does not change but the absolute value gets smaller than $|f'(\bar{v})|$.

Example 6.5 Some bacteria go quiescent (become spores) if conditions are unfavorable. Let v, w denote active and quiescent bacteria and s a substrate. Assume that the rate of going quiescent is increasing with decreasing substrate concentration. Assume further that substrate uptake is fast in comparison to reproduction and making spores. Then we have a system

$$\begin{aligned} \dot{v} &= F(s,v) - P(s)v + qw - \mu v \\ \epsilon \dot{s} &= -sv + r \\ \dot{w} &= P(s)v - qw. \end{aligned}$$

Consider the limiting case $\epsilon \to 0$. Then $s = r/v$. Define

$$p(v) = P(\frac{r}{v}), \quad f(v) = F(r/v, v) - \mu v.$$

Hence we arrive at the system (6.19) and the rate (6.22) (which determines stability) can be computed.

6.6 Slow dynamics

Rather than assuming that individuals switch between an active and a quiescent phase one can assume that they switch between a vector field f and a "slow field" κf where $\kappa \in (0,1)$. Then we have the system (with equal rates)

$$\begin{aligned} \dot{v} &= f(v) - pv + qw \\ \dot{w} &= \kappa f(w) + pv - qw. \end{aligned} \tag{6.23}$$

Suppose that $f(\bar{u}) = 0$. We look for a related stationary point of (6.23). The choice $v = w = \bar{u}$ works only for the special case $p = q$. Otherwise it is not evident how to proceed. In (Hadeler, 2008c) it has been assumed that f is homogeneous of degree 1 and that $\bar{u}\exp\{\hat{\lambda}t\}$ is an exponential solution. Then we find two related exponential solutions of the form

$$(v,w) = (\alpha_i \bar{u}, \beta_i \bar{u}) \exp\{\lambda_i t\}, \qquad i = 1,2$$

from the eigenvalue problem

$$\begin{pmatrix} \hat{\lambda} - p & q \\ p & \kappa\hat{\lambda} - q \end{pmatrix} \begin{pmatrix} \alpha_i \\ \beta_i \end{pmatrix} = \lambda_i \begin{pmatrix} \alpha_i \\ \beta_i \end{pmatrix}, \qquad i = 1,2.$$

It can be shown that both solutions are real and that the larger eigenvalue is between $\hat{\lambda}$ and 0. It can further be shown that if $\bar{u}$ is stable then the solution corresponding to the larger eigenvalue is also stable (Hadeler, 2008c). We illustrate the use of this result on a predator prey system with Holling type II functional response:

Example 6.6 (Hadeler, 2008c) Consider the homogeneous predator-prey system

$$\begin{aligned}\dot{u} &= au - b\frac{uv}{u+v} \\ \dot{v} &= c\frac{uv}{u+v} - dv,\end{aligned} \tag{6.24}$$

with $a, b, c, d > 0$. This system can be completely analyzed, e.g., in terms of the variable $\xi = u/(u+v)$ for which we get a scalar equation

$$\dot{\xi} = \xi(1-\xi)[a+d-b-(c-b)\xi]. \tag{6.25}$$

From this equation we can determine the stationary points and their stability. A stationary solution of (6.25) corresponds to an exponential solution of (6.24): If $(\bar{u}, \bar{v})\exp\{\hat{\lambda}t\}$ is an exponential solution of (6.24) then $\bar{\xi} = \bar{u}/(\bar{u}+\bar{v})$ is a stationary point of (6.25). And if $\bar{\xi}$ is a stationary point of (6.25) then there is a corresponding exponential solution of (6.24). Furthermore, the exponential solution is stable (in the sense of stability of exponential solutions) if and only if the stationary point $\bar{\xi}$ is stable. Hence the existence of exponential solutions and their stability follows from the scalar equation, but the exponents cannot be retrieved from (6.25). It turns out that for the equation (6.25) there are four orthants in parameter space with different qualitative behavior (as in a Lotka competition model).
I) $c < a + d < b$: Unstable coexistence point, attractors 0 and 1.
II) $a + d < b$ and $a + d < c$: No coexistence point. The point 0 attracts $[0, 1)$.
III) $c > a + d > b$: Coexistence point globally attracting in $(0, 1)$.
IV) $a + d > b$ and $a + d > c$: No coexistence point. The point 1 attracts $(0, 1]$.

In cases I) and III) the exponent of the coexistence solution is $\rho = (bc - ac - bd)/(b - c)$. The exponent ρ is negative in the unstable case I and positive in the stable case III.

These observations are easy to verify but a similar analysis of the problem with slow dynamics is very difficult. However, the general results guarantee that to each stable exponential solution of the two-dimensional system corresponds one stable exponential solution of the four-dimensional system. In particular, in case III, the system with slow dynamics has a stable exponential solution where prey and predator coexist.

6.7 Delay equations

In Section 6.3 we have seen that a quiescent phase and a delay have different effects. Hence it may be worthwhile to study the effect of a quiescent phase in a scalar delay equation with constant delay $\theta > 0$

$$\dot{u}(t) = f(u(t), u(t-\theta)) \tag{6.26}$$

with $f(0,0) = 0$. The system

$$\begin{aligned}\dot{v}(t) &= f(v(t), v(t-\theta)) - pv(t) + qw(t) \\ \dot{w}(t) &= pv(t) - qw(t)\end{aligned} \tag{6.27}$$

could be called the natural quiescent extension of (6.26). We linearize at $u = 0$ and at $(v, w) = (0, 0)$, respectively, and test with exponentials. Then we get the characteristic equation (β and α are the partial derivatives of f)

$$\alpha e^{-\mu\theta} + \beta - \mu = 0 \tag{6.28}$$

for (6.26) and

$$\det \begin{pmatrix} \alpha e^{-\lambda\theta} + \beta - p - \lambda & q \\ p & -q - \lambda \end{pmatrix} = 0 \tag{6.29}$$

for the system (6.27). The connection between the eigenvalues λ and μ is clearly not as simple as in (6.16). Hence the "natural extension" is not covered by Theorem 6.1. One easily understands this fact if one replaces the delay equation by a succession of ordinary differential equations representing the state at $t, t - h, t - 2h, \ldots$. Then Theorem 6.1 requires that each component, and not just the first, goes quiescent with the same rate.

Hence, in order to get the analogue of Theorem 6.1 for the delay equation, we should write the delay equation as an evolution equation in $C[-\theta, 0]$,

$$\frac{d}{dt} u_t(s) = \begin{cases} \frac{d}{ds} u_t(s) & -\theta \leq s < 0 \\ f(u_t(0), u_t(-\theta)) & s = 0 \end{cases} \tag{6.30}$$

(as usual, u_t denotes the "segment," i.e., the restriction of u to the interval $[t - \theta, t]$). Now each "component" $u_t(s)$ must go quiescent with the same rate. So we get the system

$$\begin{aligned} \frac{d}{dt} v_t(s) &= \begin{cases} \frac{d}{ds} v_t(s) - p v_t(s) + q w_t(s) & -\theta \leq s < 0 \\ f(v_t(0), v_t(-\theta)) - p v_t(0) + q w_t(0) & s = 0 \end{cases} \\ \frac{d}{ds} w_t(s) &= p v_t(s) - q w_t(s) \quad -\theta \leq s \leq 0. \end{aligned} \tag{6.31}$$

We write the equations in an elementary notation ($v(t, s) = v_t(s)$, $w(t, s) = w_t(s)$)

$$\begin{aligned} \frac{\partial}{\partial t} v(t, s) &= \frac{\partial}{\partial s} w(t, s) - p v(t, s) + q w(t, s) \\ \frac{\partial}{\partial t} w(t, s) &= p v(t, s) - q w(t, s), \quad -\theta \leq s < 0 \\ \frac{\partial}{\partial t} v(t, 0) &= f(v(t, 0), v(t, -\theta)) - p v(t, 0) + q w(t, 0) \\ \frac{\partial}{\partial t} w(t, 0) &= p v(t, 0) - q w(t, 0). \end{aligned} \tag{6.32}$$

Again we linearize at $v = w = 0$, test with exponentials, and get a differential equation and three further equations

$$\begin{aligned} \dot{v}(s) &= (\lambda + p) v(s) - q w(s) \\ (\lambda + q) w(s) &= p v(s), \quad -\theta \leq s < 0 \\ (\lambda + p) v(0) &= \alpha v(-\theta) + \beta v(t) + q w(0) \\ (\lambda + q) w(0) &= p v(0). \end{aligned}$$

We eliminate the function w and arrive at a differential equation for v and a boundary condition,

$$\dot{v}(s) = \mu v(s)$$

$$(\lambda + p - \frac{pq}{\lambda + q})v(0) = \alpha v(-\theta) + \beta v(0)$$

where at this stage

$$\mu = \lambda + p - \frac{pq}{\lambda + q} \tag{6.33}$$

is just an abbreviation. We solve this linear differential equation and insert the solution into the boundary condition. We find that λ and μ satisfy the equation

$$\alpha e^{-\mu\theta} + \beta - \mu = 0 \tag{6.34}$$

which is again (6.28). If we would insert (6.33) into (6.34) then we would get the characteristic equation for λ. Assume μ is a solution of (6.28). Then μ and λ are indeed connected by the equation (6.16) simply because (6.33) and (6.16) are equivalent.

Equation (6.27), however meaningful it may be from a modeling point of view, within the framework of quiescent phases is a system with distinct transition rates while (6.32) is a system with equal transition rates.

Example 6.7 The difference between the two approaches can be shown in the example of the blowfly equation (τ is the duration of the juvenile state)

$$\dot{u}(t) = b(u(t-\tau))e^{-\mu_0\tau} - \mu(u(t))u(t) \tag{6.35}$$

which can be derived from the Gurtin-MacCamy system (Nisbet et al., 1980; Bocharov and Hadeler, 2000; Hadeler and Bocharov, 2003; Hadeler, 2008b) with adult birth rate $b(u)$, adult death rate $\mu(u)$, and constant juvenile mortality μ_0. If there is no quiescence in the juvenile state (which amounts to $p = q = 0$ for the juvenile state and hence to different rates in the adult and in the juvenile state) then we get the "natural quiescent extension" in the form

$$\begin{aligned} \dot{u}(t) &= b(u(t-\tau))e^{-\mu_0\tau} - \mu(u(t))u(t) - pu(t) + qv(t) \\ \dot{v}(t) &= pu(t) - qv(t). \end{aligned} \tag{6.36}$$

If there is juvenile quiescence with the same rates as in the adults, then we get a similar system where the factor $\exp\{-\mu_0\tau\}$ is replaced by a larger number $\kappa = \kappa(\mu_0, p, q, \tau)$ which accounts for reduced juvenile mortality due to quiescence. For large p, q we have $\kappa \approx \exp\{-q\mu_0\tau/(p+q)\}$ (which follows immediately from the properties of Poisson processes).

6.8 Spread in space

6.8.1 *Reaction-diffusion equations*

The idea of coupled dynamics as in (6.2) can be applied to the parabolic system of two coupled scalar reaction diffusion equations:

$$\begin{aligned} v_t &= D_1\Delta v + f_1(v) - pv + qw \\ w_t &= D_2\Delta w + f_2(w) - qw + pv. \end{aligned} \tag{6.37}$$

If we imitate the procedure of (6.7), replacing $f(v)$ by $D\Delta v + f(v)$ etc., we end up with rather clumsy "viscous damped wave equations" where there are spatial derivatives within the nonlinearities, see Hadeler and Lewis (2002). If either D_1 and f_1 or D_2 and f_2 vanish, then one arrives at a single standard damped wave equation. The limiting equation of (6.37) for large p, q is

$$u_t = (\tilde{q}D_1 + \tilde{p}D_2)\Delta u + f_1(\tilde{q}u) + f_2(\tilde{p}u).$$

The following two systems have been studied in Hadeler and Lewis (2002). In the first scenario the v particles diffuse and are subject to mortality and the w particles react,

$$\begin{aligned} v_t &= D\Delta v - \mu v - pv + qw \\ w_t &= f(w) - qw + pv \end{aligned} \tag{6.38}$$

(see also Lewis and Schmitz (1996)), while in the second scenario a quiescent phase is coupled to a reaction diffusion equation,

$$\begin{aligned} v_t &= D\Delta v + f(v) - pv + qw \\ w_t &= -qw + pv. \end{aligned} \tag{6.39}$$

A single equation in the form of a damped wave equation results if one chooses to focus on one of the two variables v, w. For the system (6.38) with $\mu = 0$ we choose w and get the equation, with $\tau = 1/(p+q)$,

$$\tau w_{tt} + (1 - \tau f'(w))w_t - \tau D\Delta w_t + \tau D\Delta f(w) = \tilde{q}D\Delta w + \tilde{p}f(w), \tag{6.40}$$

while for (6.39) we choose v and get

$$\tau v_{tt} + (1 - \tau f'(v))v_t - \tau D\Delta v_t = \tilde{q}D\Delta v + \tilde{q}f(v). \tag{6.41}$$

These equations have features of damped wave equations (terms like Δw_t correspond to viscous damping) but they are parabolic because of the (almost) equivalence with (6.38) and (6.39), respectively. These systems have been studied in bounded domains with zero Dirichlet boundary conditions in Hadeler and Lewis (2002).

The problem of traveling fronts and the spread rate has been discussed in Lewis and Schmitz (1996) and Hadeler and Lewis (2002). Traveling waves are those solutions that can be expressed in terms of a single moving reference frame $z = x - ct$. The spread rate is the speed at which a locally introduced population spreads spatially. The two problems are connected. Traveling waves connecting the trivial steady state to a nontrivial steady state describe population spread with speed c. In the case that the nonlinear growth functions satisfy a convexity constraint (no Allee effect), the cooperative nature of the interaction dynamics in (6.38) and (6.39) means that the traveling wave solutions and spread rates can be fully characterized using the methods of Weinberger et al. (2002). Specifically, there exists a family of traveling wave solutions for various speeds c. Solutions exist for all speeds c greater than or equal to a minimum speed c^*. The minimum traveling wave speed is also the spread rate for a locally introduced population. Finally, the value of c^* can be determined by linear analysis about the leading edge of the invasive wave. Details are in Lewis and Schmitz (1996) and Hadeler and Lewis (2002).

6.8.2 *Reaction-transport equations*

In Hillen (2003) transport equations for spatial spread have been coupled to quiescent phases. Transport equations present alternative models to classical reaction-advection-diffusion equations, if detailed information about the movement of individuals is available. Modern tracking techniques, such as GPS data for collared mammals or birds, allow one to follow the paths of individuals and measure their mean speed, mean rate of change of direction and the distribution of turning angles. These measurements can be directly used for transport equations.

Besides moving, individuals will also stop movement to rest, to find shelter, or to forage. To model the dynamic between activity and resting the transport equation is coupled to an equation for the resting compartment, whereby the rate of stopping is spatially dependent. Let $u(t, x, s)$ denote the density of moving individuals, where $t \geq 0$ denotes time, $x \in \mathbb{R}^n$ space and $s \in V$ velocity. The set of possible velocities, V, is assumed to be a spherical shell and $|V|$ denotes its Lebesgue measure. The resting compartment is denoted by $r(t, x)$ and the total density by $N(t, x) = \int_V u(t, x, s)ds + r(t, x)$. Resting individuals that start moving can choose any velocity uniformly in V, hence a factor $|V|^{-1}$ shows up in the corresponding transition term. The stopping rate $p(x)$ is spatially dependent, while the rate q at which individuals start moving is constant. Also the turning rate $\mu > 0$ is assumed to be constant. The distribution of the newly chosen velocity is given by $T(s, s')$. The functions $l(N), g(N)$ denote loss and gain-terms, respectively. The full transport model reads

$$\begin{aligned} u_t + s \cdot \nabla u &= -\mu u + \mu \int_V T(s, s')u(.,.,s')ds' \\ &\quad -p(x)u + \frac{q}{|V|}r - l(N)u \\ r_t &= p(x) \int_V u(.,.,s)ds - qr + g(N)r - l(N)r. \end{aligned} \tag{6.42}$$

Notice that the arguments of the functions have been suppressed, except in the integrals. The turning kernel $T(s, s')$ needs to satisfy certain positivity conditions as described in detail in Hillen (2003). It is sufficient if T is positive and square integrable.

A useful tool to study transport equations is the so called "parabolic limit" (see Alt, 1980; Hillen and Othmer, 2000; Dickinson, 2000; Hillen, 2003; Chalub et al., 2004). This is in fact a scaling method for large speeds and large turning rates, or equivalently, for macroscopic time and space scales of the form

$$\tau = \epsilon^2 t, \qquad \xi = \epsilon x$$

for a small parameter $\epsilon > 0$. The details of the formal asymptotics and the corresponding convergence estimates are given in the literature cited above. Here, we only summarize the results. Up to leading order, the total population $N(\tau, \xi)$ satisfies the parabolic reaction-advection-diffusion equation

$$\begin{aligned} N_\tau &= \nabla_\xi \left(D_{pq}(\xi)\nabla_\xi N - D_{pq}(\xi)\frac{N}{q+p(\xi)}\nabla_\xi p(\xi)\right) \\ &\quad + \frac{p(\xi)}{p(\xi)+q}\tilde{g}(N)N - \tilde{l}(N)N, \end{aligned} \tag{6.43}$$

where $D_{pq}(\xi)$ denotes the *diffusion tensor*

$$D_{pq}(\xi) = \frac{q}{|V|(p(\xi)+q)}\int_V v\mathcal{F}_p(\xi)v\, dv;$$

$\mathcal{F}_p(\xi)$ is a pseudo inverse:

$$\mathcal{F}_p(\xi) = \left(\mathcal{L}_p|_{\langle 1\rangle^\perp}\right)^{-1}$$

and $\mathcal{L}_p$ denotes the *effective turning operator*

$$\begin{aligned} \mathcal{L}_p\Phi(s) &= -(\mu+p(\xi))\Phi(s) \\ &\quad +(\mu+p(\xi))\int_V \left(\frac{\mu}{\mu+p(\xi)}T(s,s') + \frac{p(\xi)}{|V|(\mu+p(\xi))}\right)\Phi(s')ds' \end{aligned}$$

and $\langle 1\rangle \subset L^2(V)$ denotes the linear subspace of functions constant in $s \in V$. The functions $\tilde{g}, \tilde{l}$ are rescaled versions of $g = \epsilon^2\tilde{g}, l = \epsilon^2\tilde{l}$, ensuring that death and reproduction occur on the macroscopic scale, and not on the scale of individual movement.

Remarks: The diffusion limit in (6.43) is remarkable in several ways:

1. The procedure quite naturally leads to nonisotropic diffusion expressed through the diffusion tensor D_{pq}. In many situations, however, the diffusion will be isotropic in which case $D_{pq} = d_{pq}I$ with a diffusion constant d_{pq} and the identity I. For example, if individuals have a constant speed $\sigma > 0$, $V = \sigma S^{n-1}$ and change of direction is uniformly distributed, $T(s,s') = |V|^{-1}$ then, as shown in Hillen and Othmer (2000), we obtain isotropic diffusion with

$$d_{pq} = \frac{\sigma^2}{n|V|}\frac{q}{(p(\xi)+q)(p(\xi)+\mu)}.$$

   More general conditions for isotropy and examples for nonisotropic diffusion are given in Hillen and Othmer (2000) and Othmer and Hillen (2002).

2. It is remarkable that (6.43) shows a taxis term including $\nabla p(\xi)$. This is a drift term in direction of higher levels of $p(\xi)$. Since the stopping rate, $p(\xi)$, is larger in favorable environments (more food, better shelter), the corresponding term describes taxis towards favorable environments. Reaction-diffusion equations with drift towards favorable environments were studied by Cosner and Lou (2003). Alternatively, the appearance of this additional taxis term can be directly motivated from a quiescent-diffusion equation, where the stopping rate is spatially dependent:

$$\begin{aligned} v_t &= D\Delta v - p(x)v + qw \\ w_t &= p(x)v - qw, \end{aligned} \tag{6.44}$$

where v describes individuals moving in space and w individuals at rest. Notice that this model corresponds to model (6.38) and model (6.39) for $f = 0$ and spatially dependent stopping rate $p(x)$.
For large transition rates p, q we obtain the limiting equation

$$\begin{aligned} u_t &= D\Delta\left(\frac{qu}{q+p(x)}\right) \\ &= D\nabla\left(\frac{q}{q+p(x)}\nabla u - \frac{qu}{(q+p(x))^2}\nabla p(x)\right), \end{aligned}$$

which shows the same taxis term as in (6.43).

3. To look at steady states that are induced by the taxis term, we assume there is no birth and death ($f = g = 0$). We consider a one-dimensional version of (6.43) on an interval $[0, l]$ with homogeneous Neumann boundary conditions. We find that for steady states we have the relation

$$N(\xi) = \kappa(q + p(\xi)),$$

with an integration constant

$$\kappa = \frac{\int_0^l N(\xi)d\xi}{ql + \int_0^l p(\xi)d\xi}.$$

This means that the shape of $N(\xi)$ follows the shape of the stopping rate, i.e., $N(\xi)$ and $p(\xi)$ have common maxima and minima.

6.9 Applications

Applications of systems with quiescent phases have been mentioned throughout the previous sections. Here we specifically discuss applications to the river drift paradox, to radiation treatment of tumors, to engineered bacteria and to infectious diseases.

6.9.1 The river drift paradox

The "river drift paradox" describes the phenomenon that various animal species persist in rapidly flowing rivers although continually individuals are drifting down the river. Apparently this problem is of a kind that showed up in Example 6.1 and also in a chemostat with washout.

Pachepsky et al. (2005) investigated the interaction of a benthic reproducing phase w and a moving phase v where individuals move (by diffusion) and can be carried away by convection. In a nondimensional form their model reads (compare (6.38))

$$\begin{aligned} v_t &= v_{xx} - \nu v_x - pv + qw \\ w_t &= w(1-w) + pv - qw, \end{aligned} \tag{6.45}$$

where ν denotes the drift velocity.

The river drift paradox can be approached in several ways. First one can consider a classical *critical domain size problem* with advection. When the link between the stationary and the mobile phases is weak ($q < 1$) then w_t remains positive for small w, and the population persists unconditionally. However, when the link is strong ($q > 1$), then persistence depends upon both the advection speed ν and the domain (river) length L. A necessary condition for persistence is that the advection speed lie below a critical threshold ($\nu < \nu^* = 2\sqrt{p/(q-1)}$). When this threshold condition is satisfied, the critical domain size approach employs the domain length L as a bifurcation parameter for existence of nontrivial solutions (i.e., persistence). Reasonable boundary conditions for the moving phase are zero flux at the top end of the stream ($x = 0$) and hostile at the bottom end of the stream ($x = L$) (Pachepsky et al., 2005). The condition for persistence is then

$$L > \frac{2}{\sqrt{\frac{4p}{q-1} - \nu^2}} \tan^{-1}\left(-\frac{1}{\nu}\sqrt{\frac{4p}{q-1} - \nu^2}\right). \tag{6.46}$$

Second, the authors consider spread in a river of infinite length, and calculate upstream and downstream traveling wave speeds. The methods for this traveling wave analysis are similar to those outlined in Lewis and Schmitz (1996) and Hadeler and Lewis (2002), but now with advection included (Pachepsky et al., 2005). The analysis can be connected to the critical domain size analysis through the threshold ν^*. A positive upstream traveling wave speed is conditional upon $\nu < \nu^*$. At $\nu = \nu^*$ the upstream invasion stalls. Thus, quite separate approaches, traveling wave speeds and critical domain size, are linked together by the critical advection speed. This approach has been extended to include generalized dispersal behavior in Lutscher et al. (2005).

Pachepsky et al. (2005) also derived the limiting equation under rapid transfer between mobile and stationary phases ($p, q \to \infty$, with $q/p = \rho$), which they call the "second Fisher approximation" for the total density of individuals,

$$(1 + \rho)u_t = u(1 - u) + \rho u_{xx} - \rho\nu u_x. \tag{6.47}$$

They use the limiting equation to find simple conditions for persistence and invasion under the assumption of strongly linked mobile and stationary populations. In agreement with our general results the authors state that "... finite residence time on the benthos ($p, q < \infty$) enhances persistence of a population." (Pachepsky et al., 2005, page 12). Also, in this problem the resting phase (immobile phase) stabilizes the dynamics.

6.9.2 Spread of genetically engineered microbes

Genetically engineered microbes (GEMs) can provide useful services in agriculture, and field trials are likely to increase in the future. Services include, for example, an extension of the growing season. This is due to prevention of ice nucleation on crops by engineered "ice-minus" bacteria (Lewis et al., 1996). However, concerns remain regarding proliferation and spread of GEMs, as well as the potential for ecosystem disruption and gene transfer.

Lewis et al. (1996) modeled spread of GEMs in the presence of competition with wild bacteria. For example, the wild counterpart to "ice-minus" bacteria is a naturally occurring "ice-plus" strain that nucleates ice crystals. While a traditional ecological approach would emphasize details of local competition, a key to modeling spread of GEMs is inclusion of a mobile compartment, describing aerosols, or surface water and groundwater suspensions, where there is rapid movement but high mortality. Here the model is

$$\begin{aligned}
\frac{\partial s_w}{\partial t} &= s_w(1 - s_w - \gamma_w s_e) + p m_w - q s_w \\
\frac{\partial s_e}{\partial t} &= r s_e(1 - s_e - \gamma_e s_w) + p m_e - q s_e \\
\frac{\partial m_w}{\partial t} &= -\mu_w m_w - p m_w + q s_w + \frac{\partial^2 m_w}{\partial x^2} \\
\frac{\partial m_e}{\partial t} &= -\mu_e m_e - p m_e + q s_e + \delta \frac{\partial^2 m_e}{\partial x^2},
\end{aligned} \tag{6.48}$$

where s and m refer to stationary and mobile compartments, and subscripts w and e denote wild and engineered strains. Note that spatial spread of strains requires linked growth and dispersal and hence nonzero transfer rates q and p.

The simplest case, which we consider here, is where wild and engineered strains are identical in all aspects but their ability to compete ($r = \delta = 1$ and $\mu_w = \mu_e = \mu$). The case with competitive exclusion of one strain by another requires one competition coefficient larger than one, and the other less than one. When one strain is only a slightly better competitor, it is reasonable to also assume $\gamma_w + \gamma_e \approx 2$. Without loss of generality we consider the case where the engineered strain is the better competitor ($\gamma_e < 1$). Although this may not always be true, it is the case of interest when it comes to the spread of GEMs.

We start by considering the limiting equation, where there are strong, balanced links between sedentary and mobile classes ($q, p \to \infty$, with $q/p = \rho$). Here the system (6.48) simplifies to a modified spatial Lotka-Volterra competition equation*

$$\begin{aligned}
(1+\rho)\frac{\partial s_w}{\partial t} &= s_w\left[1 - \rho\mu - s_w - \gamma_w s_e\right] + \rho\frac{\partial^2 s_w}{\partial x^2} \\
(1+\rho)\frac{\partial s_e}{\partial t} &= s_e\left[1 - \rho\mu - s_e - \gamma_e s_w\right] + \rho\frac{\partial^2 s_e}{\partial x^2}.
\end{aligned} \tag{6.49}$$

In this case the approach of Okubo et al. (1989) can be employed: addition of the

* Note the typo in the equivalent equations (16) and (17) from Lewis et al. (1996).

two equations and application of the condition $\gamma_w + \gamma_e = 2$ yields a single equation of Fisher form for the sedentary individuals

$$(1+\rho)\frac{\partial s}{\partial t} = s\,[1 - \rho\mu - s] + \rho\frac{\partial^2 s}{\partial x^2}. \tag{6.50}$$

Although the sedentary individuals do not actually diffuse, their behavior is consistent with the diffusion-type term in equation (6.50), because they are coupled strongly to a diffusive mobile component. This equation has a globally attracting invariant manifold $s = 1 - \rho\mu$, which is positive, providing the growth during time spent in the stationary class exceeds mortality during time spent in the mobile class. We expect initial conditions to start close to this invariant manifold, with $s_w \approx 1-\rho\mu$ and $s_e \approx 0$ except at a local perturbation which corresponds to localized introduction of the engineered strain. Hence it is reasonable to consider the case of population spread on the invariant manifold. Substitution of $s_e = 1 - \rho\mu - s_w$ into the second of equation (6.49) yields another Fisher type equation

$$\frac{\partial s_e}{\partial t} = \frac{(1-\rho\mu)(1-\gamma_e)}{(1+\rho)} s_e \left[1 - \frac{s_e}{1-\rho\mu}\right] + \frac{\rho}{1+\rho}\frac{\partial^2 s_e}{\partial x^2}, \tag{6.51}$$

with asymptotic spread rate

$$c^* = 2\frac{\sqrt{(1-\rho\mu)(1-\gamma_e)\rho}}{1+\rho}. \tag{6.52}$$

Note that spread is slowed by interstrain competition γ_e and mortality μ, but is non-monotonic with respect to the transfer rate balance $\rho = q/p$. Indeed, the worst, or speediest, invasion occurs when the mobile to stationary transfer rate slightly exceeds the stationary to mobile rate so that $p = q(1 + 2\mu)$, with speed $c^* = \sqrt{(1-\gamma_e)/(1+\mu)}$. As the mortality rate in the mobile class, μ, approaches zero, the speed simplifies to $c^* = \sqrt{1-\gamma_e}$, which is exactly half the rate calculated by Okubo et al. (1989) for the spread of a competitively superior species into another via Lotka-Volterra with simultaneous competitive growth and diffusion. The halving of the spread rate comes from differing original assumptions. Rather than allowing for simultaneous competitive growth and diffusion, equation (6.48) assumes that individuals either compete and grow, in one class, or diffuse, in another.

The case with weakly linked mobile and stationary classes can be understood using similar mathematical methods (see Lewis et al. (1996), Appendix). The invariant manifolds are found by adding the first two and second two equation of (6.48), under the assumption $\gamma_w + \gamma_e = 2$, to obtain a reduced system

$$\begin{aligned} \frac{\partial s}{\partial t} &= w(1-w) + pm - qs \\ \frac{\partial m}{\partial t} &= -\mu m - pm + qs + \frac{\partial^2 m}{\partial x^2}. \end{aligned} \tag{6.53}$$

Here the variables $s = s_w + s_e$ and $m = m_s + m_e$ represent the total number of microbes, both genetically engineered and wild, in the stationary pool and the mobile pools, respectively. Spatially homogeneous steady-state solutions to this system are

$(0, 0)$ and $(\bar{s}, \bar{m})$, where

$$\bar{s} = \frac{\mu(1-q)+p}{\mu+p} \qquad \bar{m} = \frac{q}{\mu+p}\bar{w}. \tag{6.54}$$

Contracting rectangle arguments (Smoller, 1982) show that $(\bar{s}, \bar{m})$ is a globally stable equilibrium point for (6.53) (Schmitz, 1993), and hence $s_w + s_e = \bar{s}$ and $m_s + m_e = \bar{m}$ is a globally attracting invariant manifold. On this manifold, the invading GEMs obey

$$\begin{aligned} \frac{\partial s_e}{\partial t} &= s_e(1 - s_e - \gamma_e(\bar{s} - s_e)) + pm_e - qs_e \\ \frac{\partial m_e}{\partial t} &= -\mu m_e - pm_e + qs_e + \frac{\partial^2 m_e}{\partial x^2}. \end{aligned} \tag{6.55}$$

Because $\bar{s} < 1$ (6.54) and $\gamma_e < 1$, equation (6.55) describes logistic growth in the stationary state and switching between sedentary phase and a mobile state (see Section 6.8.1). Here the spread of GEMs can be calculated as for equation (6.38). As with the strongly coupled case (above), zero mortality ($\mu = 0$) and balanced transfer rates q and p lead to a spread rate of $c^* = \sqrt{1-\gamma_e}$. Figure 2 of Lewis et al. (1996) shows spread rates for nonzero μ and unbalanced transfer rates.

6.9.3 *Tumor growth: The linear-quadratic model*

We can use the mechanism of quiescent dynamics to derive the linear quadratic model in cancer radiation treatment. There it is assumed that the surviving fraction $S(D)$ of a tumor after radiation treatment with dose $D(t)$ can be expressed as

$$S(D) = e^{-\alpha D(t) - \beta D(t)^2}, \tag{6.56}$$

where α and β are nonnegative constants. It has been shown that this model fits many data really well (Wheldon, 1988).

It is known that proliferating cells can enter a quiescent phase to eventually enter the cell cycle again. The quiescent phase is of particular interest in radiation treatment of cancer because radiation is most damaging to highly active proliferating cells. Quiescent cells are hit by radiation as well but they have time enough to repair DNA damage and recover. Hence for treatment to be successful it is important to estimate the quiescent phase. Cancer control cell cycle models were studied by Dawson and Hillen (2005) and Swierniak et al. (1996) and many others. Here we study the following model.

Let $N(t)$ denote the active tumor cells and $R(t)$ the resting tumor cells. It is assumed that cells randomly switch between the active and quiescent phases. An alternative model, where cells after proliferation directly enter the quiescent phase, has been studied in Dawson and Hillen (2005). Here we study:

$$\begin{aligned} \dot{N} &= \mu N(1 - N/K) - pN + qR - (A_1 + BD(t))\dot{D}(t)N, \\ \dot{R} &= -qR + pN - A_2\dot{D}(t)R. \end{aligned}$$

The growth of the tumor is modeled through a logistic term. Alternative models use a Gompertz law, the Bernoulli equation or a von Bertalanffy growth law (see Gyllenberg and Webb (1989), Britton (2003)). We describe the radiation damage through the *hazard function* $h(t) = (A_1 + 2BD(t))\dot{D}(t)$ (see Zaider and Minerbo (2000)), where $D(t)$ is the total dose and $\dot{D}(t)$ is the dose-rate. The parameters A_1 and A_2 describe the radiation damage caused by single hit events while the coefficient B describes double hit damage. It is assumed that quiescent cells can recover from double hit events, since they have time to repair the damage. We also assume that $A_2 < A_1$.

The limiting equation reads

$$\dot{u} = \tilde{q}u(1 - \tilde{q}u/K) - ((\tilde{q}A_1 + \tilde{p}A_2)\dot{D}(t) + \tilde{q}B\dot{D}(t)D(t))u. \tag{6.57}$$

To derive the linear-quadratic model (6.56) we assume that cell proliferation is slow on the time scale of radiation treatment. Hence we study

$$\dot{u} = -((\tilde{q}A_1 + \tilde{p}A_2)\dot{D}(t) + \tilde{q}B\dot{D}(t)D(t))u \tag{6.58}$$

which has the solution

$$u(t) = u(0)\exp(-\alpha D(t) - \beta D(t)^2),$$

with

$$\alpha = \tilde{q}A_1 + \tilde{p}A_2, \qquad \beta = \tilde{q}B.$$

The α/β-ratio is used in clinical applications to choose the best radiation protocol. It has been observed experimentally that cells in a long cell cycle have a large α/β-ratio, while cells in a short cell cycle have a low α/β-ratio. The model shows that α is a weighted mean of A_1 and A_2, while β is proportional to $\tilde{q}$ and B. Then a large α/β-ratio corresponds to small $\tilde{q}$, or small B. Small $\tilde{q}$ implies that a small fraction of the population is in the active compartment.

6.9.4 Infectious diseases

Introducing quiescent phases in the classical Kermack-McKendrick model amounts to assuming that individuals avoid contacts at random intervals (Castillo-Chavez and Hadeler, 1995; Hadeler and van den Driessche, 1997). One obtains

$$\begin{aligned}
\dot{S} &= -\beta\frac{SI}{N} - p_1 S + q_1 W \\
\dot{I} &= \beta\frac{SI}{N} - \alpha I - p_2 I + q_2 Z \\
\dot{R} &= \alpha I + \alpha Z \\
\dot{W} &= p_1 S - q_1 W \\
\dot{Z} &= -\alpha Z + p_2 I - q_2 Z \\
N &= S + I + W + Z + R
\end{aligned}$$

where S denotes active susceptible, I active infected, R the recovered, and W, Z individuals that temporally leave the risk group. The parameter β denotes the infection rate and α is the recovery rate. Here one can assume that N is constant. Hence it does not matter whether one uses mass action kinetics or standard incidence.

However, the interpretation of a quiescent phase matters. It makes a difference if people avoid social contact at all or just contacts that could cause transmission of the disease. It also matters if the total number of contacts is reduced or if it remains constant and hence the same number of contacts is distributed in the smaller then active population.

The basic reproduction number is

$$R_0 = \frac{\beta}{\alpha} \frac{q_1}{p_1 + q_1} \frac{q_2 + \alpha}{p_2 + q_2 + \alpha}. \tag{6.59}$$

In view of $d(S+I+W+Z)/dt = -\alpha(I+Z)$ it is evident that eventually $I+Z \to 0$. From $d(S+I)/dt = -\beta SI$ it follows that the total number of potential susceptibles $S+W$ is decreasing. Hence on limit sets $S+W$ is a constant and $p_1 S = q_1 W$. In contrast to the classical case there is no explicit formula or equation for the proportion of individuals which have never been infected.

Hence the model behaves essentially as the classical Kermack-McKendrick model but the quiescent phase reduces the basic reproduction number. Hadeler and van den Driessche (1997) discussed more general (and more realistic) situations where the rates depend on the prevalence of the disease.

6.9.5 Contact distributions versus migrating infective

Traditionally, the spread of epidemic diseases in space has been modeled in different ways, by contact distributions (Kendall) and by migrating individuals (Noble). A contact distribution describes the infectious force which one infectious individual at position y exerts upon a susceptible individual at position x. The contact distribution is a nonnegative symmetric convolution kernel k with $k * 1 = 1$,

$$(k * u)(x) = \int_{\mathbb{R}^n} k(x-y)u(y)dy.$$

The model assumes the form

$$\begin{aligned} S_t &= -\beta(k * I)S \\ I_t &= \beta(k * I)S - \alpha I. \end{aligned} \tag{6.60}$$

On the other hand, one can model the motion of individuals by migration processes via

$$\begin{aligned} S_t &= -\beta IS + d_S(k * S - S) \\ I_t &= \beta IS - \alpha I + d_I(k * I - I) \end{aligned} \tag{6.61}$$

where again k is a nonnegative symmetric convolution kernel with $k * 1 = 1$ and d_S, d_I are diffusion coefficients, typically different for susceptible and infected. For instance, in rabies models one assumes that only infectious individuals move, $d_S = 0$.

The contact model and the diffusion model describe different scenarios. In the contact model each individual "sits" at some location and meets other people at other locations with probability of contact decreasing with distance. The diffusion model is based on the idea that people move around and get into contact with other people. Of course this model does not imply that every person has a home base to which he/she will eventually return.

In either model, one can perform a diffusion approximation (using that the kernel is normalized and symmetric)

$$k * u \approx u + \frac{1}{2}\int_{\mathbb{R}^n} k(z) z_1^2 dz \Delta u. \tag{6.62}$$

Then the contact model (6.60) becomes Kendall's model and the diffusion model (6.61) becomes a standard system of reaction diffusion equations.

In practice the contact models and the migration models show very similar behavior. In order to compare the two approaches we consider the SIS case for both models. The contact model:

$$\begin{aligned} S_t &= -\beta S(I + \sigma I_{xx}) + \alpha I \\ I_t &= \beta S(I + \sigma I_{xx}) - \alpha I \end{aligned}$$

and thus

$$I_t = \beta(1 - I)(I + \sigma I_{xx}) - \alpha I.$$

The diffusion model:

$$\begin{aligned} S_t &= -\beta SI + \alpha I + DS_{xx} \\ I_t &= \beta SI - \alpha I + DS_{xx} \end{aligned}$$

and thus

$$I_t = \beta(1 - I)I - \alpha I + DI_{xx}.$$

Notice that this last equation is essentially the logistic equation with diffusion. We get the wave speed simply by linearizing at the leading edge (this argument can be made rigorous):

$c_0 = 2\sqrt{(\beta - \alpha)\beta\sigma}$ for the contact model.

$c_0 = 2\sqrt{(\beta - \alpha)D}$ for the diffusion model.

Hence the two formulas agree if we put $D = \beta\sigma$.

The question is whether these are just two similar but different models or whether there is some deeper connection. One connection can be made by designing a larger model for two types of stochastically moving individuals, the "quiescent" who move only in their neighborhood and the "active" who travel far. Then the two models before can be obtained as limiting cases of a larger model. Such a larger model is

$$\begin{aligned} S_t &= -S(\beta_1 I^{(1)} + \beta_2 I^{(2)}) \\ I_t^{(1)} &= \delta(\tilde{k} * I^{(1)} - I^{(1)}) - \alpha I^{(1)} + qI^{(2)} - pI^{(1)} \\ I_t^{(2)} &= S(\beta_1 I^{(1)} + \beta_2 I^{(2)}) - \alpha I^{(2)} - qI^{(2)} + pI^{(1)} \end{aligned} \tag{6.63}$$

with a nonnegative convolution kernel $\tilde{k}$, $\tilde{k} * 1 = 1$ and a coefficient $\delta > 0$.

There are susceptible S and infected individuals of two kinds, migrating $I^{(1)}$ and sedentary $I^{(2)}$. The parameters β_1 and β_2 are the transmission rates for sedentary and migrating infected individuals, respectively. A sedentary susceptible can be infected by either an infected individual residing at the same position or by a passing migrating infected. $I = I^{(1)} + I^{(2)}$ is the total number of infected individuals.

Hadeler (2003) showed that different scalings of this system lead to limiting models with contact distributions (6.60) or to limiting models with migrating infective (6.61). The migration models correspond to the situation of slow progression of the disease within the population while contact models describe spread by rapid excursions of a few highly infectious individuals.

Hence migration models and contact models can be seen as limiting cases of models with different levels of mobility.

We sketch a proof of (6.62) for normalized symmetric kernels with existing second moments. By Taylor expansion we find

$$\begin{aligned}\int_{\mathbb{R}^n} k(x-y)u(t,y)dy &= \int_{\mathbb{R}^n} k(z)u(x+z)dz \\ &= \int_{\mathbb{R}^n} k(z)(u(x) + u_x(x)z + \frac{1}{2}z^T u_{xx}(x)z + o(|z|))dz \\ &= u(x) + \frac{1}{2}\int_{\mathbb{R}^n} z^T u_{xx}(x)z + o(|z|).\end{aligned}$$

We have used $k * 1 = 1$; the u_x term goes away because of symmetry; u_{xx} is the Hessian matrix. Now

$$\int k(z)z^T u_{xx}(x)zdz = \int k(z)\sum_{ij} u_{x_i x_j}(x)z_i z_j dz$$

and

$$\int k(z)z_i z_j dz = \begin{cases} 0 & i \neq j \\ \int k(z)z_i^2 dz & i = j \end{cases}$$

and, because of the symmetry,

$$\int k(z)z_i^2 dz = \int k(x)z_1^2 dz.$$

Hence

$$\int k(z)z^T u_{xx}(x)zdz = \int k(z)z_1^2 dz\,(\Delta u)(x).$$

6.10 Discussion

Throughout this chapter, biological systems have emerged in which quiescent phases drastically change the dynamics quantitatively or even qualitatively. Generally, quiescent phases tend to slow the dynamics near equilibria, stabilize equilibria against the onset of oscillations, and enhance persistence of certain species or types.

The effect of quiescent states may be significant with respect to outcomes in specific biological systems. In fact, quiescent phases can have a quite surprising effect on the population as a whole. For example, quiescent states can induce taxis terms in movement equations. The extinction of populations (through washout) in river ecosystems can be prevented when there is a stationary phase weakly coupled to the mobile state. Cancer tumors can resist radiation treatment when cells have refuge in a quiescent state, which needs to be accounted for in radiation treatment planning. A similar effect is known for antibiotic resistance in bacteria. Balaban et al. (2004) have used a model involving a quiescent state (they called it "persisters") to fit survival data of *E. coli* bacteria which were exposed to the antibiotic *ampicillin*. They show that the existence of a persisting compartment can explain population survival and re-growth after treatment.

Moreover, our systematic approach to quiescent phases solves the longstanding discrepancy between diffusion and contact distribution models for spatial spread of epidemics, which can now be understood in terms of different scaling limits of a larger model with quiescence. It further highlights a risk of potentially erroneous conclusions about the joint effects of quiescent phases and delays.

In general, systems with quiescent phases have twice the dimension compared to systems without. Hence the mathematical analysis of such systems may become quite cumbersome (in particular in the transition from dimension two, where phase plane analysis is available, to dimension four). The methods and examples presented here provide tools to handle such systems provided the qualitative behavior of the systems without quiescent phases is well understood. However, further mathematical challenges remain, in particular to derive a solid theory for infinite dimensional systems, such as PDE's and to understand the effects of quiescent phases on global behavior, specifically the existence of compact global attractors.

6.11 Acknowledgments

This research was supported by DFG-ANumE (KPH), NSERC, MITACS (TH and MAL) and a Canada Research Chair (MAL).

6.12 References

W. Alt (1980), Biased random walk model for chemotaxis and related diffusion approximation, *J. Math. Biol.* **9**: 147-177.

N. Q. Balaban, J. Merrin, R. Chait, L. Kowali, and S. Leibler (2004), Bacterial persistence as a phenotypic switch, *Science* **305**: 1622-1625.

L. Bilinsky and K. P. Hadeler (2008), Quiescence stabilizes predator-prey relations, *J. Biol. Dynamics* accepted, open access available.

G. Bocharov and K. P. Hadeler (2000), Structured population models, conservation laws, and delay equations, *J. Differential Equations* **168**: 212-237.

N. F. Britton (2003), *Essential Mathematical Biology*, Springer, Heidelberg.

C. Castillo-Chavez and K. P. Hadeler (1995), A core group model for disease transmission, *Math. Biosc.* **128**: 41-55.

F.A.C.C. Chalub, P.A. Markovich, B. Perthame, and C. Schmeiser (2004), Kinetic models for chemotaxis and their drift-diffusion limits, *Monatsh. Math.* **142**: 123-141.

C. Cosner and Y. Lou (2003), Does movement toward better environments always benefit a population? *J. Math. Anal. Appl.* **277**: 489-503.

A. Dawson and T. Hillen (2006), Derivation of the tumor control probability (TCP) from a cell cycle model, *Comput. Math. Meth. Med.* **7**: 121-142.

R. Dickinson (2000), A generalized transport model for biased cell migration in an anisotropic environment, *J. Math. Biol.* **40**: 97-135.

L. Edelstein-Keshet, J. Watmough, and Grunbaum (1998), D. Do travelling band solutions describe cohesive swarms? An investigation for migratory locust, *J. Math. Biol.* **36**: 515-549.

W.S.C. Gurney, S.P. Blythe, and R.M. Nisbet (1980), Nicholson blowflies revisited, *Nature* **287**: 17-21.

M. Gyllenberg and G. F. Webb (1989), Quiescence as an explanation of Gompertzian tumor growth, *Growth, Development, and Aging* **53**: 25-33.

K.P. Hadeler (2003), The role of migration and contact distribution in epidemic spread, in *Biomathematical Modeling Applications in Homeland Security, Frontiers Appl. Math.,* ed. by C. Castillo-Chavez and H.T. Banks, SIAM, pp. 203-214.

K.P. Hadeler (2008a), Quiescent phases and stability, *Linear Algebra and Appl.* **428**(7): 1620-1627.

K.P. Hadeler (2008b), Neutral delay equations from and for population dyanmics, *Europ. J. Qual. Theor. Diff. Eq.* **11**: 1-18.

K.P. Hadeler (2008c), Homogeneous systems with a quiescent phase, *Math. Model. Nat. Phenom.* **3**(7): 115-125.

K.P. Hadeler and G. Bocharov (2003), Delays in population models and where to put them in particular in the neutral case, *Canadian Appl. Math. Quart.* **11**: 159-173.

K.P. Hadeler and T. Hillen (2006), Coupled dynamics and quiescent states, in *Math Everywhere,* ed. by G. Aletti, M. Burger, A. Micheletti, and D. Morale, Springer, pp 7-23.

K.P. Hadeler and M.A. Lewis (2002), Spatial dynamics of the diffusive logistic equation with sedentary compartment, *Canadian Appl. Math. Quart.* **10**: 473-499.

K.P. Hadeler and F. Lutscher (2008), Quiescent phases with distributed exit time, in preparation.

K.P. Hadeler and H.R. Thieme (2008), Monotone dependence of the spectral bound in linear compartment systems, *J. Math. Biol.* **57**: 697-712.

K.P. Hadeler and P. van den Driessche (1997), Backward bifurcation in epidemic control, *Math. Biosc.* **146**: 15-35.

T. Hillen (2003), Transport equations with resting phases, *Europ. J. Appl. Math.* **14**(5): 613-636.

T. Hillen and H. G. Othmer (2000), The diffusion limit of transport equations derived from velocity jump processes, *SIAM J. Appl. Math.* **61**(3): 751- 775.

M. A. Lewis, G. Schmitz, P. Kareiva, and J. T. Trevors (1996), Models to examine containment and spread of genetically engineered microbes, *Molecular Ecology* **5**: 165-175.

M.A. Lewis and G. Schmitz (1996), Biological invasion of an organism with separate mobile and stationary states: Modeling and analysis, *Forma* **11**: 1-25.

F. Lutscher, E. Pachepsky, and M.A. Lewis (2005), The effect of dispersal patterns on stream populations, *SIAM Rev.* **47**: 749-772.

T. Malik and H.L. Smith (2006), A resource-based model of microbial quiescence, *J. Math. Biol.* **53**: 231-252.

T. Malik and H.L. Smith (2008), Does dormancy increase fitness of bacterial populations in time-varying environments, *Bull. Math. Biol.* **70**: 1140-1162.

M.G. Neubert, P. Klepac, and P. van den Driessche (2002), Stabilizing dispersal delays in predator-prey metapopulation models, *Theor. Popul. Biol.* **61**: 339-347.

A. Okubo, P. K. Maini, M. H. Williamson, and J. D. Murray (1989), On the spatial spread of the grey squirrel in Britain, *Proc. R. Soc. Lond. B* **238**: 113-125.

H.G. Othmer and T. Hillen (2002), The diffusion limit of transport equations II: Chemotaxis equations, *SIAM J. Appl. Math.* **62**(4): 1122-1250.

E. Pachepsky, F. Lutscher, R.M. Nisbet, and M.A. Lewis (2005), Persistence, spread and the drift paradox, *Theor. Pop. Biol.* **67**: 61-73.

G. Schmitz (1993), *A Model for the Spread of Genetically Engineered Microbes*, Master thesis, University of Utah.

J. Smoller (1982), *Shock Waves and Reaction-Diffusion Equations,* Springer-Verlag, Berlin.

A. Swierniak, A. Polanski, and M. Kimmel (1996), Optimal control problems arising in cell-cycle-specific cancer chemotherapy, *Cell Prolif.* **29**: 117- 139.

H.F. Weinberger, M.A. Lewis, and B. Li (2002), Analysis of linear determinacy for spread in cooperative models, *J. Math. Biol.* **45**: 183-218.

T.E. Wheldon (1988), *Mathematical Modelling in Cancer Research,* Adam Hilger, Bristol.

M. Zaider and G.N. Minerbo (2000), Tumor control probability: a formulation applicable to any temporal protocol of dose delivery, *Phys. Med. Biol.* **45**: 279-293.

CHAPTER 7

Spatial scale and population dynamics in advective media

Roger M. Nisbet
University of California at Santa Barbara

Kurt E. Anderson
University of California at Riverside

Edward McCauley
University of Calgary

Ulrike Feudel
Carl von Ossietzky University Oldenburg

Abstract. We review recent research on mathematical models of populations that disperse in media with net unidirectional flow. Examples include drifting invertebrates in rivers and streams, marine organisms whose larvae are dispersed in local longshore currents, and plants with wind or waterborne seeds. We focus on theory relating to two issues: conditions for population persistence and biotic responses to abiotic forcing. For both issues, we identify key length scales that impact qualitative dynamics. Population persistence in an idealized, spatially homogeneous, infinitely long system is commonly guaranteed if invasion waves can advance upstream. Demographic and dispersal characteristics of organisms in finite systems determine the "critical domain size," i.e., the minimum system size for population viability. Spatial heterogeneity introduces a number of scenarios where population persistence involves source-sink dynamics. Interpretation of many aspects of steady state and transient responses to environmental forcing involves the "response length," a measure of the distance over which the impact of a point-source disturbance is felt. The recent advances have implications for future theoretical, experimental and field work, along with policy development in the areas of conservation biology and environmental assessments.

7.1 Introduction

Many organisms live in *advective* media, defined as media possessing either a net unidirectional flow or a complex velocity field with local unidirectional flow due to features such as vortices. Examples include drifting macroinvertebrates in rivers and streams, marine organisms whose larvae are dispersed in local longshore currents, and plants with wind or waterborne seeds. If the directional bias in dispersal is strong, the theory underpinning many fundamental questions in population ecology may no longer hold, or at least requires careful re-interpretation. Our aim in this essay is to review recent developments concerning population dynamics in advective media, with particular focus on two issues: *conditions for population persistence* and *biotic responses to abiotic forcing*. Although we hope that the new theory will have broad applicability, our primary emphasis is on population dynamics in rivers and streams.

We pay particular attention to the interrelationship between processes operating at multiple spatial and temporal scales. Establishing such links empirically is commonly impossible, even with large quantities of data and sophisticated statistical approaches (Diehl et al. 2008). Our essay rests on the premises that determining the underlying ecological mechanisms is an essential prerequisite to understanding these links, and that simple mathematical models can help elucidate the broader implications of mechanisms found to occur at one particular scale in space or time.

One recurring theme in the essay will be *characteristic lengths* that influence population dynamics in advective systems. Each of our main themes introduces a key length scale. Population persistence commonly requires that a river be longer than the *critical domain size* (defined in Section 7.3), or that there is a "source" region that exceeds a critical length. Many aspects of the population response to forcing can be related to the *response length* (defined in Section 7.4).

We use two relatively simple, single-species models to illustrate most of our points. These are introduced in Section 7.2. Section 7.3 introduces the so-called "drift paradox" and reviews requirements for population persistence. Section 7.4 focuses on responses to abiotic forcing for both steady-state and transient situations. We end with a brief discussion, where we suggest directions for future theoretical and empirical studies.

7.2 Models

Speirs-Gurney model

Many properties of populations in advective systems can be conveniently described using "strategic," single species models. The simplest of these are *reaction-advection-diffusion models*, the formulation of which is well covered in many texts (Cantrell and Cosner 2003; Edelstein-Keshet 1988; Gurney and Nisbet 1998; Kot 2001; Murray 1989; Nisbet and Gurney 1982; Okubo 1980; Shigesada and Kawasaki 1997). A key study of population dynamics in rivers and estuaries by Speirs and Gurney (2001), subsequently referred to as SG, assumed logistic population growth with advection and diffusion. This model was generalized by Lutscher *et al.* (2006), who added some simple hydraulics, and allowed for spatial variability in both the environment and in biotic parameters.

The SG model is a one-dimensional representation of a river (the x-axis) with $x = 0$ representing the "source," and the point $x = L$ the "mouth". Water in the river flows at a uniform speed v. A population of organisms, with density $n(x,t)$ at location x at time t, has a local per capita growth rate (births - deaths) $rf(n)$ where $f(0) = 1$ and $f'(n) \leq 0$ for all $n \geq 0$. Thus r represents the intrinsic growth rate of a small population and $f(n)$ characterizes density dependence, if present. Organisms are advected by the stream at a uniform velocity v, and, in addition, move randomly (due to individual behavior and/or small-scale turbulence) with diffusion coefficient D. The local population at any point changes due to the combined effects of reproduction and mortality, advection, and diffusion, the mathematical representation of this combination being the partial differential equation

$$\frac{\partial n}{\partial t} = \underbrace{rnf(n)}_{\text{population growth rate}} - \underbrace{v\frac{\partial n}{\partial x}}_{\text{advection}} + \underbrace{D\frac{\partial^2 n}{\partial x^2}}_{\text{diffusion}} \quad \text{for } 0 < x < L. \tag{7.1}$$

A "zero-flux" boundary condition is assumed at the point $x = 0$. Individuals that leave the river at its mouth are assumed not to return. Speirs and Gurney implemented this by assuming that the boundary $x = L$ is "hostile" or "absorbing," so that

$$vn(0,t) - D\left[\frac{\partial n}{\partial x}\right]_{x=0} = 0, \text{ and } n(L,t) = 0, \text{ for all } t \geq 0. \tag{7.2}$$

There are other representations of the boundary condition at the river mouth (Ballyk and Smith 1999; Fagan et al. 1999); these typically make quantitative rather than qualitative changes to the results discussed in this essay.

Drift-Benthos model

Many of the organisms that have motivated recent research spend all or part of their lives on the benthos. Furthermore, in many riverine systems, "storage zones" of essentially stationary water are important (Allen 1995). A model that retains much of the simplicity of SG, yet recognizes these factors, is the "drift-benthos" (DB) model, that describes a population of benthic organisms that occasionally enter the drift where they disperse passively by advection and diffusion (Lutscher et al. 2005; Pachepsky et al. 2005). This is an appropriate idealization for many aquatic insects. The model shares many assumptions with a previous model, motivated by the dynamics of microbial populations in the gut (Ballyk and Smith 1999).

The DB model uses one partial differential equation (PDE) to describe the dynamics of the sub-population in the drift, while an ordinary differential equation (ODE) is used to model the sub-population on the benthos (Table 7.1) because individuals on the benthos are assumed to be immobile. Reproduction, if included, only occurs on the benthos. "Emigration" from the benthos occurs at a density-independent, per capita rate μ, and settlement from the drift to benthos occurs at a rate σ. There are a number of options for describing recruitment - see Table 7.1 for examples.

Of particular ecological interest is the situation where an organism spends a very small fraction of its life in the drift, but moves a significant distance with each jump. For the DB model (Table 7.1), this corresponds mathematically to the situation $\sigma \to \infty$; $v \to \infty$; with the ratio v/σ and the quantity $\sqrt{D/\sigma}$ remaining finite. In that limiting situation, Lutscher et al. (2005) showed that the population dynamics are well approximated by the integro-differential equation shown in Table 7.1. The average distance traveled in a jump, the *dispersal length*, is typically close to the larger of the two quantities v/σ and $\sqrt{D/\sigma}$, each of which has the units of length.

7.3 Population persistence and the drift paradox

One long-running concern in stream ecology is the so-called "drift paradox," according to which extinction is inevitable in a closed population subject only to downstream drift. A physical analogy is the extinguishing of a candle flame in a strong wind (Straube and Pikovsky 2007). A variety of hypotheses involving some compensatory upstream movement have been proposed as resolutions of the drift paradox. An early hypothesis (Muller 1954; Muller 1982) is that adult insects compensate for downward drift of the insect larvae through upstream flight before oviposition. This hypothesis can be generalized to cover a diverse range of upstream movement mechanisms without invoking two life stages with opposing dispersal biases. In particular, sufficiently strong diffusion can lead to persistence (Ballyk et al. 1998; Ballyk and Smith 1999; Speirs and Gurney 2001), a theme we emphasize in this section of the essay. A contrasting hypothesis (Waters 1972) was that the paradox would be resolved if insects were to reside mainly on the benthos, and only the surplus over the local carrying capacity would drift downstream. One of our models (DB) points to a variant on this hypothesis.

Population persistence in an infinitely long river

For an infinitely long river ($L \to \infty$), a necessary and sufficient condition for population persistence in the SG is $v < c_{SG}$, where $c_{SG} = 2\sqrt{Dr}$. For a proof see Appendix A of Speirs and Gurney (2001). This result is intuitively appealing, since c_{SG} is the speed at which an invasion of the species would propagate upstream in an infinitely long system in the absence of advection (Fisher 1937); thus *a necessary condition for a population to persist is that its tendency to propagate upstream "wins" over washout by advection.* In the absence of diffusion ($D = 0$), persistence is thus impossible.

Pachepsky et al. (2005) studied persistence for a closed population described by the DB model. They identified two situations. First, a sufficient condition for population persistence is that $r > \mu$, i.e., the local growth rate of a small population exceeds the emigration rate. This condition has some loose similarity to the Waters hypothesis (see above), but the analysis shows that persistence depends on the local population growth rate on the benthos rather than on the carrying capacity. Second, Pachepsky et al. showed that if this inequality is reversed, population persistence is possible if

$v < c_{DB}$ where c_{DB} is again the speed of propagation of an invasion wave in a system with no advection. For the DB model, this speed is given by

$$c_{DB} = 2\sqrt{Dr}\sqrt{\frac{\sigma}{\mu - r}} = c_{SG}\sqrt{\frac{\sigma}{\mu - r}} \approx c_{SG}\sqrt{\frac{\sigma}{\mu}} \text{ if } r \ll \sigma, \mu. \quad (7.3)$$

The final approximation in the above equations refers to a biologically plausible situation where an organism enters and exits the drift many times per generation. In these circumstances, the ratio σ/μ represents the proportion of time that an animal spends on the benthos versus the drift, and the persistence condition differs from that in the SG model only by an easily interpreted numerical factor.

Critical domain size

Several authors (Ballyk et al. 1998; Ballyk and Smith 1999; Speirs and Gurney 2001) have demonstrated a link between the drift paradox and the classic problem of determining the minimum size for maintenance of a viable population in a region of space with absorbing boundaries. It is well known that a population with dynamics given by eq. (7.1), but with zero advection and with absorbing boundaries at *both* ends of the system, can only persist if its size exceeds a critical value, proportional to $\sqrt{D/r}$ and commonly called the *critical domain size* (Kierstead and Slobodkin 1953; Okubo 1984; Skellam 1951). The intuitive ecological interpretation is that for population persistence, the system must be large enough that, on average, an individual replaces itself before reaching the absorbing boundary. For a viable population with a specified intrinsic growth rate that is confined to a region of fixed size, the implication is that there is an upper bound to the diffusion coefficient.

For both the SG and DB models, advection increases the critical domain size, and for population persistence in a system of known size, then there is an upper bound to the diffusion coefficient D that increases with river velocity v. There is also a lower bound, related to the requirement described above that the population must be able to propagate upstream. Together, these bounds define a range of diffusivities that permits population persistence.

Spatial heterogeneity

For an infinitely long system, the work described above leads to powerful, intuitive insight relating persistence to the possibility of upstream invasion. Lutscher et al. (2006) proved that this intuition remains valid for spatially heterogeneous versions of the SG and DB model. There is some literature on invasion wave speeds in non-advective 1-D systems. Shigesada et al. (1986) showed that fine-grained, periodic spatial variation in growth and diffusive dispersal modifies the Fisher wave speed to $c = \sqrt{r_a D_h}$, where r_a is the *arithmetic* mean of the intrinsic growth rate, and D_h is the *harmonic* mean of the diffusion coefficient. Thus fine scale variation in growth rates has only a small effect on wave speed whereas fine scale variation in dispersal can produce "dispersal bottlenecks" that can slow or even stop the advance (see also Shigesada and Kawasaki (1997)).

The concept of critical domain size can inform studies of heterogeneous environments. Van Kirk and Lewis (1997) showed that populations in patchy environments can persist, and invasion waves can advance, even when the spatially averaged growth rates predict extinction. Their model involves "bad" (smaller than critical domain size) patches alternating with "good" (larger than critical domain size) patches; persistence was produced by source-sink dynamics.

More recent work points to another important role for source-sink dynamics; persistence may be possible, *even without upstream dispersal*, at points downstream of a population "source". Straube and Pikovsky (2007) noted that for the SG model with zero diffusion, an infinitely long tail can persist downstream of a point source (δ - function). The mechanism is simple - population exported from the source is advected downstream while simultaneously growing at the intrinsic growth rate. This mechanism has previously attracted the attention of ecologists (e.g., Holmes 2002; Reynolds and Glaister 1993). Of greater interest, Straube and Pikovsky also showed that if some finite region of a stream has sufficiently high diffusion, then population persistence is guaranteed at all downstream points. The size of this region must exceed a critical size, the formula for which has intriguing similarity to that for critical domain size. It can be shown that the same applies for the DB model as well (U. Feudel, unpublished). In short, the combination of advective motion and spatial heterogeneity can produce source-sink dynamics. The implication is that for systems where diffusion is insufficiently strong to explain population persistence (as in several ecological examples discussed by Speirs and Gurney), very distant upstream sources may be implicated. Guaranteed persistence of the downstream population can also be achieved when the flow velocity is changing along the river. The simplest case would be a piecewise change of the velocity which again leads to a critical size of a region with sufficiently slow flow. This can be related to the concept of "storage zones" mentioned above.

A contrasting approach to modeling population dynamics in advective systems with strong heterogeneity follows an approach originally developed for studies of interspecific competition (Chesson 2000). The growth rate of a small population is partitioned into terms involving nonlinearities and the (spatial) co-variance of vital rates with the environment (Melbourne and Chesson 2005; Melbourne and Chesson 2006). This work yields predictions of the differences between local versus regional scale dynamics (appropriately defined), but in its present form does not give an estimate of the length at which the transition occurs. There is an obvious opening for theoretical studies that explore the ideas of Melbourne and Chesson in models with explicit representations of advective movement.

7.4 Response to abiotic forcing

Population response to variable environments is a classic theme in theoretical ecology, with large literatures relating to both density-independent growth (e.g., Caswell (2001) and references therein), and to fluctuations near an equilibrium (e.g., Nisbet and Gurney (2003) and references therein). There is a similarly large body of literature on the exotic spatio-temporal dynamics that can exist in population models where a spatially homogeneous, steady state is unstable. However, in advective systems, there are plausible arguments suggesting that local dynamics will commonly have a stable equilibrium (Nisbet et al. 1997); for this reason we focus here on dynamics close to a stable equilibrium state in a spatially homogeneous, or near-homogeneous, system. We describe the key concepts, using the integro-differential variant of the DB model with "open recruitment" (Table 7.1).

Steady-state response

In the absence of spatial heterogeneity, a population described by the DB model would have a stable, spatially homogeneous equilibrium value. Downstream of a localized perturbation at some point x_0, deviations from this steady state population decay exponentially (proportional to $\exp[-(x - x_0)/L_R]$). For illustrations and discussion see Anderson et al. (2005), Nisbet et al. (2007), and Diehl et al. (2008). The parameter L_R, which we call the *response length*, is a measure of the distance over which the population perturbation is spread. For the integro-differential version of the DB model, the response length is well approximated as the mean distance traveled in the organism's lifetime.

The steady-state response to arbitrary, spatially extended, environmental variation is most easily calculated using Fourier analysis. This approach allows us to represent *any* arbitrary pattern of spatial variation in the environment as a sum (or integral) of simple sinusoids with different (spatial) wavelengths L_E, and was first applied to a spatial ecological model by Roughgarden (1974). Roughgarden demonstrated that for systems with random (diffusive) dispersal, spatial patterns in population density greatly amplify fine scale environmental variability when dispersal is low, with increasing attenuation as dispersal rates increase. Roughgarden named these responses *tracking* and *averaging*. Tracking and averaging were subsequently studied for other models (Engen et al. 2002; Gurney and Nisbet 1976). Anderson et al. (2005) studied tracking and averaging for the DB model with open recruitment, by calculating population distributions for the idealized circumstance where spatial variation in demographic (i.e., recruitment and mortality) rates or emigration rate is represented as a simple sinusoid with wavelength L_E. In the case of recruitment variation,

$$R(x) = \bar{R}\left(1 + a\cos\left(\frac{2\pi x}{L_E}\right)\right) \tag{7.4}$$

where $\bar{R}$ represents the spatial mean of the recruitment rate and the *amplitude*, a, is

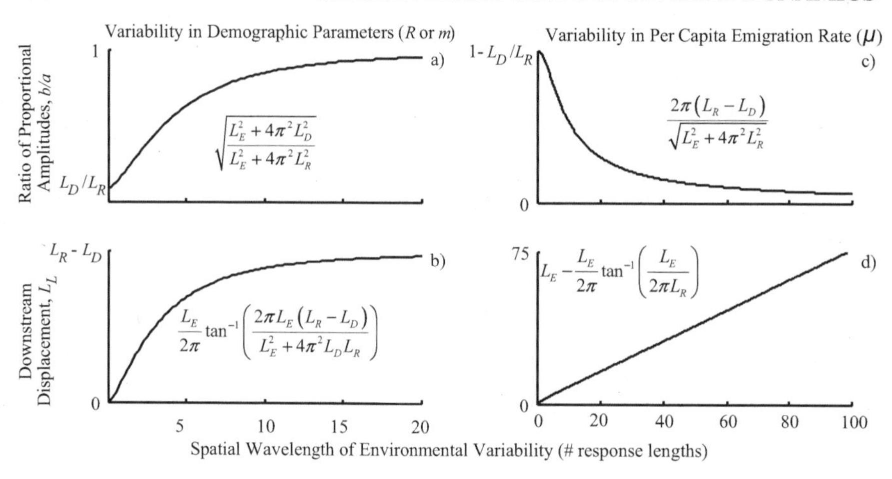

Figure 7.1 Properties of the steady state population response to environmental forcing at different spatial wavelengths derived for the integro-differential equation version of the DB model (Table 7.1). Panels (a) and (b) describe the response to variability in demographic parameters; panels (c) and (d) describe the response to variability in dispersal rate. The top panels (a,c) show the ratio of proportional amplitudes (b/a) - defined in the text. The bottom panels (b,d) show the downstream displacement (L_L). The distance units shown on the horizontal axes scale the wavelengths of environmental variability relative to the response length of the population. Also shown are the formulae (derived in Anderson et al. (2005) to which the reader is referred for further details) used to construct the figures. The quantity L_D in the equations is the average distance traveled by an organism in a single jump and is equal to $Q/(\sigma A)$ (parameters defined in Table 7.1). Adapted from Fig. 3 of Anderson et al. (2005).

the maximum proportional deviation of the recruitment rate from this mean value. The spatial wavelength L_E is a measure of the spatial scale of environmental variation. They showed that, far from the source, the population distribution is another sinusoid, but with a different amplitude, b, and displaced by a "lag," L_L from the original sinusoid (see Fig. 7.1a,b). Explicit expressions, also displayed in Fig. 7.1a,b, for the amplification (b/a) and lag show that the small scale variations in recruitment are averaged and the large scale variations are tracked. The pattern is reversed with spatial variation in the per capita emigration rate from the benthos (Fig. 7.1c,d); in that situation, small scale variation is tracked and large scale variation averaged. With both patterns the transition from tracking to averaging occurs at a spatial scale related to the response length. The analysis can be made more formal by the use of transfer functions (Nisbet et al. 2007).

Anderson et al. (2006a) and Nisbet et al. (2007) extended this analysis by adding, in turn, to the basic DB model density-dependence of demographic and dispersal rates, an immobile resource, and predation by an "ideal-free" predator that disperses towards regions with higher resource levels. In each case, the response length is readily calculated, and it determines the spatial scale at the transition from tracking to averaging. The implication is that the *response length plays a central role in deter-*

mining the population response to environmental variability. It is thus encouraging that much of the information required to estimate its value (longevity, local emigration and demographic rates and their functional dependences) is precisely the information that is available from traditional small scale experiments (Diehl et al. 2008). The missing link is typically the distribution of dispersal distances, which is likely to vary strongly with flow conditions. For one case, a small stream in the Sierra Nevada mountains in California, Diehl et al. estimated response lengths for many taxa of aquatic insects to range from meters to hundreds of meters. For stoneflies in a creek in southern England, Anderson et al. (2005) estimated response lengths ranging from hundreds of meters to a few kilometers, depending on flow rate.

Transient dynamics

Describing transients is challenging because there is a different transient for every initial disturbance. Traditional local stability analyses describe the asymptotic dynamics of a perturbed system, i.e., dynamics that occur if we wait long enough. A long-established metric characterizing asymptotic dynamics in a system with a stable equilibrium is *resilience*, the ultimate rate of approach to equilibrium (Holling 1973) - for a recent review see Botton et al. (2006). There is a need for an analogous metric describing the early time course of the transient (Neubert and Caswell 1997). Thus *reactivity* is defined as the maximum possible growth rate that could occur *immediately* following a perturbation; Neubert and Caswell provided a recipe for its calculation without explicit reference to particular initial conditions. The *amplification envelope* characterizes the complete time course of transients, including how big they can get and how much time might elapse before asymptotic behavior sets in. A system with a stable equilibrium may have positive reactivity, implying that an initial perturbation may grow in magnitude even if it is ultimately dissipated (Neubert and Caswell 1997; Neubert et al. 2004). Such systems are called *reactive*.

Anderson et al. (2008) studied these metrics for a number of models of advective systems, paying particular attention to models incorporating multiple processes that operate on different time scales (e.g., behavioral versus demographic; nutrient uptake versus utilization). As in the studies of steady-state response, analysis is facilitated by using spatial Fourier analysis. For the DB model, each Fourier component obeys a set of *ordinary* differential equations (ODEs) to which the original theory of Neubert and Caswell for nonspatial models can apply (Neubert et al. 2002). Thus, Anderson et al. were able to investigate how reactivity and the amplification envelope vary with the spatial wavelength - a powerful way of characterizing "scale dependence" of these quantities.*

A number of interesting results emerged. First, a system may be reactive at some

* For models with an advective component, the Fourier transformed equations are complex (meaning they involve $\sqrt{-1}$). However, all ecologically relevant quantities of course turn out to be real. Most of the theory of Neubert et al. (2002) remains applicable if transposed matrices are replaced by Hermitian conjugate matrices.

spatial scales and not at others. Second, the duration of amplification varies strongly with spatial scale. For the DB model, there was a striking relationship between characteristics of the amplification envelope and the response length - a long response length implying that peak amplification was small and occurred fast. Indeed, each (spatial) Fourier component of the transient response decayed over time at a rate largely determined by the openness number, defined as the ratio L_R/L_E (Nisbet et al. 2007). This clean relationship did not always hold with the consumer-resource models, probably because interactions in this model can lead to *flow-induced instability* (Malchow 1995; Malchow 2000; Rovinsky and Menzinger 1993) - including nonequilibrium spatial patterns with their own characteristic scale set by model parameters and unrelated to the response length.

7.5 Directions for future research

In the preceding sections, we surveyed recent theoretical advances relating to persistence of populations in advective media and their response to abiotic forcing. Our survey highlighted two characteristic length scales - the critical domain size and the response length - that can guide ecologists in the design of future studies. These length scales complement the analogous quantities in use in hydrology and in studies of the transport and turnover of nutrients - for example the "processing length" and "retention length" (Fisher et al. 1998; Newbold 1992; Newbold et al. 1981). Further examples are in Table 2 of Anderson et al. (2006b).

These advances have implications for future theoretical, experimental, and field work. We discuss each of these in turn.

The theoretical work described here is all based on 1-D models, whereas the fluid environment in the natural environment has obvious 3-D structure in space and time. The implications of this are poorly understood. There have been a few "strategic" studies of population persistence in 2-D systems, (e.g., Speirs and Gurney (2001), Holmes (2002), and Straube and Pikovsky (2007)), including applications motivated by estuarine dynamics where tidal cycles in the advective components may be strong. Other studies have been targeted at particular systems, for example a series of studies of copepod dynamics in the North Atlantic (Speirs et al. 2005; Speirs et al. 2004). The theoretical and practical issues that arise in such studies are common to many studies that involve the interplay of physical forcing and population dynamics (Freund et al. 2006; Oschlies and Garcon 1998; Sandulescu et al. 2007), but strong advection may induce new dynamic patterns.

A bigger challenge is "testing" the theory. Many hypotheses relating to population dynamics can be tested using carefully designed experiments in microcosms or mesocosms, but strong advection introduces major problems relating to spatial scale. We noted above that even with rather small organisms (aquatic insects), response lengths are likely to be of order tens or hundreds of meters. This is comparable with, or longer than all current systems of experimental stream channels. One way around this problem is to identify a small system that has some "functional similarity" (Petersen and Englund 2005) to the system of interest. The experimental system need not involve the same, or even similar, organisms, if the aim is to test or develop "generic," process-based models. The challenge is to design these small experiments so as to be relevant at relevant larger scales, one technique that can help such design being dimensional analysis (Petersen and Englund 2005; Petersen and Hastings 2001). What matters is that important dimensionless combinations of model parameters take similar values in the experimental system and in its larger counterpart. One recent study (Simpson et al. 2008) illustrates the potential for this approach. Simpson et al. studies the effects of spatial heterogeneity on upstream and downstream invasion speeds for periphyton in small experimental streams (a few meters in length). From data in that paper, we estimate the response length to be of order 10cm. It is tempting to conjecture that dynamics of invasion waves of periphyton in the small system will help elucidate processes that are implicated in the dynamics of aquatic invertebrates (response length of order tens or hundreds of meters) in stretches of streams as long as a few kilometers.

Direct experimental testing of the models in the field is of course the ultimate challenge. There are exciting possibilities that involve exploiting the growing body of data on the effects of stable isotope enrichment in streams. Modeling the dynamics of a stable isotope within the framework used in this essay is particularly simple if we assume that the isotope is added to a system *at equilibrium*, and that the rare isotope does not disturb this equilibrium. Concentrations of the rare isotope bound in different system components then obey *linear* equations with coefficients proportional to the *fluxes*, a property that has previously been used to help flux estimation in biochemical networks (Yangimachi et al. 2001) and in ecosystem ecology (Wollheim et al. 1999). One encouraging precedent (Wollheim et al. 1999) used a discrete time model, very similar in spirit to those discussed above to interpret data from a study of the Kuparuk river in Alaska (Peterson et al. 1997). The assumed food web is complex (Fig. 1 of Wollheim et al.), but we can reasonably conjecture that simpler models are possible as some inferred fluxes are very small.

The theory reviewed in this essay has potential applicability to many aspects of stream, river, and watershed management. One particular problem, assessment of in-stream flow needs in anthropogenically impacted rivers and streams, has been discussed in detail elsewhere (Anderson et al. 2006b). Current approaches often utilize simple hydrological or habitat-association methods which lack feedback relationships among biological components and/or physical biological coupling. Another major problem concerns understanding "urban footprints" and their environmental impact on flowing water systems. Here, recent work on the determination of nutrient uptake lengths has provided some insight (e.g., Gibson and Meyer (2007); Haggard et al. (2005)). However, the relationship among spatial scales identified by theory needs to be examined along with a more complete examination of the implication of increasing or decreasing these "lengths" for environmental quality (e.g., oxygen profiles, eutrophication - algal blooms or increased macrophyte density clogging waterways and adversely affecting other biological populations (Gucker et al. 2006;

Scrimgeour and Chambers 2000)). Our hope is that the recent theoretical advances may facilitate development of new, and better, methodology for investigating population viability or ecosystem function in streams and rivers.

7.6 Acknowledgments

We thank Sebastian Diehl, Frank Hilker, Mark Lewis, Frithjof Lutscher, Mike Neubert and Elizaveta Pachepsky for many valuable discussions. The research was supported by the National Science Foundation (Grant DEB-0717259), NSERC (Canada), and the Alberta Ingenuity Centre for Water Research (T2A-4).

Table 7.1: Equations for the "drift-benthos" (DB) model. The form presented here is an adaptation of the formalism used by Lutscher et al. (2006).

Variables

$n_B(x,t)$ = population density (number per unit length) in benthos
$n_D(x,t)$ = population density (number per unit volume) in the drift
$Q(x,t)$ = discharge rate (volume/time)
$A(x,t)$ = cross-sectional area of drift
$F(x,t)$ = net flux of water into river (volume/time) - often set to zero

Balance equations

$$A\frac{\partial n_D}{\partial t} = \underbrace{\mu n_B}_{\text{emigration}} - \underbrace{\sigma A n_D}_{\text{settlement}} - \underbrace{Q\frac{\partial n_D}{\partial x}}_{\text{advection}} + \underbrace{\frac{\partial}{\partial x}\left(DA\frac{\partial n_D}{\partial x}\right)}_{\text{diffusion}} \qquad \text{DRIFT}$$

$$\frac{\partial n_B}{\partial t} = \underbrace{R}_{\text{recruitment}} - \underbrace{m n_B}_{\text{mortality}} - \underbrace{\mu n_B}_{\text{emigration}} + \underbrace{\sigma A n_D}_{\text{settlement}} \qquad \text{BENTHOS}$$

$$\frac{\partial A}{\partial t} = -\frac{\partial Q}{\partial x} + F \qquad \text{WATER}$$

Integro-differential equation

$$\frac{\partial n_B}{\partial t} = \underbrace{R}_{\text{recruitment}} - \underbrace{m n_B}_{\text{mortality}} - \underbrace{\mu n_B}_{\text{emigration}} + \underbrace{\sigma \int_0^L h(x,u) n_B(u) du}_{\text{settlement}}$$

with "kernel," $h(x,u)$ derived from balance equations

Hydrology Specification of dynamics of $A(x,t)$ - often assumed constant

Boundary conditions

$$Q n_D(0,t) - DA\left(\frac{\partial n_D}{\partial x}\right)_{x=0} = 0 \qquad \text{Zero flux at } x=0.$$

$$n_D(L,t) = 0. \qquad \text{Absorbing boundary at } x = L.$$

Options

$$R = \begin{cases} \text{constant} & \text{open recruitment} \\ r n_B & \text{local reproduction} \\ r n_B(1 - n_B/K) & \text{logistic reproduction} \end{cases}$$

Parameters

L River length
D Diffusion coefficient
μ Emigration rate from benthos
σ Settlement rate

7.7 References

J.D. Allen (1995), *Stream Ecology*, London, Chapman and Hall.

K.E. Anderson, R.M. Nisbet and S. Diehl (2006a), Spatial scaling of consumer-resource interactions in advection-dominated systems, *American Naturalist* **168**:358-372.

K.E. Anderson, R.M. Nisbet, S. Diehl and S.D. Cooper (2005), Scaling population responses to spatial environmental variability in advection-dominated systems, *Ecology Letters* **8**: 933-943.

K.E. Anderson, R.M. Nisbet and E. McCauley (2008), Transient responses to spatial perturbations in advective systems, *Bulletin of Mathematical Biology* **70**: 1480-1502.

K.E. Anderson, A.J. Paul, E. McCauley, L. Jackson, J.R. Post and R.M. Nisbet (2006b), Instream flow needs in streams and rivers: The importance of understanding ecological dynamics, *Frontiers in Ecology and Environment* **4**: 309-318.

M. Ballyk, L. Dung, D.A. Jones and H.L. Smith (1998), Effects of random motility on microbial growth and competition in a flow reactor, *SIAM Journal on Applied Mathematics* **59**: 573-596.

M. Ballyk and H. Smith (1999), A model of microbial growth in a plug flow reactor with wall attachment, *Mathematical Biosciences* **158**: 95-126.

S. Botton, M. van Heusden, J.R. Parsons, H. Smidt and N. van Straalen (2006), Resilience of microbial systems towards disturbances, *Critical Reviews in Microbiology* **32**: 101-112.

R.S. Cantrell and Cosner (2003), *Spatial Ecology via Reaction-Diffusion Equations*, Wiley Series in Mathematical and Computational Biology, John Wiley and Sons, Chichester, UK.

H. Caswell (2001), *Matrix Population Models - Construction, Analysis and Interpretation*, Sunderland, Massachusetts, Sinauer Associates, Inc. Publishers.

P. Chesson (2000), Mechansisms of maintenace of species diversity, *Annual Review of Ecology and Systematics* **31**: 343-366.

S. Diehl, K.E. Anderson and R.M. Nisbet (2008), Population responses of drifting stream invertebrates to spatial environmental variability: An emerging conceptual framework, in *"Aquatic Insects: Challenges to Populations,"* ed. by J. Lancaster and R. A. Briers, CABI Publishing, pp. 158-183.

L. Edelstein-Keshet (1988), *Mathematical Models in Biology*, New York, Random House.

S. Engen, R. Lande and B.-E. Saether (2002), Migration and spatio-temporal variation in population dynamics in a heterogeneous environment, *Ecology* **83**: 570-579.

W.F. Fagan, R.S. Cantrell and C. Cosner (1999), How habitat edges change species interactions, *The American Naturalist* **153**: 165-182.

R.A. Fisher (1937), The wave of advance of advantageous genes, *Annals of Eugenics* **7**: 355-369.

S.G. Fisher, N.B. Grimm, E. Marti, R.M. Holmes and J.B. Jones, Jr. (1998), Material spiralling in stream corridors: a telescoping ecosystem model, *Ecosystems* **1**: 19-34.

J.A. Freund, S. Mieruch, B. Scholze, K. Wiltshire and U. Feudel (2006), Bloom dynamics in a seasonally forced phytoplankton-zooplankton model: Trigger mechanisms and timing effects, *Ecological Complexity* **3**: 129-139.

C.A. Gibson and J.L. Meyer (2007), Nutrient uptake in a large urban river, *Journal of the American Water Resources Association* **43**: 576-587.

B. Gucker, M. Brauns and M.T. Pusch (2006), Effects of wastewater treatment plant discharge on ecosystem structure and function of lowland streams, *Journal of the North American Benthological Society* **25**: 313-329.

W.S.C. Gurney and R.M. Nisbet (1976), Spatial pattern and the mechanism of population regulation, *Journal of Theoretical Biology* **59**: 361-370.

W.S.C. Gurney and R.M. Nisbet (1998), *Ecological Dynamics*, New York, Oxford University Press.

B.E. Haggard, E.H. Stanley and D.E. Storm (2005), Nutrient retention in a point-source-enriched stream, *Journal of the North American Benthological Society* **24**: 29-47.

C.S. Holling (1973), Resilience and stability of ecological systems, *Annual Review of Ecology and Systematics* **4**: 1-23.

S.J. Holmes (2002), Turbulent flows and simple behaviors: Their effects on strategic determinations of population persistence, University of Strathclyde, Glasgow, UK.

H. Kierstead and L.B. Slobodkin (1953), The size of water masses containing plankton blooms, *Journal of Marine Research* **12**: 141-147.

M. Kot (2001), *Elements of Mathematical Ecology*, New York, Cambridge University Press.

F. Lutscher, M.A. Lewis and E. McCauley (2006), The effects of heterogeneity on spread and persistence in rivers, *Bulletin of Mathematical Biology* **68**: 2129-2160.

F. Lutscher, E. Pachepsky and M.A. Lewis (2005), The effect of dispersal patterns on stream populations, *SIAM Journal of Applied Mathematics* **65**: 1305-1327.

H. Malchow (1995), Flow-and locomotion-induced pattern formation in nonlinear population dynamics, *Ecological Modelling* **82**: 257-264.

H. Malchow (2000), Motional instabilities in prey-predator systems, *Journal of Theoretical Biology* **204**:639-647.

B.A. Melbourne and P. Chesson (2005), Scaling up in population dynamics: integrating theory and data, *Oecologia* **145**: 179-187.

B.A. Melbourne and P. Chesson (2006), The scale transition: scaling up population dynamics with field data, *Ecology* **87**: 1478-1488.

K. Muller (1954), Investigations on the organic drift in North Swedish streams, *Report of the Institute of Freshwater Research, Drottningholm* **34**: 133-148.

K. Muller (1982), The colonization cycle of freshwater insects, *Oecologia* **53**: 202-207.

J.D. Murray (1989), *Mathematical Biology*, Heidelberg, Germany, Springer-Verlag.

M.G. Neubert and H. Caswell (1997), Alternatives to resilience for measuring the responses of ecological systems to perturbations, *Ecology* **78**: 653-665.

M.G. Neubert, H. Caswell and J.D. Murray (2002), Transient dynamics and pattern formation: reactivity is necessary for Turing instabilities, *Mathematical Biosciences* **175**: 1-11.

M.G. Neubert, T. Klanjscek and H. Caswell (2004), Reactivity and transient dynamics of predator-prey and foodweb models, *Ecological Modelling* **178**: 29-38.

J.D. Newbold (1992), Cycles and spirals of nutrients, in *"River Flows and Channel Forms,"* ed. by G. Petts and P. Calow, Blackwell Science, pp. 130-159.

J.D. Newbold, J.W. Elwood, R.V. O'Neill and E.W. Van Winkle (1981), Measuring nutrient spiralling in streams, *Canadian Journal of Fisheries and Aquatic Sciences* **38**: 860-863.

R.M. Nisbet, K.E. Anderson, E. McCauley and M.A. Lewis (2007), Response of equilibrium states to spatial environmental heterogeneity in advective systems, *Mathematical Biosciences and Engineering* **4**: 1-13.

R.M. Nisbet, S. Diehl, W.G. Wilson, S.D. Cooper, D.D. Donalson and K. Kratz (1997), Primary productivity gradients and short-term population dynamics in open systems, *Ecological Monographs* **67**: 535-553.

R.M. Nisbet and W.S.C. Gurney (1982), *Modelling Fluctuating Populations*, Chichester, Wiley.

R.M. Nisbet and W.S.C. Gurney (2003), *Modelling Fluctuating Populations*, Princeton, NJ, USA, Blackburn Press.

A. Okubo (1980), *Diffusion and Ecological Problems: Mathematical Models*, Biomathematics **10**, Berlin, Springer-Verlag.

A. Okubo (1984), Critical patch size for plankton patchiness, in *"Mathematical Ecology,"* ed. by S. A. Levin and T. G. Hallam, Lecture Notes in Biomathematics **54**, Springer-Verlag, New York, pp. 456-477.

A. Oschlies and V. Garcon (1998), Eddy-induced enhancement of primary production in a model of the north Atlantic Ocean, *Nature* **394**: 266-269.

E. Pachepsky, F. Lutscher, R.M. Nisbet and M.A. Lewis (2005), Persistence, spread and the drift paradox, *Theoretical Population Biology* **67**: 61-73.

J.E. Petersen and G. Englund (2005), Dimensional approaches to designing better experimental ecosystems: a practitioners guide with examples, *Oecologia* **145**: 216-224.

J.E. Petersen and A. Hastings (2001), Dimensional approaches to scaling experimental ecosystems: designing mousetraps to catch elephants, *American Naturalist* **157**: 324-333.

B.J. Peterson, M. Bahr and G.W. Kling (1997), A tracer investigation of nitrogen cycling in a pristine tundra river, *Canadian Journal of Fisheries and Aquatic Sciences* **54**: 2361-2367.

C.S. Reynolds and M.S. Glaister (1993), Spatial and temporal changes in phytoplankton abundance in the upper and middle reaches of the River Severn, *Archiv Fur Hydrobiologie* Suppl. 101: 1-22.

J. Roughgarden (1974), Population dynamics in a spatially varying environment: how population size "tracks" spatial variation in carrying capacity, *The American Naturalist* **108**: 649-664.

A.B. Rovinsky and M. Menzinger (1993), Self-organization induced by the differential flow of activator and inhibitor, *Physical Review Letters* **70**: 778-781.

M. Sandulescu, C. Lopez, E. Hernandez-Garcia and U. Feudel (2007), Plankton blooms in vortices: the role of biological and hydrodynamic timescales, *Nonlinear Processes in Geophysics* **14**: 443-454.

G.J. Scrimgeour and P.A. Chambers (2000), Cumulative effects of pulp mill and municipal effluents on epilithic biomass and nutrient limitation in a large northern river ecosystem, *Canadian Journal of Fisheries and Aquatic Sciences* **57**: 1342-1354.

N. Shigesada and K. Kawasaki (1997), *Biological Invasions: Theory and Practice,* Oxford; New York; Tokyo, Oxford University Press.

N. Shigesada, K. Kawasaki and E. Teramoto (1986), Traveling periodic-waves in heterogeneous environments, *Theoretical Population Biology* **30**: 143-160.

K. Simpson, E. McCauley and W.A. Nelson (2008), Spatial heterogeneity has differential impacts on upstream versus downstream rates of spread in experimental streams, *Oikos* **117**:1491-1499.

J.G. Skellam (1951), Random dispersal in theoretical populations, *Biometrika* **38**:196-218.

D.C. Speirs and W.S.C. Gurney (2001), Population persistence in rivers and estuaries, *Ecology* **82**: 1219-1237.

D.C. Speirs, W.S.C. Gurney, M.R. Heath and S.N. Wood (2005) Modelling the basin-scale demography of Calanus finmarchicus in the north-east Atlantic, *Fisheries Oceanography* **14**: 333-358.

D.C. Speirs, W.S.C. Gurney, S.J. Holmes, M.R. Heath, S.N. Wood, E.D. Clarke, I.H. Harms et al. (2004), Understanding demography in an advective environment: modelling Calanus finmarchicus in the Norwegian Sea, *Journal of Animal Ecology* **73**: 897-910.

A.V. Straube and A. Pikovsky (2007), Mixing-induced global models in open active flow, *Physical Review Letters* **99**: 184503.

T.F. Waters (1972), The drift of stream insects, *Annual Review of Entomology* **17**: 253-272.

W.M. Wollheim, B. Peterson, L. Deegan, M. Bahr, J.E. Hobbie, D. Jones, W.B. Bowden et al. (1999), A coupled field and modeling approach for the analysis of nitrogen cycling in streams, *Journal of the North American Benthological Society* **18**: 199-221.

K.S. Yangimachi, D.E. Stafford, A.F. Dexter, A.J. Sinskey, S. Drew and G. Stephanopoulos (2001), Application of radiolabeled tracers to biocatalytic flux analysis, *European Journal of Biochemistry* **268**: 4950-4960.

CHAPTER 8

Using multivariate state-space models to study spatial structure and dynamics

Richard A. Hinrichsen
Hinrichsen Environmental Consulting

Elizabeth E. Holmes
Northwest Fisheries Science Center

Abstract. Routine estimation of stochastic growth rates from population count data can be badly biased when measurement error is present. For univariate count time series—that is a time series of population counts from a single site or population—a number of solutions have been developed for this problem. These solutions use a state-space approach that incorporates a process model and a measurement model. In this chapter, we extend this approach in order to analyze count data from multiple sites or subpopulations. We show how a multivariate state-space model can be used in a maximum likelihood framework to estimate the stochastic growth rates from multisite data. This modeling framework allows one to take into account the spatial correlation in growth rates or process variability across multiple sites. We also show how model selection, specifically Aikake's information criteria (AIC) and a bootstrap variant designed for state-space models, can be used to make inferences about the underlying population structure. This allows researchers to measure the data support for alternative models of growth rate and covariance structure within the population. We apply multivariate state-space modeling to a multisite data set from endangered salmon populations in the Snake River basin.

8.1 Introduction

Populations in nature are rarely unstructured, that is acting as a single, well-mixed, and random-mating unit. Instead populations are structured by various mechanisms. One ubiquitous mechanism that structures populations is spatial subdivision—spread out across multiple sites, populations naturally form subpopulations that covary to a restricted degree.

In our work as population analysts, we are primarily concerned with risk assessment and forecasting of imperiled populations. Understanding the spatial structure within a population is important in this context, because spatial structure has a strong effect on extinction risk and the degree to which a population is buffered from environmental fluctuations. Monitoring data collected for populations of concern reflect this ubiquitous spatial structure. Such data are typically collected from multiple census sites, sites which are often thought to represent subpopulations that are at least partly independent. Modeling multisite data, however, presents serious challenges for population analysts. In many cases, monitoring data are limited to simple abundance counts, and data necessary to specify spatial structure—movement patterns or common environmental drivers—are missing. This hinders the use of mechanistic spatial models which require knowledge of the movement and the covariance of environmental drivers throughout a landscape (such as the models used in Lahaye et al. 1994, Dunning et al. 1995, and Schumaker et al. 2004).

Recently, statistical approaches based on time-series analysis and maximum likelihood estimation* have been developed to analyze population count data and infer underlying dynamics (Lindley 2002, Holmes and Fagan 2002, Holmes 2004, Staples et al. 2004, Dennis et al. 2006). These methods are based on research concerning the asymptotic distributions of abundance that evolve from stochastic population processes (Tuljapurkar and Orzack 1980, Dennis et al. 1991, Holmes and Semmens 2004, Holmes et al. 2007). To date, this research has focused on the analysis of single population time series and how to deal with multiple sources of variability, specifically variability from environmental fluctuations and from measurement errors. These approaches use univariate state-space models that incorporate both variance in population growth due to process error and variance in the observations due to measurement error.

In this chapter, we extend this theory and present an analytical framework for the analysis of multisite count data based on *multivariate* state-space models. This work has two related objectives. The first objective is to estimate the stochastic growth rates and variances that drive the dynamics of the population given a *known* or hypothesized spatial structure within the population. The second objective is to infer the spatial pattern of synchrony and correlation across sites. The latter objective allows us to make statistical inferences about which groups of sites act as independent subpopulations with uncorrelated changes in population abundance, which groups of sites act as independent but correlated subpopulations, and whether the sites appear to be independent observations of a single population. Figure 8.1 illustrates some of the different structures that a group of five sites might have: independent with different growth rates, independent with a shared growth rate and uncorrelated or correlated variability, and fully synchronized such that they appear to be observations of a single population. By synchronized, we mean that the sites not only have correlated changes in abundance but the sites also track each other over time (without diverging).

The methods described in this chapter are designed to infer the spatial patterns of synchrony and correlation by disentangling the variability due to measurement error,

* Bayesian state-space approaches have also been developed; however, this chapter focuses exclusively on frequentist approaches and research.

which causes the appearance of asynchrony and uncorrelation, from the underlying variability in population counts due to temporal variability in growth rates. However, we do not model the mechanisms causing these patterns explicitly—rather the methods look for the consequences: synchrony and correlation. For example, dispersal is one mechanism that can synchronize population dynamics. We do not model movement rather we model the resulting synchrony. Similarly common abiotic environmental drivers, common exposure to diseases, and a common prey base can cause correlated growth rates, but we do not model drivers explicitly only the resultant correlation. Determining what mechanisms drive patterns of synchronization and correlation revealed by the analysis would require separate studies and different types of data; however the patterns that are revealed may suggest which mechanisms are more likely and guide further data collection and analysis.

8.2 Multivariate state-space models for multisite population processes

The stochastic exponential model with Gaussian errors is the asymptotic approximation for a wide-variety of density-independent population processes, including complex age-structured and spatially-structured processes (Holmes et al. 2007). As such, this model is the foundation of much research on stochastic population dynamics. Written in log space, this model is

$$x_t = x_{t-1} + \mu + e_t, \tag{8.1}$$

where x_t represents the log-population abundance at time t, and μ is the mean rate of population growth per time step. The process-error term, e_t, represents the stochastic deviations in population-growth rate through time. The process errors are assumed to have a normal distribution[†] with a mean of zero and constant variance. The stochastic exponential model is closely related to the stochastic Gompertz model, $x_t = bx_{t-1} + \mu + e_t$, which is the stochastic approximation for a variety of density-dependent processes (Ives et al. 2003, Dennis et al. 2006). Although we use the stochastic exponential process in this chapter, the framework we present can be used for a stochastic Gompertz process also.

Suppose that instead of a single population, there are m subpopulations, which together comprise the total population. We can model the dynamics of this type of population using a multivariate stochastic exponential model:

$$\mathbf{X}_t = \mathbf{X}_{t-1} + \mathbf{B} + \mathbf{E}_t, \tag{8.2}$$

where $\mathbf{X}_t$ is an $m \times 1$ vector of log abundance in each of the m subpopulations at time t. $\mathbf{B}$ is an $m \times 1$ vector of the underlying stochastic growth rates, $\mu_1, \mu_2, \ldots, \mu_m$, in each of the m subpopulations. The process-error term, $\mathbf{E}_t$, is an $m \times 1$ vector of the serially uncorrelated stochastic deviations in each subpopulation's growth rate at time t. We assume that the process errors can be correlated between subpopulations

[†] The normality assumption arises not from convenience but from the multiplicative nature of population growth. As a result, the error terms in a process become normal (in log space) over multiple time steps.

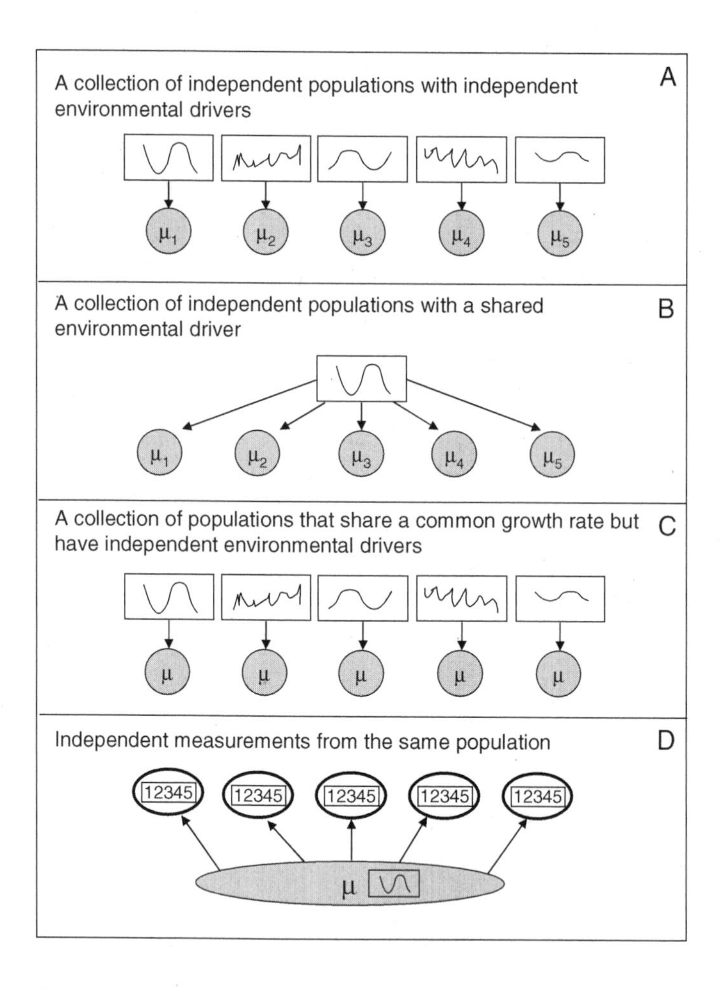

Figure 8.1 Some of the different spatial structures possible for multisite data. The μ in the figures refers to the subpopulation's stochastic growth.

by specifying that $\mathbf{E}_t$ has a multivariate normal distribution with a mean of zero and an $m \times m$ covariance matrix $\mathbf{Q}$.

Monitoring data also contain variance due to measurement error, and this will confound the estimation of $\mathbf{Q}$—the variance due to process error. Recent methods for addressing measurement error in population data use state-space models, which combine a model for the hidden true abundances with a model for the observations of the true abundance (deValpine and Hastings 2002, Lindley 2002, Dennis et al. 2006, Holmes et al. 2007). We use the same approach to address measurement error in multisite data by using a multivariate state-space model. This is achieved by combining equation (8.2) with a measurement equation that relates the observed values of log abundance at time t to the true abundances at time t:

$$\mathbf{Y}_t = \mathbf{Z}\mathbf{X}_t + \mathbf{D} + \boldsymbol{\Upsilon}_t. \tag{8.3}$$

$\mathbf{Z}$ is an $n \times m$ matrix that defines how the n observations relate to the m true abundances; in general, the n observations could be any additive combination of the m true abundances. The $n \times 1$ vector $\mathbf{D}$ specifies the bias between the observations and the true abundances. The measurement errors at time t are denoted by $\boldsymbol{\Upsilon}_t$, which is an $n \times 1$ vector of serially uncorrelated disturbances with a mean of zero and an $n \times n$ covariance matrix $\mathbf{R}$. It is important to note that $\mathbf{R}$ and $\mathbf{D}$ are not the same as the variance and bias in the sampling process—for example, the errors resulting from counting animals from, say, a plane or the errors resulting from only sampling along a transect. The sampling process is but one source, and probably a minor source, of measurement variability and bias in many population data sets. Bigger sources of measurement variability come from temporal changes in sightability due to effects of age-structure, effects of environmental conditions, and changes in the fraction of the population contained in a site-specific census. These other sources of measurement variance are usually unknown and unknowable (for all practical purposes).

Equation (8.3) permits many different relationships between the measurements and the true abundances. For this chapter, we consider only two cases. In case 1, there are m subpopulations, and each is associated with one measurement time series. In this case, m equals n, $\mathbf{Z}$ is an $m \times m$ identity matrix, and $\mathbf{D}$ equals 0[‡]. In case 2, there is one subpopulation, $m = 1$, that has been measured at n different sites. In this case, $\mathbf{Z}$ is an $n \times 1$ vector of ones, and $\mathbf{X}_t$ is a scalar (since $m = 1$). In this case, we allow $\mathbf{D}$ to be an $n \times 1$ vector with the first element equal to zero and the other elements estimated. This allows for the possibility that differences in the mean log abundance between sites are due to different biases in the measurement errors at each site.

Equations (8.2) and (8.3) form the multisite state-space model. The objective is to use this model to estimate the parameters $\{\mathbf{B}, \mathbf{Q}, \mathbf{R}, \mathbf{D}\}$: $\mathbf{B}$ gives the mean population growth rate in each subpopulation, $\mathbf{Q}$ gives the variance in the population growth between time steps, and $\mathbf{R}$ and $\mathbf{D}$ give the measurement-error variance and bias for each site. We assume for this chapter that the process errors and measurement errors are Gaussian and uncorrelated. These assumptions allow us to use estimation methods designed for linear Gaussian state-space models. The assumption of uncorrelated errors can easily be relaxed (see for example Shumway and Stoffer 2000). The assumption of Gaussian errors can also be relaxed (Durbin and Koopman 2000); however parameter estimation would be considerably more involved.

8.3 Specification of the spatial structure among the subpopulations

In its unconstrained form, the multisite state-space model allows each subpopulation to have its own population growth rate, its own process-error variance, and any level

‡ $\mathbf{D}$ cannot be estimated in this case because it is a scaling factor that drops out during estimation.

of correlation in the process errors between subpopulations. It allows a similar level of flexibility in the measurement errors. We can incorporate spatial structure by imposing constraints on the $\mathbf{B}$, $\mathbf{Q}$, and $\mathbf{R}$ terms. For example, we might specify that the process-error variances are the same across subpopulations, or that the measurement errors are independent.

We will denote the alternative model structures by the triplet $\{f_B, f_Q, f_R\}$, where f_B denotes the constraint used for $\mathbf{B}$, f_Q denotes the constraint used for $\mathbf{Q}$, and f_R denotes the constraint used for $\mathbf{R}$. In all cases, $f = 1$ will denote the unconstrained form.

8.3.1 Structure of the population growth rates (f_B)

1. $f_B = 1$ *Each subpopulation has an independent and different mean growth rate.* In this case, $\mathbf{B} = \boldsymbol{\mu}$, where $\boldsymbol{\mu}$ is an $m \times 1$ vector of subpopulation-specific stochastic growth rates μ_i. This is the unconstrained form for $\mathbf{B}$.
2. $f_B = 2$ *Each subpopulation has the same mean growth rate.* $\mathbf{B} = \boldsymbol{\mu}$, where $\boldsymbol{\mu}$ is an $m \times 1$ vector of μ's, all of which are equal.
3. $f_B = 3$ *There is one population ($m = 1$) that has been measured at n different sites.* Consequently, there is only one population growth rate. In this case, $\mathbf{B}$, $\mathbf{Q}$, and $\mathbf{X}_t$ are scalars: $\mathbf{B} = \mu$, $\mathbf{Q} = \sigma_{var}$, and $\mathbf{X}_t = x_t$. This case also affects the structure in the measurement errors: $\mathbf{Z}$ is an $n \times 1$ vector of ones and $\mathbf{D}$ is an $n \times 1$ vector of biases, the first of which is set equal to 0.[§]

8.3.2 Structure of the process-error variances (f_Q)

1. $f_Q = 1$ *An unconstrained covariance matrix.* Each subpopulation has a different level of process-error variance, and each pair of subpopulations has a different level of covariance between their process errors. In this case, $\mathbf{Q}$ is an $m \times m$ covariance matrix with terms on the diagonal and off-diagonals.
2. $f_Q = 2$ *A diagonal covariance matrix with unequal diagonal entries.* In this case, each subpopulation has a different level of process-error variance, but the process errors between subpopulations are independent. Thus, the off-diagonal terms in the covariance matrix are 0. $\mathbf{Q}$ is an $m \times m$ covariance matrix with terms on the diagonal and zeros on the off-diagonals.
3. $f_Q = 3$ *A diagonal covariance matrix with equal diagonal entries.* Each subpopulation has the same level of process-error variance, but the errors are independent. $\mathbf{Q} = \sigma_{var}\mathbf{I}$, where σ_{var} is the common process-error variance term and $\mathbf{I}$ is an $m \times m$ identity matrix. This gives a covariance matrix with all terms on the diagonal equal to σ_{var} and the off-diagonal terms equal to 0.

[§] Effectively we are estimating the biases relative to the bias for the first measurement time series. This should be kept in mind when interpreting the estimated true abundances.

4. $f_Q = 4$ *A covariance matrix with equal variances and covariances.* Each subpopulation has the same level of process-error variance, and the covariances between process errors are equal between any two subpopulations. $\mathbf{Q} = \sigma_{var}\mathbf{I} + \sigma_{cov}(\boldsymbol{U} - \mathbf{I})$, where σ_{var} is the common variance term and σ_{cov} is the common covariance term. $\mathbf{I}$ is an $m \times m$ identity matrix and $\boldsymbol{U}$ is an $m \times m$ unit matrix. This gives a covariance matrix with all terms on the diagonal equal to σ_{var} and the off-diagonal terms equal to σ_{cov}.

8.3.3 Structure of the measurement errors (f_R)

The constraints on the measurement-error variances are the same as for the process-error variance. The different measurement-error models are denoted $f_R = 1, 2, 3$ or 4, where the constraints are defined as in Section 8.3.2 with references to 'process-error' replaced with 'measurement-error' and with references to $\mathbf{Q}$ replaced with $\mathbf{R}$.

8.4 Estimation of the population parameters using maximum likelihood

Equations (8.2) and (8.3), along with the model constraints specified by $\{f_B, f_Q, f_R\}$, form the constrained model for the multisite data. Using this model, we can estimate the parameters that describe the population dynamics, $\mathbf{B}$ and $\mathbf{Q}$, and the parameters that describe the measurement error, $\mathbf{R}$ and $\mathbf{D}$. There are two main approaches to parameter estimation: maximum likelihood estimation and Bayesian estimation. In this chapter, we focus on maximum likelihood estimation. However, the likelihood functions specified in this chapter are also used in Bayesian estimation. Thus, this chapter provides the building blocks needed for a Bayesian approach as well.

8.4.1 The likelihood function

The first step of maximum likelihood estimation is to specify the likelihood of the parameters, $\Theta = \{\mathbf{B}, \mathbf{Q}, \mathbf{R}, \mathbf{D}\}$, given the observed data. In our case, the data are n time series of observations for time 1 to T. We denote the n observations at time t as $\mathbf{Y}_t$, and the set of all observations for time t to T as $\mathbf{Y}_1^T \equiv \mathbf{Y}_1, \mathbf{Y}_2, \ldots, \mathbf{Y}_T$.

The observations at time t, $\mathbf{Y}_t$, are dependent on the past observations, $\mathbf{Y}_1^{t-1}$. Thus we cannot write the likelihood simply as $L(\Theta) = \prod_{t=1}^{T} p(\mathbf{Y}_t)$. Instead we write the likelihood as a product of the conditional probabilities¶:

$$L(\Theta|\mathbf{Y}_1^T) = p(\mathbf{Y}_1) \prod_{t=2}^{T} p(\mathbf{Y}_t|\mathbf{Y}_1^{t-1}), \tag{8.4}$$

¶ For more background on the derivation of the likelihood see, for example, Harvey (1989) section 3.4.

where $p(\mathbf{Y}_t|\mathbf{Y}_1^{t-1})$ is the probability density function for $\mathbf{Y}_t$ given all of the observations up to time $t-1$. Note that $\mathbf{Y}$ is not Markov (only $\mathbf{X}$ is), thus we must condition on $\mathbf{Y}_1^{t-1}$ rather than $\mathbf{Y}_{t-1}$. The distribution of $p(\mathbf{Y}_t|\mathbf{Y}_1^{t-1})$ is multivariate normal, and we denote the mean of this distribution as $\tilde{\mathbf{Y}}_{t|t-1}$ and its covariance matrix as $\mathbf{F}_t$. $\tilde{\mathbf{Y}}_{t|t-1}$ is defined as $\mathrm{E}(\mathbf{Y}_t|\mathbf{Y}_1^{t-1})$, the expected value of $\mathbf{Y}_t$ conditioned on $\mathbf{Y}_1^{t-1}$. $\mathbf{F}_t$ is defined as $\mathrm{E}((\mathbf{Y}_t - \tilde{\mathbf{Y}}_{t|t-1})(\mathbf{Y}_t - \tilde{\mathbf{Y}}_{t|t-1})')$, also conditioned on $\mathbf{Y}_1^{t-1}$. The initial conditions are specified by $p(\mathbf{Y}_1)$, and will be treated as either an estimated or a nuisance parameter (see Section 8.4.2).

Using the probability density for a multivariate normal, we can write out the likelihood function given in equation (8.4) as

$$L(\Theta|\mathbf{Y}_1^T) = \prod_{t=1}^{T} \frac{\exp\left\{-\frac{1}{2}(\mathbf{Y}_t - \tilde{\mathbf{Y}}_{t|t-1})'\mathbf{F}_t^{-1}(\mathbf{Y}_t - \tilde{\mathbf{Y}}_{t|t-1})\right\}}{\left((2\pi)^n|\mathbf{F}_t|\right)^{1/2}}. \tag{8.5}$$

To calculate the likelihood, we need estimates of $\tilde{\mathbf{Y}}_{t|t-1}$ and $\mathbf{F}_t$. We do not have these directly, but we can solve for them indirectly by rewriting them in terms of $\mathbf{X}_t$ and the deviations in $\mathbf{X}_t$ from the predicted values. First, we define $\mathbf{x}_{t|t-1} \equiv \mathrm{E}(\mathbf{X}_t|\mathbf{Y}_1^{t-1})$. Then, using the measurement equation (equation (8.3)), we have:

$$\begin{aligned} \tilde{\mathbf{Y}}_{t|t-1} &= \mathrm{E}(\mathbf{Y}_t|\mathbf{Y}_1^{t-1}) = \mathrm{E}(\mathbf{Z}\mathbf{X}_t + \mathbf{D} + \boldsymbol{\Upsilon}_t|\mathbf{Y}_1^{t-1}) \\ &= \mathbf{Z}\mathbf{x}_{t|t-1} + \mathbf{D}. \end{aligned} \tag{8.6}$$

Next, we define $\mathbf{P}_{t|t-1} \equiv \mathrm{E}((\mathbf{X}_t - \mathbf{x}_{t|t-1})(\mathbf{X}_t - \mathbf{x}_{t|t-1})')$. Then, using the measurement equation (equation (8.3)) again, we have:

$$\begin{aligned} \mathbf{F}_t &= \mathrm{E}\left(\left[\mathbf{Y}_t - \tilde{\mathbf{Y}}_{t|t-1}\right]\left[\mathbf{Y}_t - \tilde{\mathbf{Y}}_{t|t-1}\right]'\right) \\ &= \mathrm{E}\left(\left[\mathbf{Z}(\mathbf{X}_t - \mathbf{x}_{t|t-1}) + \boldsymbol{\Upsilon}_t\right]\left[\mathbf{Z}(\mathbf{X}_t - \mathbf{x}_{t|t-1}) + \boldsymbol{\Upsilon}_t\right]'\right) \\ &= \mathbf{Z}\mathbf{P}_{t|t-1}\mathbf{Z}' + \mathbf{R}. \end{aligned} \tag{8.7}$$

Thus using equations (8.6) and (8.7), we can solve for the likelihood if we have estimates of $\mathbf{x}_{t|t-1}$, the expected value of $\mathbf{X}_t$ given the observed data up to time $t-1$, and $\mathbf{P}_{t|t-1}$, the deviations between $\mathbf{X}_t$ and $\mathbf{x}_{t|t-1}$.

8.4.2 *Estimation of* $\mathbf{x}_{t|t-1}$ *and* $\mathbf{P}_{t|t-1}$ *using the Kalman filter*

The multisite state-space model is a linear dynamical system with discrete time and Gaussian errors. This type of problem is extremely important in many engineering fields. In 1960, Rudolf Kalman published an algorithm that solves for the optimal (lowest mean square error) estimate of the hidden $\mathbf{X}_t$ based on the observed data up to time t for this class of linear dynamical system. This algorithm, now known as the Kalman filter, gives an estimate of $\mathrm{E}(\mathbf{X}_t|\mathbf{Y}_1^t)$, which we will denote as $\mathbf{x}_{t|t}$, and the covariance, $\mathrm{E}((\mathbf{X}_t - \mathbf{x}_{t|t})(\mathbf{X}_t - \mathbf{x}_{t|t})')$, which we will denote as $\mathbf{P}_{t|t}$. The Kalman filter also provides the optimal estimates of $\mathbf{X}_t$ conditioned on the data up to time $t-1$, i.e., $\mathbf{x}_{t|t-1}$ and its covariance, $\mathbf{P}_{t|t-1}$. These are the estimates that are needed to calculate $\tilde{\mathbf{Y}}_{t|t-1}$ and $\mathbf{F}_t$ in equations (8.6) and (8.7). These in turn are used to calculate the likelihood. The Kalman filter is widely used in time-series analysis, and there are many textbooks covering it and its applications. The books by Harvey (1989) and Shumway and Stoffer (2000) are particularly useful for ecologists because they are geared towards physical, biological, and economics applications.

The Kalman filter is a recursion that consists of a set of prediction equations followed by a set of updating equations. The prediction equations are so named because they predict the states at time t given information up to and including time $t-1$:

$$\mathbf{x}_{t|t-1} = \mathbf{x}_{t-1|t-1} + \mathbf{B} \quad (8.8)$$

$$\mathbf{P}_{t|t-1} = \mathbf{P}_{t-1|t-1} + \mathbf{Q}. \quad (8.9)$$

Using the output from the prediction equations, new estimates conditioned on the data up to time t are calculated using the updating equations:

$$\mathbf{x}_{t|t} = \mathbf{x}_{t|t-1} + \mathbf{P}_{t|t-1}\mathbf{Z}'\mathbf{F}_t^{-1}(\mathbf{Y}_t - \mathbf{Z}\mathbf{x}_{t|t-1} - \mathbf{D}) \quad (8.10)$$

$$\mathbf{P}_{t|t} = \mathbf{P}_{t|t-1} - \mathbf{P}_{t|t-1}\mathbf{Z}'\mathbf{F}_t^{-1}\mathbf{Z}\mathbf{P}_{t|t-1}. \quad (8.11)$$

This recursive algorithm is started with initial values $\mathbf{x}_{0|0}$ and $\mathbf{P}_{0|0}$, which are the mean and variance of the population abundance at time $t=0$. Using those initial values, one iterates through the prediction and updating equations for $t = 1, 2, 3, \ldots, T$. This provides the time series, $\mathbf{x}_{t|t-1}$ and $\mathbf{F}_t$, that are needed to calculate the likelihood.

Typically, there is no prior information for the abundances at time $t=0$. One solution is to estimate $\mathbf{x}_{0|0}$ and $\mathbf{P}_{0|0}$ as extra free parameters. Alternatively, the initial conditions can be specified using a diffuse prior distribution for $\mathbf{X}_0$. This is done by setting $\mathbf{P}_{0|0} = \kappa\mathbf{I}$ (where $\mathbf{I}$ is an $m \times m$ identity matrix), substituting this into the Kalman filter equations and allowing κ to grow arbitrarily large. Since $\mathbf{X}_0$ is defined as normal with a mean of $\mathbf{x}_{0|0}$ and variance $\mathbf{P}_{0|0}$, this has the effect of setting a diffuse prior on $\mathbf{X}_0$. When $f_B = 1$ or 2, this diffuse prior leads to $\mathbf{x}_{1|1} = \mathbf{Y}_1$ and $\mathbf{P}_{1|1} = \mathbf{R}$. When $f_B = 3$, $\mathbf{x}_{1|1}$ and $\mathbf{P}_{1|1}$ are scalars, and the diffuse prior leads to $\mathbf{x}_{1|1} = \left(\mathbf{O}'\mathbf{R}^{-1}\mathbf{Y}_1\right) / \left(\mathbf{O}'\mathbf{R}^{-1}\mathbf{O}\right)$ and $\mathbf{P}_{1|1} = 1/\left(\mathbf{O}'\mathbf{R}^{-1}\mathbf{O}\right)$, where $\mathbf{O}$ indicates an $m \times 1$ vector of ones.

For simplicity, we presented the likelihood calculation as if there were no missing values in the data. However, one of the strengths of state-space approaches is that missing values are easy to accommodate; if some values within $\mathbf{Y}_t$ are missing, those values become a place-holder that will be filled with the optimal estimate for the missing data point. Harvey (1989), section 3.4, shows how the Kalman filter equations are modified when there are missing values.

8.4.3 Maximization of the likelihood function

The Kalman filter provides estimates of $\mathbf{x}_{t|t-1}$ and $\mathbf{P}_{t|t-1}$ that together with equations (8.5), (8.6), and (8.7) allow us to calculate the likelihood of the parameters, Θ.

Our objective is to find the Θ that maximizes the likelihood. There are a variety of approaches to the maximization problem. One standard approach is a Nelder-Mead algorithm, which is available as a pre-packaged routine for most computing software. However, for the multisite state-space model, we found that this algorithm did not always converge. Another approach, which we found to always converge, is the estimation-measurement (EM) algorithm presented in Shumway and Stoffer (1982) and Shumway and Stoffer (2000, section 4.3). The EM algorithm involves iteratively estimating the true, hidden, abundances conditioned on all of the data, using that to re-estimate the parameters and then using the updated parameters to re-estimate the true abundances. This is repeated until the likelihood converges.[‖] Another wrinkle that can be added is restricted maximum likelihood (REML). Because of the measurement errors, there is a negative temporal correlation in the data. This negative correlation provides additional information which can be used to improve the estimates (Staples et al. 2004, Dennis et al. 2006). In Section 8.6, we will discuss bootstrap methods for specifying the confidence intervals, standard errors, and bias of the maximum likelihood parameters.

One issue of concern for all of these maximization methods is that when the time series are short (T is small) or contain many missing values, the likelihood surface can become multimodal. The problem in this case is that the likelihood surface has its largest peak with either the $\mathbf{Q}$ or $\mathbf{R}$ diagonal terms set at zero, and there is a smaller peak at the correct value where all $\mathbf{Q}$ and $\mathbf{R}$ diagonal terms are nonzero. The result is that all of the variance in the data is put into process-error or measurement-error variance. Intuitively, what is happening is that there is not enough information in the data to partition the variance. If this is discovered to be a problem, which will be apparent by either of the $\mathbf{Q}$ or $\mathbf{R}$ diagonal terms going to zero, there are two general solutions. First, the size of the model can be constrained such that it is commensurate with the information in the data. For example, the population structure can be constrained (e.g., by setting $f_B = 3$ or $f_Q = 4$) so that there are fewer parameters to estimate. The second general approach is add an informative prior on the variance parameters using a Bayesian approach. In this case, the prior will affect the posterior estimates. This is the objective in this case, since the data do not contain enough information in and of themselves to partition the variance. Obviously, the use of an informative prior should be done with caution, but there are situations where researchers have external information on the plausible range of measurement-error or process-error variance.

8.5 Investigation of the population structure using model-selection criteria

In Section 8.4, we specified a particular population structure by putting constraints on $\mathbf{B}$, $\mathbf{Q}$, and $\mathbf{R}$. We can also use the multisite state-space framework to measure the data support for different population structures (Figure 8.1) rather than specifying a

‖ The EM algorithm is a hill-climbing algorithm. Thus steps must be taken to ensure that it does not get stuck on local maxima (Biernacki et al. 2003)

structure *a priori*. The different structures are denoted by the triplet $\{f_B, f_Q, f_R\}$ presented in Section 8.3. These form a nested set of models varying from unstructured (a single population but measured with multiple time series) to fully structured (different stochastic growth rates and process-error variances in each subpopulation and correlations in the process errors between subpopulations).

Using model-selection criteria (Burnham and Anderson 2002, Johnson and Omland 2004), we can measure the data support for the different models. The basic idea is that different models are fit to the data, the fit of the model to the data is measured using the likelihood function, and the fit is penalized for the number of parameters estimated by the model. The latter corrects for the fact that more complex models will tend to fit data better, simply because there is more flexibility in the model. The function that specifies how the likelihood is penalized for complexity is the model-selection criterion, and it gives a relative measure of data support. There are a variety of different model-selection criteria used in model selection. The most commonly used are Akaike's information criterion (AIC) (Akaike 1973, Burnham and Anderson 2002), Bayesian or Schwarz information criterion (BIC) (Mcquarrie and Tsai 1998), and deviance information criterion (DIC) (Spiegelhalter et al. 2002). Mcquarrie and Tsai (1998) is a good reference for model selection approaches specific to time-series data, and Burnham and Anderson (2002) is a good reference for model-selection approaches for the ecological sciences. In the example below, we illustrate the use of AIC for measuring the data support for different structures within a group of chinook salmon subpopulations.

8.6 Analysis of Snake River chinook salmon dynamics and structure

The Snake River is one of the major tributaries of the Columbia River, and historically it produced a large proportion of the chinook salmon within the Columbia basin. However, anthropogenic impacts such as the construction of hydropower dams on the Columbia and Snake Rivers, habitat destruction, and over-fishing led to large declines in the chinook populations within the Snake River and its tributaries. In 1992, the Snake River spring/summer chinook Evolutionary Significant Unit** (ESU) was listed as threatened under the U.S. Endangered Species Act, along with other salmonid ESUs in the Columbia River basin. This Snake River ESU includes all wild (not hatchery-released) chinook salmon that spawn in the spring and summer in the Snake River and its tributaries: the Tucannon, Grande Ronde, Imnaha, and Salmon Rivers (Figure 8.2). Chinook that spawn in the spring and summer spend their first year in freshwater near their natal streams and migrate to the ocean as yearlings.

To illustrate the use of the multisite state-space model, we analyzed time series from six distinct chinook subpopulations in the upper reaches of the Snake River basin (Figure 8.2). Chinook salmon show strong fidelity to their natal streams, thus the fish spawning within a specific stream are most likely to have been spawned in that stream

** Evolutionary Significant Unit is the term for a population segment that is considered distinct for the purpose of conservation under the U.S. Endangered Species Act (Waples 1991).

Figure 8.2 Map of the Snake River spring/summer chinook ESU. The location of the subpopulations are shown with the gray ovals: 1) Bear Valley/Elk Creek , 2) Sulphur Creek, 3) Marsh Creek, 4) Upper Valley Creek, 5) Big Creek, 6) Lemhi River. To reach these spawning areas, fish must pass through (or be barged around) eleven hydropower dams.

or nearby. The six subpopulations we analyzed were Bear Valley/Elk Creek, Sulphur Creek, Marsh Creek, Valley Creek, Big Creek, and Lemhi River (Figure 8.2). The time-series data for each subpopulation represent estimates of the spawning salmon abundances within each subpopulation from 1980 to 2001 (Figure 8.3). Our analysis focused on two questions: 1) Given a particular structure for the six subpopulations, what are the maximum likelihood estimates of the stochastic growth rates and the true abundances? and 2) What population structures are most supported by the data?

8.6.1 Estimation of the stochastic growth rates and true abundances

To separate out the measurement errors and provide estimates of the true abundances within each spawning site, we used the Kalman smoother. The Kalman smoother provides the optimal estimates of $\mathbf{X}_t$ given all the data, $\mathbf{Y}_1^T$. The Kalman smoother

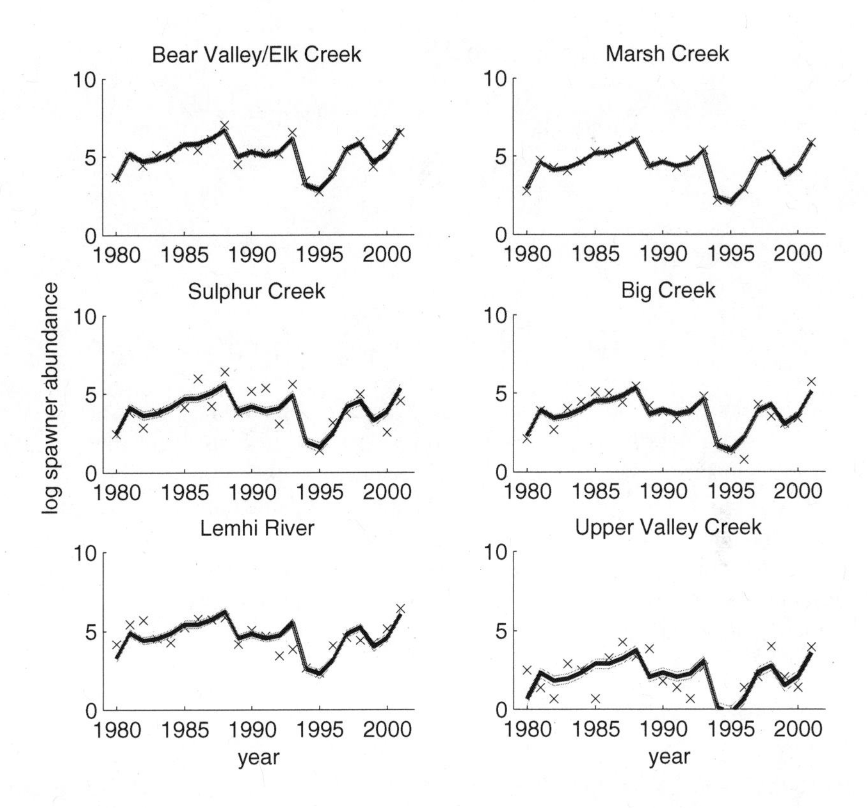

Figure 8.3 The 22-year time series of spawning abundance and smoothed estimates for 1) Bear Valley/Elk Creek, 2) Sulphur Creek, 3) Marsh Creek, 4) Upper Valley Creek, 5) Big Creek, and 6) Lemhi River. The $\times$'s are the actual spawner counts, the black line is the smoothed estimates of spawner abundance from the Kalman smoother, and the gray lines give the upper and lower 95% CIs for the smoothed estimates (the estimates with the measurement errors removed).

starts with the $\mathbf{x}_{T|T}$ estimates from the Kalman filter[††] and works backwards from T to 1 using the following updating equations:

$$\mathbf{x}_{t-1|T} = \mathbf{x}_{t-1|t-1} + \mathbf{J}_{t-1}\left(\mathbf{x}_{t|T} - \mathbf{x}_{t|t-1}\right) \tag{8.12}$$

$$\mathbf{P}_{t-1|T} = \mathbf{P}_{t-1|t-1} + \mathbf{J}_{t-1}\left(\mathbf{P}_{t|T} - \mathbf{P}_{t|t-1}\right)\mathbf{J}'_{t-1}, \tag{8.13}$$

where $\mathbf{J}_{t-1} \equiv \mathbf{P}_{t-1|t-1}\mathbf{P}^{-1}_{t|t-1}$. At the end of the recursion, we have the smoothed estimates of $\mathbf{X}_t$ conditioned on all the data. The smoothed estimates are denoted $\mathbf{x}_{t|T}$.

[††] The Kalman filter is a forward recursion and provides an optimal prediction of $\mathbf{X}_t$ given the past, $\mathbf{Y}_1^{t-1}$. The Kalman smoother is a backwards recursion that provides optimal estimates of the past given the future, in this case $\mathbf{X}_t$ given $\mathbf{Y}_1^T$.

The smoothed estimates have a simple relationship to the maximum likelihood estimates for the stochastic growth rates:

$$\mathbf{B} = \frac{1}{T-1}\left(\mathbf{x}_{T|T} - \mathbf{x}_{1|T}\right), \quad \text{if } f_B = 1 \text{ or } 3, \tag{8.14}$$

$$\mathbf{B} = \frac{1}{T-1}\frac{\mathbf{O}'\mathbf{Q}^{-1}\left(\mathbf{x}_{T|T} - \mathbf{x}_{1|T}\right)}{\mathbf{O}'\mathbf{Q}^{-1}\mathbf{O}}, \quad \text{if } f_B = 2, \tag{8.15}$$

where $\mathbf{O}$ is an $m \times 1$ matrix of ones.

8.6.2 Investigation of the subpopulation structure using AIC and AICb

To determine which population structure was best supported by the data, we used Akaike's Information Criteria (AIC) (Akaike 1973, Burnham and Anderson 2002) and the bootstrap AICb (Cavanaugh and Shumway 1997). In particular, we were interested in whether the six sites should be treated as one population sampled with six independent time series or as six separate subpopulations with correlated process errors. AIC and AICb measure the data support for models with different population structure. Models with lower AIC and AICb scores have better data support relative to models with higher AIC or AICb scores.

For model q, where q specifies a unique $\{f_B, f_Q, f_R\}$ triplet, the AIC is defined as:

$$\text{AIC}_q = -2\log L(\hat{\Theta}_q) + 2p, \tag{8.16}$$

where $\hat{\Theta}_q$ is the parameter set $\{\mathbf{B}, \mathbf{Q}, \mathbf{R}, \mathbf{D}\}$ that maximizes the likelihood of the observed data $\mathbf{Y}_1^T$ given model q, and p is the number of effective parameters in model q. We calculated the maximum likelihood estimates $\hat{\Theta}_q$ using the EM algorithm (Shumway and Stoffer 1982, 2000).

AICb is a variant of AIC that corrects for AIC's bias towards overly complex models when the sample size is small. AICb has the same objective as the more familiar AICc—the small sample-size corrected AIC (Burnham and Anderson 2002)—but AICb is designed for state-space models. AICb for model q is defined as:

$$\text{AICb}_q = -2\log L(\hat{\Theta}_q) + 2\left\{\frac{1}{N}\sum_{b=1}^{N} -2\log\frac{L(\hat{\Theta}_q(b))}{L(\hat{\Theta}_q)}\right\}, \tag{8.17}$$

where $\hat{\Theta}_q(b)$, $b = 1, \ldots, N$, represents a set of N bootstrap replicates of $\hat{\Theta}_q$. The bootstrap replicates are generated using the following procedure (Stoffer and Wall 1991; Shumway and Stoffer 2000, section 2.6). Using the parameters $\hat{\Theta}_q$, the model q is fit to the data. This provides a time series of the innovations ϵ for $t = 1$ to T, where $\epsilon_t \equiv \mathbf{Y}_t - \tilde{\mathbf{Y}}_{t|t-1}$. The bootstrap replicates of the innovations time series, $\epsilon(b)$, are generated by taking T samples with replacement from ϵ. The bootstrap-generated $\epsilon(b)$ are then used in what is termed the innovations form of the state-space model to generate a bootstrapped $\mathbf{Y}(b)$ time series. This process is repeated N times to produce N bootstrapped $\mathbf{Y}(b)$ time series. For each of these $\mathbf{Y}(b)$ time series, the parameters that maximize the likelihood of the bootstrapped data $\mathbf{Y}(b)$ are found. This produces the N bootstrapped estimates: $\hat{\Theta}_q(b)$, $b = 1, \ldots, N$.

8.6.3 Confidence intervals and diagnostics

To determine the accuracy of the parameter estimates, we used a parametric bootstrap approach (Shumway and Stoffer 2000, section 2.6). The N bootstrap replicates of $\hat{\Theta}_q(b)$ were used to estimate confidence intervals, standard errors, and bias for each of the estimated model parameters. To construct 95% confidence intervals for the k-th parameter within $\hat{\Theta}_q$, the 2.5 and 97.5 percentiles for the k-th parameter in the bootstrap replicates $\hat{\Theta}_q(b)$ were used as the lower and upper confidence limits.

The bootstrap standard error for the k-th parameter was defined as the mean squared difference between the k-th parameter in the bootstrapped samples and the mean value of the k-th parameter in the bootstrapped samples:

$$\mathrm{SE} = \frac{1}{N-1} \sum_{i=1}^{N} \left(\hat{k}_q(b) - \frac{1}{N} \sum_{b=1}^{N} \hat{k}_q(b) \right)^2, \tag{8.18}$$

where $\hat{k}_q(b)$ denotes the k-th parameter in $\hat{\Theta}_q(b)$. The bootstrap bias was calculated as:

$$\mathrm{bias} = \frac{1}{N} \sum_{b=1}^{N} \hat{k}_q(b) - \hat{k}_q, \tag{8.19}$$

where $\hat{k}_q$ is the maximum likelihood estimate of the k-th parameter in $\hat{\Theta}_q$. As a rule of thumb, bias is considered a potential problem when it exceeds 5% of the SE.

Diagnostics were applied to the fitted state-space models to determine whether they were appropriate for analyzing the salmon data. When running model diagnostics, prediction errors in state-space models play a similar role to that of residuals in an ordinary least-squares regression. Like residuals, the prediction errors are assumed to be independent and normally distributed. The normality assumption was examined using quantile-quantile (QQ) plots (Chambers et al. 1983) of the standardized prediction errors. Jarque-Bera tests for normality were also applied to the standardized prediction errors (Cromwell et al. 1994). The assumption of serially independent prediction errors was examined using autocorrelation-function plots and the Box-Pierce test for independence (Box and Pierce 1970).

8.6.4 Results

The estimates of the mean stochastic growth rate were positive across all the top models. This indicates a population that is increasing. The stochastic growth rate estimate for the best model was 0.14, suggesting a robust mean growth rate of 14% per year. However, the estimated growth rates had large standard errors and correspondingly wide confidence intervals that included negative values (Table 8.1). This indicates that even with six 22-year time series, the data are insufficient for confidently estimating whether the population is increasing or decreasing.

Results on the population structure however are more informative. Out of the 36 state-space models considered, the model with $f_B = 3$, $f_Q = 1$,[‡‡] and $f_R = 2$ had the lowest AICb and AIC values (Table 8.1). This is the model with a single population that is measured with six different time series, each with independent and different measurement errors. This model had considerably more support (ΔAIC $>$ 10) than the next competitor. The model equivalent to six independent salmon subpopulations each measured independently, $\{f_B, f_Q, f_R\} = \{1, 2, 2\}$, fit extremely poorly. This model ranked 30 out of the 36 models with a ΔAICb score of 170.7 compared to the best model (Table 8.1).

The result that the model with a single population fit the data best indicates that the six subpopulations were highly correlated. Surporting this, we also found that the model $\{f_B, f_Q, f_R\} = \{1, 1, 2\}$, which has an unrestricted process-error covariance matrix, also indicated that the process-error correlations were high. The correlation coefficients for this model ranged from 0.86 to 1.0 (Table 8.2). In contrast, the measurement errors were found to be uncorrelated. The best-fitting model had a diagonal measurement-error covariance matrix, $\mathbf{R}$, with unequal variances (f_R = 2) and zero correlation between all subpopulations. The measurement-error variances differed greatly between the six subpopulations (Table 8.3). Marsh Creek had an estimated measurement-error variance of 0.03 (SE = 0.03) compared to Upper Valley Creek, which had an estimated measurement-error variance of 1.14 (SE = 0.36). Changing from a diagonal $\mathbf{R}$ matrix with unequal variances to a diagonal matrix with equal variances produced the second best model with a ΔAICb of 14.5 above the best model. Bias in the measurement-error estimates ranged from 4% to 63% of SE.

The methods are designed to look for correlation and synchrony across sites. In the case of these salmon time series, we see both strong correlation and strong synchrony. From the time-series data alone, we cannot infer what mechanism is driving this pattern in the salmon data. These six salmon stocks are exposed to a similar ocean environment and river-migration environment, and this would lead to correlated process errors. However, the process errors would need to be perfectly correlated in order to produce synchrony because without perfect correlation, the time series across the six sites would eventually diverge. This suggests that there is another mechanism that is causing synchrony. Dispersal, in this case straying of spawners to nonnatal streams, is known to occur and is a possible mechanism for the synchrony.

Diagnostics were run on the model with the lowest AICb score. The QQ plots indicated no deviation from normality in the prediction errors except for the Marsh Creek subpopulation. For Marsh Creek, the normal QQ plot showed large deviations from a straight line in the tails of the prediction errors, and the Jarque-Bera test indicated that the distribution was not normal (p-value = 0.003). This deviation from normality, however, was driven by a single prediction error (from the year 1994), which

[‡‡] Note that when $f_B = 3$, $m = 1$ so $\mathbf{Q}$ is a scalar.

Table 8.1 *Results of the model-selection analyses which fit models with different population structures to the salmon data. Models are by ranked by ΔAICb scores. Lower ΔAICb indicates more data support for that model. Generally, a ΔAICb > 10 indicates low data support.* p *indicates the number of parameters in each model.*

	Model form			Mean	Lower 95%	Upper 95%			
Rank	f_B	f_Q	f_R	μ	limit	limit	p	ΔAIC	ΔAICb
1	3	1	2	0.14	-0.32	0.59	14	0	0
2	3	1	3	0.11	-0.27	0.54	9	19	14.5
3	3	1	4	0.09	-0.23	0.44	10	20.2	18.9
4	2	4	2	0.14	-0.37	0.60	15	22.9	31.6
5	1	4	2	0.13	-0.34	0.61	20	28.5	36
6	2	4	3	0.11	-0.27	0.52	10	44.4	46.4
7	2	2	4	0.00	-0.07	0.08	15	49.2	50.5
8	2	3	4	-0.01	-0.08	0.07	10	47	51.5
9	1	4	3	0.11	-0.31	0.51	15	52.5	53.2
10	2	4	4	0.08	-0.20	0.34	11	45.5	57.5
11	1	3	4	-0.01	-0.08	0.06	15	55.3	58.8
12	1	2	4	-0.02	-0.13	0.09	20	57.8	73
13	1	4	4	0.08	-0.18	0.35	16	53.6	73.2
14	3	1	1	0.10	-0.19	0.40	29	8.3	88.8
15	2	3	3	0.06	-0.05	0.16	9	109.1	104.3
16	2	3	2	0.07	-0.04	0.17	14	113.1	110.1
17	2	2	3	0.06	-0.07	0.17	14	117.9	118
18	2	2	2	0.05	-0.06	0.16	19	119.9	121.8
19	2	3	1	-0.01	-0.07	0.06	29	41.7	122.4
20	2	4	1	0.10	-0.19	0.39	30	28.3	123.7
21	2	1	2	0.14	-0.20	0.48	34	44.5	124.2
22	2	2	1	0.01	-0.06	0.08	34	37.1	125.5
23	1	3	1	-0.01	-0.08	0.05	34	45.6	132.8
24	1	3	3	0.06	-0.05	0.15	14	119	135.2
25	1	3	2	0.07	-0.05	0.18	19	123	144.2
26	1	1	2	0.12	-0.29	0.53	39	52.6	145.2
27	1	4	1	0.11	-0.18	0.38	35	32.6	150.2
28	2	1	3	0.17	-0.11	0.47	29	53.1	157.1
29	1	2	3	0.06	-0.04	0.17	19	127.8	161.5
30	1	2	2	0.06	-0.05	0.16	24	129.4	170.7
31	1	2	1	0.00	-0.07	0.07	39	42.4	172.3
32	1	1	3	0.12	-0.32	0.57	34	61.7	180.2
33	2	1	4	0.17	-0.12	0.47	30	54.9	213.2
34	1	1	4	0.12	-0.33	0.54	35	63.6	231.5
35	2	1	1	0.15	-0.10	0.40	49	52.3	274.9
36	1	1	1	0.09	-0.15	0.33	54	61.1	294.7

Table 8.2 *Estimated correlation coefficients for the process errors. Model* $\{f_B, f_Q, f_R\} = \{1, 1, 2\}$ *was used to estimate the unconstrained covariance matrix* $\mathbf{Q}$ *which was then used to calculate the correlation matrix.*

	Bear Valley/ Elk Cr.	Marsh Cr.	Sulphur Cr.	Big Cr.	Lemhi R.
Marsh Cr.	1.00				
Sulphur Cr.	0.99	0.99			
Big Cr.	1.00	1.00	0.99		
Lemhi R.	0.86	0.87	0.79	0.85	
Up. Valley Cr.	0.98	0.99	0.95	0.98	0.93

Table 8.3 *Measurement-error variances using the best-fitting model with lowest AICb.*

	Estimate	SE	Lower 95% CI	Upper 95% CI	Bias
Bear Valley/Elk Cr.	0.13	0.05	0.04	0.23	-0.20
Marsh Cr.	0.03	0.03	0.00	0.10	0.14
Sulphur Cr.	0.66	0.22	0.26	1.11	-0.17
Big Cr.	0.28	0.09	0.11	0.47	-0.22
Lemhi R.	0.53	0.16	0.24	0.87	-0.21
Up. Valley Cr.	1.14	0.36	0.49	1.83	-0.20

was 2.74 standard deviations below zero. When this prediction error was deleted, the Jarque-Bera test indicated no significant deviation from normality (p-value = 0.59). The Box-Pierce tests, based on lags up to five years, indicated that the prediction errors were serially uncorrelated. The autocorrelation functions, however, did indicate a relatively large negative lag-1 autocorrelation for Sulphur Creek (r = -0.53) and a relatively large positive lag-5 autocorrelation (r = 0.45) for the Bear Valley/Elk Creek subpopulation.

8.7 Discussion

The analysis of the salmon data suggests that the population dynamics within the upper Snake River basin are highly synchronized. The best fitting models indicated very high correlations in the year-to-year fluctuations in subpopulation growth rates and a common stochastic growth rate for all of the six subpopulations. This implies that these subpopulations tend to act as a single population. Biologically, this is not surprising; a certain amount of straying of spawners into nonnatal streams is known to occur and, in addition, the salmon from the different spawning sites are exposed to a similar environment after they leave their spawning stream. They migrate down the same river corridor to the ocean and then spend two to four years in the ocean. In contrast, the modeling suggests that measurement errors are uncorrelated among the six subpopulations, with variances that differ. This is not surprising given that site differences can greatly affect the accuracy of spawning-abundance counts and given that counts at different subpopulations are made on different days.

Aside from revealing these important patterns in the data, does the multivariate technique improve the accuracy of the stochastic growth rate estimates—relative to simply fitting a univariate model to each subpopulation time series independently, then taking the average? At first glance, the answer appears to be no. The univariate stochastic growth rates can be obtained by using the model $\{f_B, f_Q, f_R\} = \{1, 2, 2\}$. This is the model that specifies an independent stochastic growth rate and variance for each subpopulation, and treats the data as if there is no correlation between subpopulations or measurements. This model gives an SE of 0.052 for the average stochastic growth rate, while the best multivariate model, $\{f_B, f_Q, f_R\} = \{3, 1, 2\}$, gives a much larger SE of 0.23. Shouldn't we expect the model with the lowest AICb to produce lower standard errors? The answer is no, because standard error estimates of models with poor AICb are unreliable. The standard errors are largely a function of the estimated variance matrices (Harvey 1989). Therefore, poor estimates of the variance matrices mean poor standard error estimates and poor confidence intervals. The model $\{f_B, f_Q, f_R\} = \{1, 2, 2\}$, which gives low standard errors, has one of the worst ΔAICb scores (170.7 in Table 8.1), and therefore inferences on precision are not as reliable as those from the top model.

Bias is another part of accuracy that must be considered. The stochastic growth rate estimates were not biased, but variance estimates were. For example, the model with smallest AICb had a process-error variance that was biased downward by 20% SE. It also had biases in the measurement-error variance that ranged from 14% to 22%. Lindley (2003) found that when time series are short, the Kalman filter (used in this chapter) tends to lead to underestimates of the true process error. This suggests that some bias correction procedure ought to be investigated for the variance estimates. Another possibility is using restricted maximum likelihood, which was found to generate unbiased estimates of process- and measurement-error variance in a univariate setting (Staples et al. 2004). Currently, however, this method does not handle multivariate data or missing values, it sometimes fails to converge, and it can generate negative estimates of measurement-error variance. The slope method (Holmes 2001) can also reduce process-error bias, but it also does not handle multivariate data and may generate negative variance estimates.

Multivariate state-space modeling has a long, rich history in the engineering and economics literature and has proved a powerful tool for modeling and forecasting dynamical systems. This approach allows analysts to deal with data from multiple sites simultaneously, handle missing values, and impose different assumptions concerning the spatial structure within the population dynamics and within the measurement process. Although we have assumed a linear model with Gaussian and uncorrelated errors, these assumptions can be relaxed and the same framework could be used but with the parameters estimated via alternate estimation algorithms. In summary, the multivariate state-space approach provides a formal framework for incorporating spatial structure into the analysis of multisite time series data and can reveal important relationships among subpopulations—relationships that would remain concealed with a single-site or nonspatial approach.

8.8 References

H. Akaike (1973), Information theory and an extension of the maximum likelihood principle, in *"Second international symposium on information theory,"* ed. by B.N. Petrov and F. Csaki, Akademiai Kiado, Budapest, Hungary, pp. 267–281.

C. Biernacki, G. Celeux, and G. Govaert (2003), Choosing starting values for the EM algorithm for getting the highest likelihood in multivariate Gaussian mixture models, *Comp. Stat. Data Anal.* **41**: 561–575.

G. E. P. Box and D. A. Pierce (1970), Distribution of residual correlations in autoregressive-integrated moving average time series models, *J. Am. Stat. Assoc.* **65**: 1509–1526.

K. P. Burnham and D. R. Anderson (2002), *Model Selection and Multi-model Inference,* New York, Springer.

J. E. Cavanaugh and R. H. Shumway (1997), A bootstrap variant of AIC for state-space model selection, *Stat. Sinica* **7**: 473–496.

J. Chambers, W. Cleveland, B. Kleiner, and P. Tukey (1983), *Graphical Methods of Data Analysis,* The Wadsworth Statistics/Probability Series, Belmont, Calif, Wadsworth International Group.

J. B. Cromwell, W. C. Labys, and M. Terraza (1994), *Univariate Tests for Time Series Models,* Sage Publications, Inc., Thousand Oaks, CA.

B. Dennis, P. L. Munholland, and J. M. Scott (1991), Estimation of growth and extinction parameters for endangered species, *Ecol. Monogr.* **61**: 115–143.

B. Dennis, J. M. Ponciano, S. R. Lele, M. L. Taper, and D. F. Staples (2006), Estimating density dependence, process noise, and observation error, *Ecol. Monogr.* **76**: 323-341.

P. deValpine and A. Hastings (2002), Fitting population models incorporating process noise and observation error, *Ecol. Monog.* **72**: 57–76.

J. B. Dunning, Jr., D. J. Stewart, B. J. Danielson, B. R. Noon, T. L. Root, R. H. Lamberson, and E. E. Stevens (1995), Spatially explicit population models: current forms and future uses, *Ecol. Appl.* **5**: 3–11.

J. Durbin and S. J. Koopman (2000), Time series analysis of non-Gaussian observations based on state space models from both classical and Bayesian perspectives, *J. R. Statist. Soc. B* **62**: 3–56.

A. C. Harvey (1989), *Forecasting, Structural Time Series Models and the Kalman Filter,* Cambridge University Press, Cambridge, UK.

E. E. Holmes (2001), Estimation risks in declining populations with poor data, *Proc. Natl. Acad. Sci. USA* **98**: 5072–5077.

E. E. Holmes (2004), Beyond theory to application and evaluation: diffusion approximations for population viability analysis, *Ecol. Appl.* **14**: 1272–1293.

E. E. Holmes and W. F. Fagan (2002), Validating population viability analysis for corrupted data sets, *Ecology* **83**: 2379–2386.

E. E. Holmes, J. L. Sabo, S. V. Viscido, and W. Fagan (2007), A statistical approach to quasi-extinction forecasting, *Ecol. Lett.* **10**: 1182–1198.

E. E. Holmes and B. Semmens (2004), Population viability analysis for metapopulations: a diffusion approximation approach, in *"Ecology, Genetics, and Evolution of Metapopulations,"* ed. by I. Hanski and O. E. Gaggiotti, Elsevier Press, pp. 565–598.

A. R. Ives, B. Dennis, K. L. Cottingham, and S. R. Carpenter (2003), Estimating community stability and ecological interactions from time-series data, *Ecol. Monogr.* **73**: 301–330.

J. B. Johnson and K. S. Omland (2004), Model selection in ecology and evolution, *Trends Ecol. Evol.* **19**: 101–108.

R. E. Kalman (1960), A new approach to linear filtering and prediction problems, *Transactions of the ASME Journal of Basic Engineering D* **82**: 35–45.

W. S. Lahaye, R. J. Gutierrez, and H. R. Akcakaya (1994), Spotted owl metapopulation dynamics in southern California, *J. Anim. Ecol.* **63**: 775–785.

S. T. Lindley (2003), Estimation of population growth and extinction parameters from noisy data, *Ecol. Appl.* **13**: 806–813.

A. D. R. Mcquarrie and C.-L. Tsai (1998), *Regression and Time Series Model Selection,* World Scientific Publishing Company, Singapore.

N. H. Schumaker, T. Ernst, D. White, J. Baker, and P. Haggerty (2004), Projecting wildlife responses to alternative future landscapes in Oregon's Willamette basin, *Ecol. Appl.* **14**: 381–400.

R. H. Shumway and D. S. Stoffer (1982), An approach to time series smoothing and forecasting using the EM algorithm, *J. Time Ser. Anal.* **3**: 253–264.

R. H. Shumway and D. S. Stoffer (2000), *Time Series Analysis and its Applications,* Springer-Verlag, New York.

D. J. Spiegelhalter, N. G. Best, B. P. Carlin, and A. van der Linde (2002), Bayesian measures of model complexity and fit, *J. Roy. Stat. Soc. B* **64**: 583–639.

D. F. Staples, M. L. Taper, and B. Dennis (2004), Estimating population trend and process variation for PVA in the presence of sampling error, *Ecology* **85**: 923–929.

D. S. Stoffer and K. D. Wall (1991), Bootstrapping state-space models: Gaussian maximum likelihood estimation and the Kalman filter, *J. Am. Stat. Assoc.* **86**: 1024–1033.

S. D. Tuljapurkar and S. H. Orzack (1980), Population dynamics in variable environments. I. Long-run growth rates and extinction, *Theor. Popul. Biol.* **18**: 314–342.

R. S. Waples (1991), Pacific salmon, *Oncorhynchus* spp., and the definition of species under the Endangered Species Act, *Mar. Fish. Rev.* **53**: 11–22.

CHAPTER 9

Incorporating the spatial configuration of the habitat into ecology and evolutionary biology

Ilkka Hanski
University of Helsinki

Abstract. Though ecologists and population biologists have appreciated the role of spatial processes in natural populations for a long time, there has been relatively little work towards integrating a realistic description of the landscape structure into population biological theories, models, and empirical research. Landscape structure is studied by landscape ecologists and geographers, but generally with limited or no reference to population processes. This chapter outlines the spatially realistic approach to metapopulation dynamics, which has facilitated the conceptual unification of population ecology and landscape ecology, and extends the spatially realistic approach to evolutionary biology.

9.1 Introduction

The classic population concept that can be traced back to Malthus (1798) and which was developed in ecology and evolutionary biology in the early part of the 20th century (McIntosh 1985) assumes that all individuals interact equally and share the same environment. This is clearly not true for real populations at spatial scales greater than the daily movement range of individuals, for the simple reason that interactions become restricted by the physical distance. At larger spatial scales, habitat heterogeneity is typically another factor, apart from just long distances, that influences population processes and thereby population structures. Viewed from the perspective of a particular species, a landscape may be heterogeneous in many different ways. It may consist of several different types of habitat that may be used for foraging and reproduction, but typically in this case there is spatial variation in habitat quality: not all habitat types are the same. The classic metapopulation concept (Levins 1969; Hanski 1999) assumes that there is just one type of habitat, but it is fragmented into discrete patches. Incorporating spatial structure into population studies and into population models has been a major goal in ecology and evolutionary biology for the past 20 years (Tilman and Kareiva 1997; Hanski 1999), with sporadic early contributions published since the 1930s (Wright 1931; Nicholson 1933; Andrewartha and Birch 1954; Huffaker 1958).

This chapter is concerned with the ecology and evolutionary biology of metapopulations, that is, species occurring as a network of local populations in a network of discrete habitat patches (Hanski and Gaggiotti 2004). Viewed from the perspective of a particular local population in a particular habitat patch, there are two processes to consider. First, the performance of the local population: its average size and temporal variability, expected life-time, genetic composition, interactions with other species, and so forth. Second, how well the population is connected in terms of migration (dispersal) and gene flow to other local populations and possibly to suitable but presently unoccupied habitat in the fragmented landscape. In theoretical studies, the structure of the patch network is often simplified by assuming that all local populations are equally connected and all habitat patches are identical, in which case there is no consideration for the actual spatial configuration of the habitat in the landscape. Levins's (1969; 1970) metapopulation model is the archetypal example. In contrast, landscape ecologists interested in population dynamics have had a particular interest in developing an explicit account of the influence of the spatial configuration of the habitat and the structure of the landscape on population processes as well as on other processes (Turner, Gardner et al. 2001). For instance, recent studies have examined how the landscape structure influences movements of individuals (Schippers, Verboom et al. 1996; Pither and Taylor 1998; Haddad 1999; Bunn, Urban et al. 2000; Jonsen and Taylor 2000; Byers 2001) and population persistence (Hill and Caswell 1999; With and King 1999), and researchers have examined the significance of the spatial configuration of the habitat as opposed to just the amount of habitat for population persistence (Fahrig 2002).

Merging of metapopulation ecology and spatial ecology more generally with landscape ecology has been anticipated for a long time (Hanski and Gilpin 1991), but the two disciplines still largely adhere to their own research traditions and own literatures (see Figure 1.2 in Hanski and Gaggiotti 2004). A reflection of disciplinary differences is that shared key concepts such as connectivity have different meanings (Tischendorf and Fahrig 2000; Moilanen and Hanski 2001). In landscape ecology, connectivity is usually seen as a property of an entire landscape, and is defined in an *ad hoc* manner via simulations. In metapopulation ecology, connectivity is viewed as a property of discrete habitat patches and local populations, and is defined as the expected rate of migration or gene flow to or from a local population or a habitat patch in the patch network. The following measure of connectivity has been used as a surrogate for the rate of immigration to the focal habitat patch i (Hanski 1994; Hanski 1999; Moilanen and Nieminen 2002)

$$S_i = A_i^{\zeta_{im}} \sum_{j \neq i} p_j A_j^{\zeta_{em}} \frac{\alpha^2}{2\pi} e^{-\alpha d_{ij}}. \tag{9.1}$$

Here, S_i is the connectivity and A_i is the area of patch i, p_j is the incidence (probability) of occupancy of patch j, d_{ij} is the distance between patches i and j, $1/\alpha$ is the average migration distance, and ζ_{im} and ζ_{em} are two parameters describing the scaling of immigration and emigration rates with patch area. The factor $\alpha^2/2\pi$ ensures that the exponential dispersal kernel integrates to one over the two-dimensional space (some other dispersal kernel could be used instead of the exponential; see Ovaskainen and Hanski 2004).

In this chapter, I review research approaches to ecology and evolutionary biology that I have dubbed as spatially realistic metapopulation approaches and models. These approaches take the classic metapopulation concept of a network of local populations as the starting point, but extend it by incorporating the influence of the spatial configuration of the habitat on individual and population processes. The concept and measure of connectivity described above plays a key role here, as it provides a practical way of addressing questions about the spatial interactions of populations both in empirical and modelling studies. I address issues from the level of individual movements to the dynamics of populations and metacommunities of competing species and to the evolution of migration rate, with a focus on the influence of the spatial configuration of habitat.

9.2 Modeling migration in fragmented landscapes

Migration (dispersal) is the key additional process that needs to be considered in spatial ecology in addition to the familiar demographic rates describing temporal dynamics. All species migrate in one way or another, and there is a plethora of good evolutionary reasons to expect them to do so, from kin competition and inbreeding avoidance to coping with temporal variability in environmental conditions (Clobert, Wolff et al. 2001; Ronce and Olivieri 2004; Ronce 2007). I return to questions about the evolution of migration rate in Section 9.5; here I address the estimation of migration parameters for species living in fragmented landscapes, which is an essential task in empirical studies.

In the case of large-bodied mammals and birds, it is feasible to track individuals using radio telemetry, satellite tracking, and other comparable technologies. In these cases, the limiting factor is often the number of individuals that can be studied, while the complex and plastic movement behavior of vertebrates makes analyses challenging (see Ovaskainen and Crone, this volume). In the case of small animals, direct study of movements is occasionally possible by other tracking technologies, such as the harmonic radar (Ovaskainen, Smith et al. 2008), but in practice it is more feasible to mark and recapture individuals within an appropriate study area and to infer the parameters of migration based on individual capture histories. Analyzing such data is complicated by the fact that the pattern of recaptures is influenced by three factors, intrinsic migration behavior of the species, the structure of the landscape (which necessarily influences the movement patterns), and the spatio-temporal distribution of recapture effort (Ovaskainen 2004). For instance, just plotting the frequency distribution of movement distances, which is still a common practice in empirical studies, is clearly inadequate, because that distribution is affected by all the three factors.

To tease apart the different components, one needs to employ models, and furthermore, to take into account the influence of landscape structure, one needs to employ a spatially realistic model.

Mark-recapture studies conducted in metapopulations living in highly fragmented landscapes may involve tens of local populations; hence it is out of question to parameterize a model for all pair-wise connections among the populations, which is a possibility when there are a few populations only (Hestbeck 1991; Hilborn 1991). However, assuming isotropic migration, one may further assume that emigrants leaving population i are distributed among the surrounding populations and habitat patches in relation to their distances from population i, with the area of the receiving habitat patch potentially modifying the numbers of immigrants, as larger patches are larger targets for immigrants. One may distinguish between mortality within populations and mortality during migration on the reasonable assumption that the former is not influenced by the connectivity of the patch, while mortality during migration is influenced by connectivity, as the probability of surviving migration can be expected to increase with the connectivity of the population of origin.

Table 9.1. Simulation results for a cohort of butterflies in a fragmented landscape using parameter values estimated with the VM model (Hanski, Alho et al. 2000) for *Proclossiana eunomia* (Petit, Moilanen et al. 2001).

	Males	Females
Initial number of butterflies (input)	295	306
Successful migration events from one patch to another	333	543
Number of butterflies dying during migration	34	14
Pooled number of butterfly-days in the system	1833	3068
Percentage of butterfly-days spent outside the natal patch	43%	53%

These ideas have been implemented in a statistical model of mark-recapture data (Hanski, Alho et al. 2000) with 6 parameters, namely mortality within habitat patches and in the landscape matrix, scaling of emigration and immigration with habitat patch area, emigration constant, and distance dependence of migration. Having estimated the parameter values, one may calculate quantities such as shown in Table 9.1 for the butterfly *Proclossiana eunomia* (Petit, Moilanen et al. 2001), including the total number of individual-days spent in the metapopulation, number of successful migration events, percentage of time spent outside the natal population, and the number of deaths during migration. In the example in Table 9.1, there was considerable migration among the populations, and the migration was inferred to be mostly successful (only 10% of individuals died during migration), apparently because in this metapopulation local populations were located close to each other. Mortality during migration is a key cost of migration with important consequences for the evolution of migration (Section 9.5). Deviations from the model fit due to specific populations may reveal important biological factors and processes, such as, e.g., migration depending on local sex ratio (Wahlberg, Klemetti et al. 2002). Ovaskainen and Crone (this volume; see also Ovaskainen 2004, Ovaskainen et al. 2008) describe an alternative and more mechanistic way of modelling mark-recapture data based on a diffusion approximation of correlated random walk with habitat selection at patch boundaries. This model assumes that movements obey a Markov process, which is probably a good approximation for many though not all organisms. In the present context, the important advantage of both models is that they allow a detailed study of how the spatial configuration of the habitat influences movements of individuals. Both models can and have been used as sub-models describing movements in individual-based demographic or evolutionary models, with the great advantage that the movement component can be rigorously parameterized. I return to this point in Section 9.5.

9.3 Metapopulation dynamics

I now turn to population dynamics in fragmented landscapes, where the habitat patches are generally small and hence also the respective local populations tend to be small and extinction-prone. In this case, long-term persistence is possible only at the level of the patch network, with recolonization of currently unoccupied patches compensating for local extinctions. This is the scenario assumed in the classic metapopulation theory (Hanski 1999). The best-developed modeling approach to metapopulation ecology is based on stochastic patch occupancy models (SPOM; Hanski 1994, Day and Possingham 1995, Frank 1998, Moilanen 1999, Hanski and Ovaskainen 2000), which simplify the modelling task by considering only the presence or absence of the focal species in the habitat patches. This is an acceptable simplification for many highly fragmented landscapes.

Mathematically, SPOMs are formulated as Markov chains (discrete time) or Markov processes (continuous time) with 2^n possible states in a network of n habitat patches (Hanski and Ovaskainen 2003; Ovaskainen and Hanski 2004). The qualitative behavior of a SPOM is simple, as the metapopulation will eventually enter the absorbing state in which all the patches are empty, and which corresponds to metapopulation extinction. In large patch networks time to metapopulation extinction may however be very long in comparison with time to local extinction, and the dynamics settle to a quasi-stationary distribution, the stationary probability distribution conditioned on nonextinction. This is an important concept for ecologists, implying that at the time scale of interest the size of the metapopulation fluctuates due to ongoing local extinctions and recolonizations but there is no temporal trend in metapopulation size. Metapopulation models formulated as SPOMs have been reviewed by, e.g., Hanski and Ovaskainen (2003) and Ovaskainen and Hanski (2004). Here I draw attention to four issues and results that have particular relevance for this chapter on the influence of the spatial configuration of habitat on population processes.

First, a useful distinction may be made between homogeneous and heterogeneous SPOMs. The familiar Levins (1969) metapopulation model is a deterministic approximation of a simple SPOM with an infinite number of identical patches. A stochastic version of the Levins model with a finite number of patches, which is often called the stochastic logistic model, is an example of homogeneous SPOMs with identical and equally connected patches. Homogeneous SPOMs with identical transition probabilities may be solved numerically even for large n, and the models are tractable for mathematical analysis (Kryscio and Lefèvre 1989; Jacquez and Simon 1993; Nåsell 1996; Ovaskainen 2001). These models have been applied widely in population biology (Norden 1982), epidemiology (Weiss and Dishon 1971), chemistry (Oppenheim, Shuler et al. 1977), and even sociology (Bartholomew 1976). In contrast, heterogeneous SPOMs allow for variation in patch sizes and connectivities, and are clearly more appropriate for real metapopulations. Unfortunately, as the size of the state space with n patches is now 2^n, the analysis of large patch networks presents computational challenges.

Second, helpful insight to metapopulation dynamics in heterogeneous networks can be gained via deterministic approximations of the stochastic models. A deterministic continuous-time approximation of a heterogeneous SPOM describing the rate of change in the probability of patch i being occupied, p_i, is given by a system of n equations for a network of n patches (Hanski and Gyllenberg 1997; Hanski and Ovaskainen 2000; Ovaskainen and Hanski 2001),

$$dp_i/dt = C_i(\boldsymbol{p})(1 - p_i) - E_i(\boldsymbol{p})p_i, \tag{9.2}$$

where $C_i(\boldsymbol{p})$ gives the colonization rate of patch i when it is empty and $E_i(\boldsymbol{p})$ gives the extinction rate of the population in patch i when it is occupied. $\boldsymbol{p}$ is the vector of the n occupancy probabilities. The deterministic approximation given by Eq. (9.2) is derived from the full stochastic model by accounting only for the deterministic drift and ignoring stochastic fluctuations. To develop this model further, it is customary to make biologically plausible assumptions about patch area dependence of the extinction rate $E_i(\boldsymbol{p})$ and connectivity dependence of the colonization rate $C_i(\boldsymbol{p})$. Making the simple assumptions $E_i = e/A_i$ and $C_i = cS_i$, where S_i is the connectivity measure given by Eq. (9.1) (with some simplification), leads to the spatially realistic Levins model (Ovaskainen and Hanski 2004). A key point is that the essential behavior of the original n-dimensional model (Eq. (9.2) for n patches) is well approximated by a one-dimensional equation (Ovaskainen and Hanski 2001; 2002). The equilibrium size of the metapopulation is given by

$$\tilde{p}^*_\lambda = 1 - \delta/\lambda_M, \tag{9.3}$$

where $\delta = e/c$ and λ_M is the leading eigenvalue of a matrix constructed with the extinction and colonization terms. λ_M is called the metapopulation capacity of a fragmented landscape (Hanski and Ovaskainen 2000). Equation (9.3) may be compared with the equilibrium in the Levins model that has been modified with the assumption that only fraction h of the patches is suitable for colonization (Lande 1987),

$p^* = 1 - \delta/h$. This comparison shows that λ_M is a quantity that plays in the spatially realistic model the same role as the amount of suitable habitat in the original nonspatial model. λ_M thus measures the influence of both the amount and the spatial configuration of suitable habitat on metapopulation growth and size at equilibrium. λ_M is a convenient parameter of landscape structure from the viewpoint of metapopulation viability, and better justified for that purpose than many other parameters commonly used in landscape ecology (Turner, Gardner et al. 2001). For instance, λ_M allows one to examine the relative roles of habitat loss and fragmentation on metapopulation persistence (Hanski and Ovaskainen 2000; Ovaskainen, Sato et al. 2002), an important issue in landscape ecology (Hill and Caswell 1999; With and King 1999; Fahrig 2001), and in conservation biology (McCarthy, Lindenmayer et al. 1997).

Third, though the deterministic analysis provides helpful insight to the dynamics of metapopulations, it is inadequate for quantitative analysis in the case of small patch networks, in which extinction-colonization stochasticity inevitably playes a big role, and in, e.g., cases where the dynamics are spatially correlated. To model metapopulation dynamics in such situations, one may turn to stochastic models, but the difficulty here is that the mathematical analysis of heterogeneous SPOMs for all but tiny networks is hampered by the huge size of the state space. Ovaskainen (2002) has described a useful approximation to overcome these problems. His approach is based on the same idea as the concept of effective population size in population genetics: construct a homogeneous SPOM which behaves in the same manner as the heterogeneous SPOM with respect to some model properties of interest, and which homogeneous SPOM can be analyzed mathematically (for technical details see the original paper). In the case of the spatially realistic Levins model, the transformed model has three parameters, which are the effective patch number, the effective colonization rate, and the effective extinction rate.

My fourth point concerns the fundamental quantity of time to metapopulation extinction. The "effective metapopulation" approach may be extended to a variety of biologically interesting situations, such as correlated local dynamics or temporally varying environmental conditions (Ovaskainen 2002). For example, Ovaskainen and Hanski (2003) used this approach to show that if extinctions and colonizations are correlated, with a correlation coefficient ρ, the effective number of habitat patches is reduced to $n_e = n/((n-1)\rho + 1)$. The correlation changes the qualitative behavior of the model in several ways. Most importantly, for $\rho > 0$ the mean time to extinction does not increase exponentially with the number of habitat patches, as predicted by a homogeneous SPOM (Ovaskainen 2001), but it grows according to the power law $T \sim n^{1/\rho}$ (Ovaskainen and Hanski 2003). This contrast between spatially uncorrelated and correlated metapopulation dynamics parallels the difference in extinction time of local populations under demographic versus environmental stochasticities (Lande 1993; Foley 1994). In both cases, the correlation can be the dominant factor determining the life-time of a population or a metapopulation, respectively.

Empirical studies

Ecologists and conservation biologists have conducted hundreds of empirical studies on the occurrence of species in highly fragmented landscapes. Many researchers have set out to test what is often called the area-isolation paradigm, namely that the spatial distribution of species is largely determined by the areas and isolations (connectivities) of habitat patches. This "paradigm," which is often seen as an integral part of the metapopulation theory, is contrasted with the view that what really matters for the occurrence of species is not habitat area and isolation but habitat quality and spatial variation in habitat quality from one patch to another. An extensive literature has grown around this issue (reviewed by Fahrig 1997; 2003, Hanski 2005, Pellet et al. 2007). Incidentally, the literatures on the species-area relationship and the island biogeographic theory (MacArthur and Wilson 1967) contain a parallel debate about the importance of island area versus habitat heterogeneity in explaining the increasing number of species on islands with increasing area (Williamson 1981; Rosenzweig 1995; Whittaker 1998).

It is apparent from the previous section that metapopulation models for fragmented landscapes have been constructed that incorporate the influence of habitat patch area and connectivity on the dynamics; this is what I call the spatially realistic approach. The rationale for this approach is that the area and isolation effects, stemming from the general extinction proneness of small populations and general distance dependence of migration, can be assumed to operate in many species in many landscapes. The intention is not to argue that habitat quality would not matter, which would be a biologically naive argument, but one should realize that variation in habitat quality is very species and habitat specific and it is difficult to extract anything generally comparable to habitat area and connectivity; there is no general answer. It is often critically important to know what really determines the occurrence of species in particular cases, not least for conservation and management, but how much habitat area, quality, and isolation matter must depend on the specific circumstances. Furthermore, each empirical study is necessarily based on a limited number of habitat patches and variables that are measured, and exactly which patches are included in the study makes a difference. Including more patches of very low quality will most likely increase the statistical significance of habitat quality in explaining habitat occupancy; adding tiny patches (which an ecologist might be tempted to exclude because they do not often support a local population) would increase the significance of patch area; and including some very isolated patches might do the same for the significance of connectivity. The point is that there is no general answer, and one should not be misled to assume that 10 studies demonstrating the importance of habitat quality have somehow demonstrated the general unimportance of the spatial configuration of habitat for the dynamics of species living in fragmented landscapes.

It should also be recognized that if a species occurs as a metapopulation in a fragmented landscape and occupies, at the stochastic quasi-equilibrium, only a fraction of the patches, there cannot be explanatory variables that could explain well in which particular patches a species happens to occur at one point in time. The effects of patch area, quality, and connectivity are expected to become more evident if more data are available and the long-term incidence (probability) of patch occupancy is analyzed (the variable p_i in Eq. (9.2)) rather than presence or absence at one point in time. In a similar manner, island area typically explains a large fraction of variation in the number of species on islands, because summing up the occurrences of many species in a set of islands averages out much of the stochastic variation. There are also several other biological reasons why patch area and isolation do not often explain a large fraction of variation in the occurrence of species in empirical studies (Hanski 2005; Hanski and Pöyry 2007), including the measure of isolation used. Most empirical studies still employ the distance to the nearest occupied habitat patch, or even to the nearest patch regardless of occupancy, though such measures are known to underestimate the effect of connectivity (Moilanen and Nieminen 2002). Biologically, Eq. (9.1) is a well justified measure of connectivity for empirical studies.

Spatially realistic models predict that the viability of the entire metapopulation depends on the amount and spatial configuration of habitat in the network. This prediction is clearly of great significance for conservation and management, but conducting empirical studies at the network level rather than at the level of individual patches takes more time and resources and hence the former are much less numerous than the latter. Nevertheless, a few studies have addressed the question about the extinction threshold in empirical studies. The most rigorous work has been done on the Glanville fritillary butterfly (*Melitaea cinxia*) in Finland, for which data are available from tens of patch networks. The spatially realistic model fits well to these data and the results provide convincing evidence for the extinction threshold, absence of the species from networks with $\delta > \lambda_M$ (see Eq. (9.3)). Qualitatively similar results have been reported for the Marsh fritillary (*Euphydryas aurinia*) in England (Bulman, Wilson et al. 2007), for the white-backed woodpecker (*Dendrocopos leucotos*) in Sweden (Carlson 2000), the three-toed woodpecker (*Picoides tridactylus*) in Finland (Pakkala, Hanski et al. 2002), and insects and fungi living in decomposing logs in Finland (Hanski 2005).

9.4 Metacommunity dynamics of competing species

The dynamic theory of island biogeography of MacArthur and Wilson (1963; 1967) explains the number of species on islands by their areas and isolations from the mainland. Hence the island model is a spatially realistic model - it incorporates the consequences of the spatial configuration of the habitat on population processes. As the basic island model assumes noninteractive species, it may be constructed by simply adding up models for single-species metapopulation dynamics: the expected number of species on an island, or in a habitat fragment, is given by the sum of the species-specific long-term probabilities of occurrence on the island, in other words the p^* values described in the previous section. As a matter of fact, it can be shown that the island model is just a limiting case of the spatially realistic metapopulation model (Hanski 2009). Viewing the island theory from the perspective of single-species metapopulation theory leads to potentially helpful models, such as the species-area relationship derived from single-species incidence functions (Ovaskainen and Hanski 2003).

In reality, all species interact with a smaller or larger number of other species. Often the interactions are spread out so thinly across many other species that no particular species appears strongly dynamically coupled with the focal species. For instance, most insectivorous birds feed on hundreds or even thousands of insect species, and most are preyed upon by a common set of natural enemies. In such situations, models ignoring interspecific interactions may successfully describe many features of multispecies communities. The island theory is an example. But not all communities are like this.

The single-species metapopulation model has been extended to multiple competing species. In this context, models have typically assumed that competition increases extinction rate, decreases colonization rate, or both (Levins and Culver 1971; Hanski 1983; Hastings 1987; Nee and May 1992), thus reducing the chances of landscape-level coexistence. These models have not incorporated any description of landscape structure, as all habitat patches and local populations have been assumed to be equally large and equally connected. On the other hand, a particular concern has been to properly account for the spatial correlation structure in the occurrence of species. Evidently, if interspecific competition increases the extinction rates of competing species, they are less likely to co-occur locally than expected by random and independent distribution of species. Indeed, much of the theoretical literature on the spatial dynamics of interacting species, whether they are competitors or engaged in predator-prey interactions, has been focused on elucidating spatial pattern formation (nonrandom spatial patterns) due to interspecific interactions (Dieckmann, Law et al. 2000; Ovaskainen and Cornell 2006). This is a striking and interesting phenomenon, but it should be recognized that spatio-temporal correlations maintained by interspecific interactions are strongest when the fixed spatial structure of the environment is simplest; most theoretical models assume completely homogeneous environment. When the habitat occurs in discrete patches that differ in size and isolation, landscape structure constrains the spatial dynamics and greatly influences spatial patterns in the occurrence of species. The situation is analogous with respect to ecological differences among the species. In the case of two or more equal competitors, interspecific interactions generate and maintain spatio-temporal correlations in species abundances that may critically influence metacommunity dynamics. But in real communities, competing species typically exhibit differences in their ecologies that constrain their dynamics, just like spatial heterogeneity constrains the spatial dynamics in a metacommunity.

With these considerations in mind, I have constructed a model for competitors in a fragmented landscape that includes a description of the spatial configuration of the habitat but ignores spatial correlations induced by interspecific interactions - because such correlations are not expected to be very influential in communities of unequal competitors in heterogeneous landscapes (Hanski 2008). The model is essentially an extension of the single-species spatially realistic metapopulation model (Eq. (9.2)) to any number of competing species, constructed on the assumption that local competition reduces the effective patch areas (local carrying capacities) as experienced by the co-occurring species. Reduced patch areas implicitly correspond to smaller populations and hence translate to an increased rate of local extinction and reduced rate of colonization in the model. Empirical studies on interspecific resource competition have demonstrated that competition typically reduces the sizes of coexisting populations (Connell 1983; Schoener 1983; Goldberg and Barton 1992). One good example of competitive metacommunities is zooplankton living in networks of rock pools and other small water bodies (Cottenie and De Meester 2005; Kolasa and Romanuk 2005). In a well-studied metacommunity of three species of *Daphnia* water fleas (Hanski and Ranta 1983), interspecific competition increased annual extinction rate of local populations by 64% (Bengtsson 1991). Other examples are discussed by Hanski (1999) and many chapters in Holyoak et al. (2005).

The model allows the calculation of deterministic equilibria with fast iteration. The model predicts that, not surprisingly, the number of coexisting species decreases with increasing strength of competition (Hanski 2008). The number of coexisting species also decreases with decreasing average rate of colonization, for a given average rate of extinction, essentially because poor colonizers have a more fragile occurrence in a fragmented landscape to start with and hence their viability is more sensitive to competition than that of good colonizers. The number of coexisting species also decreases with increasing range of migration, modelled with the help of the connectivity measure described by Eq. (9.1). In the case of species that have limited range of migration, strong competition may lead to spatially restricted ranges of the species, such as shown in Fig. 9.1a. This may happen when the spatial distribution of habitat patches is random or aggregated, which leads to spatial variation in the strength of migration. Two species may settle into a stable equilibrium in which their spatial distributions are complementary, the boundary occurring in parts of the network where patch density is low: competition may then make it impossible for a species to cross such a relative barrier to migration (the same result has been obtained in a different kind of competition model; Goldberg and Lande 2006). On the other hand, in the case of many species that compete somewhat less strongly, the spatially restricted stable distributions may become nested, as shown in Fig. 9.1b. Several species are now restricted to the most favorable part of the network, where the connectivity of the patches is high due to high patch density and/or large patch areas (Hanski 2008).

This model demonstrates how the spatial configuration of habitat may lead to qualitatively new outcomes of competition, such as partly nested distributions of competing species. This counterintuitive result is caused by the complex web of direct and indirect interactions among the species: in the case of three or more species, a shared competitor has a negative direct effect, decreasing the patch carrying capacity for the focal species, but a positive indirect effect, in reducing the abundance (incidence of patch occupancy) of a shared competitor. Another interesting result concerns the influence of spatial variation in patch areas: other things being equal, the number of coexisting species is greater when there is spatial variation in patch areas (Hanski 2008). This result relates to competition-colonization trade-off, poor competitors persisting well in small habitat patches if they are good colonizers.

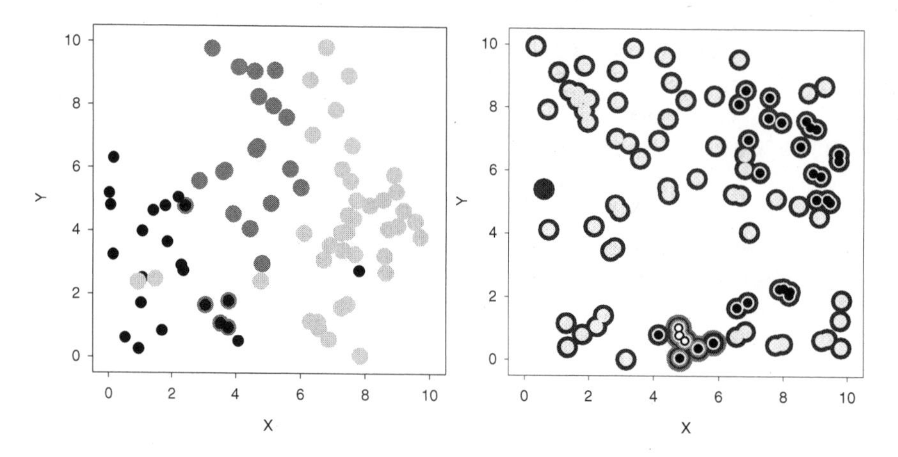

Figure 9.1 Two examples of limited spatial distributions of many similar competing species with short range of migration. The colors indicate the presence of different species in different habitat patches with probability of occupancy (p) greater than 0.2. (a) Very strong competitors exhibit practically exclusive spatial distributions, while (b) moderately strong competitors may end up with partly nested distributions. (See color insert following page 202.)

The modelling work that I have described in this section belongs to a rapidly expanding literature on metacommunity ecology and dynamics, reviewed by Holyoak et al. (2005) and Urban et al. (2008). Metacommunity dynamics have been studied within several different frameworks, of which the above model belongs to the patch dynamics framework. So far, metacommunity ecology has largely ignored the actual spatial configuration of the habitat, with some notable exceptions, for instance contributions that relate the species composition in local communities to the degree of isolation of the respective habitat patch in the landscape (Kruess 2003; Gripenberg and Roslin 2005; Tscharntke, Rand et al. 2005; van Nouhuys 2005).

9.5 Genetic and evolutionary dynamics

The spatial configuration of habitat in landscapes influences ecological dynamics, and knowing about the population dynamic consequences of spatial structure improves our understanding of and capacity to predict population dynamics. In the same manner, one could expect that the spatial configuration of habitat influences and possibly interacts with evolutionary dynamics. This is best understood in the case of heterogeneous environments consisting of dissimilar habitats, for which there is a vast body of evolutionary theory including processes of speciation and adaptive radiation. Turning back to the theme of competing species in metacommunities, just like a mixture of dissimilar habitats may allow a score of specialist species to coexist – each having its stronghold in a different type of habitat – natural selection may favor different genotypes in different habitats, and thus landscape heterogeneity is one of the factors that may maintain genetic diversity (Levene 1953; Hedrick 1986) and thereby the potential for evolutionary change. My concern here is however more circumscribed and the same as in the previous sections: assuming that there is just one type of habitat, what difference might the actual spatial configuration of that habitat make to genetic and evolutionary dynamics? I start with an example from our research on the Glanville fritillary butterfly (Hanski 1999; Ehrlich and Hanski 2004). This example is particularly instructive, because the relevant genetic information was available for the study populations for a long time, yet its significance became apparent only when analyzed in the spatially realistic context.

The enzyme phosphoglucose isomerase (PGI) catalyzes the second step in glycolysis. It is highly polymorphic in most animals and plants (e.g., Katz and Harrison 1997, Filatov and Charlesworth 1999, Dahlhoff and Rank 2000), which is the reason why it was one of the most commonly studied molecules in the enzyme electophoretic studies of genetic polymorphism in the 1970s and 1980s. It was, and still is, a convenient genetic marker for many population genetic studies. This is how we used it in our studies of the Glanville fritillary in the 1990s, for instance to characterize the level of inbreeding in small populations and to study how inbreeding depression would influence local dynamics and population extinction (Saccheri, Kuussaari et al. 1998). It was known from previous studies on e.g., *Colias* butterflies that PGI may not be a neutral locus. Watt (Watt, Chew et al. 1977; Watt 1983) had demonstrated differences in the performance and fitness components of individuals with different genotypes. Unfortunately, we had only samples of allele frequencies from many natural populations, no data on the performance or fitness of individuals, and there did not seem to be any way of assessing possible selection with such samples. Some additional information would have been needed.

In our case, progress was made when we took into account knowledge about the spatial configuration of the landscape from which the samples originated, the areas and the connectivities of the habitat patches, and the ages of the local populations inhabiting these patches, which are known thanks to long-term monitoring of the large metapopulation of the Glanville fritillary in the Åland Islands in Finland (Hanski 1999; Nieminen, Siljander et al. 2004). Several other genetic loci served as helpful controls, as the same samples had been genotyped for multiple loci, which share the same demographic history as the candidate gene *Pgi*. We found that a particular allozyme allele was significantly more frequent in newly-established isolated populations than in old and well-connected populations (Haag, Saastamoinen et al. 2005), and that the *Pgi* allelic composition of local populations influenced their growth rate, which became apparent when the effects of habitat patch area and isolation were also taken into account (Hanski and Saccheri 2006). These conclusions have been subsequently supported by studies on individual butterflies, which have documented significant associations between molecular variation in *Pgi* and variation in the flight metabolic rate and dispersal rate in the field (Niitepõld, Smith et al. 2009), body temperature of butterflies at flight in low ambient temperatures (Saastamoinen and Hanski 2008), egg clutch size (Saastamoinen 2007), and even lifespan (Saastamoinen, Ikonen et al. 2009).

The emerging discipline of landscape genetics addresses the influence of landscape structure on population genetics (Manel, Schwartz et al. 2003; Storfer, Murphy et al. 2007). However, landscape genetic studies appear to be primarily concerned with neutral genetic variation, with descriptions of genetic variation and spatial genetic structures, and with the application of spatial statistics and landscape ecological approaches to genetic data. In contrast, the example that I have just described is focused on functionally important molecular variation and attempts to develop a spatially realistic approach for metapopulation genetic studies (Gaggiotti 2004; Whitlock 2004). Ultimately, all these approaches are likely to merge, just like metapopulation ecology and landscape ecology will hopefully become increasingly integrated.

Modeling the evolution of migration

The evolution of migration rate is a classic topic in evolutionary biology and life-history ecology (Ronce, Olivieri et al. 2001; Ronce and Olivieri 2004). Evolution of migration rate is gaining additional importance from the threats that habitat loss and fragmentation pose to the viability of very large numbers of species (Hanski 2005) and from the possibility that evolutionary changes in migration rate (as well as in other life history traits) might ameliorate the impact of fragmentation. In metapopulations consisting of small extinction-prone local populations some migration is clearly necessary for long-term persistence. On the other hand, "too much" migration may elevate mortality during migration so greatly, and it may lead to such an excessive loss of time, that persistence is again compromised (Comins, Hamilton et al. 1980; Hanski and Zhang 1993; Olivieri and Gouyon 1997). Though natural selection does not operate to produce the optimal migration rate for the long-term survival of species or metapopulations (Comins, Hamilton et al. 1980), it is nonetheless possible that an evolutionary change in migration rate following habitat loss and fragmentation might reduce the risk of extinction (Leimar and Nordberg 1997).

But what is the likely change in migration rate in response to habitat loss and fragmentation? There are so many different selective forces affecting the evolution of migration rate (Ronce, Olivieri et al. 2001; Ronce and Olivieri 2004) that there is no simple answer to this question. For instance, habitat loss and fragmentation increase mortality during migration, because it becomes increasingly difficult for migrants to locate another fragment of habitat, which should select for reduced migration (van Valen 1971). Increasing genetic relatedness of individuals in increasingly isolated local populations should select for increased migration (Hamilton and May 1977; Gandon and Rousset 1999), and so should the opportunity to recolonize habitat patches that have become unoccupied following local extinction, and more generally the chance to move to a low-density population (Gadgil 1971; Roff 1975). Given the multitude of often opposing selection pressures, it is perhaps not surprising that researchers have come up with conflicting suggestions as to what might be the net effect of habitat fragmentation on the evolution of migration rate. For instance, Dempster (1991) expected evolution to reduce migration rate in butterflies living in increasingly fragmented habitats (see also Thomas et al. 1998, Hill et al. 1999), whereas

Hanski (1999) suggested that fragmentation would generally select for increased migration rate. The matter cannot be settled without having a means of considering all the major selective forces, and their interactions, at the same time. And this cannot be done without employing appropriate models.

Heino and Hanski (2001) constructed a spatially realistic evolutionary model to investigate the evolution of migration rate in fragmented habitats, using the Glanville fritillary butterfly as a model species. The model parameters were estimated with independent data whenever possible. Migration was modelled using the movement model outlined in Section 9.2, which allowed the estimation of parameter values with empirical data. The remaining parameter values were selected in such a manner that the model produced realistic short-term and long-term metapopulation dynamics. Reassuringly, when the migration rate was allowed to evolve in the model, it settled to a value close to the empirically observed one (Heino and Hanski 2001). This analysis suggested that the dominant selective forces were mortality and time lost during migration as well as the opportunity to establish new local populations in currently unoccupied patches. Starting from a little-fragmented landscape, with increasing fragmentation the predicted migration rate first declined, apparently due to increased cost of migration, but with further fragmentation migration rate increased when an increasing number of habitat patches became available for recolonization.

More detailed theoretical and empirical studies of the same butterfly metapopulation have revealed how the migration rate of butterflies in particular local populations depends on their ages and population dynamic connectivities to other populations (Hanski, Erälahti et al. 2004). The migration rate is predicted and was observed to be higher in new than in old populations, apparently because new populations are likely to be established by exceptionally mobile individuals and because the relevant traits exhibit high heritability in the Glanville fritillary (Saastamoinen 2008) as migration-related traits do also more generally (Roff and Fairbairn 1991). The results on *Pgi* polymorphism referred to above match nicely these results and suggest that *Pgi* is functionally involved in the evolution of migration rate in the Glanville fritillary. Among the new populations, the migration rate increased with decreasing connectivity (increasing isolation), whereas among old populations the opposite was both predicted and observed. A reduced migration rate in old isolated populations is largely due to emigration of the more mobile individuals away from the population and limited immigration due to great isolation. These results resolve the two opposing verbal predictions that have been put forward about the impact of increasing habitat fragmentation on migration rate. Dempster (1991) emphasized increasing emigration losses and expected migration rate to become reduced with fragmentation (increasing isolation of habitat patches); in the Glanville fritillary, this was observed for old populations. Hanski (1999) was primarily thinking of improved colonization opportunities with increasing fragmentation, hence expecting increased migration rate with increasing fragmentation - which was observed for new populations. Because both effects operate simultaneously in a metapopulation, one has to use a model to work out the overall consequences of fragmentation. In the case of the Glanville fritillary, and assuming realistic parameter values for this species and its natural landscape, the overall effect has been increasing migration rate when the habitat becomes highly fragmented (Heino and Hanski 2001; Hanski, Erälahti et al. 2004), though the quantitative result will depend on the spatial configuration of the landscape.

To return to the question whether evolutionary changes may make a difference to the long-term survival of species in changing environments, Heino and Hanski (2001) showed that an evolutionary rescue is theoretically possible: natural selection may change a migration rate to such an extent that a metapopulation will persist in a landscape in which it would go extinct without the evolutionary change. However, the calculations also indicated that in practice such a rescue is unlikely in the Glanville fritillary, largely because a change in migration rate has both positive and negative consequences for population sizes and hence cannot much compensate for habitat loss and fragmentation. Only when the contrast is between relatively uniform and highly fragmented habitats is the level of migration likely to make a truly significant difference for population persistence. Clearly, conservationists should not count on evolution to solve the extinction crisis caused by habitat loss and fragmentation!

9.6 Conclusion

I have outlined in this essay a research approach to spatial ecology - the spatially realistic metapopulation approach - that is helpful for bringing theoretical and empirical studies closer to each other, and which may facilitate the integration of studies from individual movement behavior to population and community dynamics to evolutionary dynamics. Much of the theory in spatial ecology has remained detached from empirical research, to a large extent because the theory does not take into account the influence of realistic landscape structure on population processes. The spatially realistic metapopulation approach specifically addresses the influence of the spatial configuration of the habitat on the ecological and evolutionary processes affecting individuals, populations, and communities. Just like the island biogeographic theory of MacArthur and Wilson (1967) in the 1970s, the spatially realistic approach prescribes research tasks for ecologists engaged in empirical work, in terms of testing model assumptions and predictions. A key component of this research is the measure of connectivity described in the introduction (or some comparable measure), which can be employed both in empirical and modelling studies.

9.7 References

H.G. Andrewartha and L.C. Birch (1954), *The Distribution and Abundance of Animals*, University of Chicago Press, Chicago.

D.J. Bartholomew (1976), Continuous time diffusion models with random duration of interest, *J. Math. Sociol.* **4**: 187-199.

J. Bengtsson (1991), Interspecific competition in metapopulations, in *Metapopulation Dynamics: Empirical and Theoretical Investigations*, ed. by M. E. Gilpin and I. Hanski, Academic Press, London, pp. 219-237.

C.R. Bulman, R.J. Wilson, A.R. Holt, L.G. Bravo, R.I. Early, M.S. Warren and C.D. Thomas

(2007), Minimum viable metapopulation size, extinction debt, and the conservation of a declining species, *Ecological Applications* **17**: 1460-1473.

A.G. Bunn, D.L. Urban and T.H. Keitt (2000), Landscape connectivity: A conservation application of graph theory, *Journal of Environmental Management* **59**: 265-278.

J.A. Byers (2001), Correlated random walk equations of animal dispersal resolved by simulation, *Ecology* **82**: 1680-1690.

A. Carlson (2000), The effect of habitat loss on a deciduous forest specialist species: The White-backed Woodpecker (*Dendrocopos leucotos*), *Forest Ecology and Management* **131**: 215-221.

J. Clobert, J.O. Wolff, J.D. Nichols, E. Danchin and A.A. Dhondt (2001), Introduction, in *Dispersal*, ed. by J. Clobert, E. Danchin, A.A. Dhondt and J.D. Nichols, Oxford University Press, Oxford, pp. 17-21.

H. N. Comins, W.D. Hamilton and R.M. May (1980), Evolutionary stable dispersal strategies, *Journal of Theoretical Biology* **82**: 205-230.

J.H. Connell (1983), On the prevalence and relative importance of interspecific competition: evidence from field experiments, *American Naturalist* **122**: 661-696.

K. Cottenie and L. De Meester (2005), Local interactions and local dispersal in a zooplankton metacommunity, in *Metacommunities: Spatial Dynamics and Ecological Communities*, ed. by M. Holyoak, M.A. Leibold and R.D. Holt, Chicago University Press, Chicago, pp. 189-212.

E.P. Dahlhoff and N.E. Rank (2000), Functional and physiological consequences of genetic variation at phosphoglucose isomerase: Heat shock protein expression is related to enzyme genotype in a montane beetle, *Proceedings of the National Academy of Sciences of the United States of America* **97**: 10056-10061.

J.R. Day and H.P. Possingham (1995), A stochastic metapopulation model with variability in patch size and position, *Theoretical Population Biology* **48**: 333-360.

J.P. Dempster (1991), Fragmentation, isolation and mobility of insect populations, in *Conservation of Insects and their Habitats*, ed. by N.M. Collins and J.A. Thomas, Academic Press, London, pp. 143-154.

U. Dieckmann, R. Law and J.A.J. Metz (2000), *The Geometry of Ecological Interaction: Simplifying Spatial Complexity*, Cambridge University Press, Cambridge.

P. Ehrlich and I. Hanski, Eds. (2004), *On the Wings of Checkerspots: A Model System for Population Biology*, Oxford University Press, New York.

L. Fahrig (1997), Relative effects of habitat loss and fragmentation on population extinction, *Journal of Wildlife Management* **61**: 603-610.

L. Fahrig (2001), How much habitat is enough?, *Biological Conservation* **100**: 65-74.

L. Fahrig (2002), Effect of habitat fragmentation on the extinction threshold: A synthesis, *Ecological Applications* **12**: 346-353.

L. Fahrig (2003), Effects of habitat fragmentation on biodiversity, *Annual Review of Ecology, Evolution, and Systematics* **34**: 487-515.

D.A. Filatov and D. Charlesworth (1999), DNA polymorphism, haplotype structure and balancing selection in the leavenworthia PgiC locus, *Genetics* **153**: 1423-1434.

P. Foley (1994), Predicting extinction times from environmental stochasticity and carrying-capacity, *Conservation Biology* **8**: 124-137.

K. Frank and C. Wissel (1998), Spatial aspects of metapopulation survival - from model results to rules of thumb for landscape management, *Landscape Ecology* **13**: 363-379.

M. Gadgil (1971), Dispersal: population consequences and evolution, *Ecology* **52**: 253-261.

O.E. Gaggiotti (2004), Multilocus genotype methods for the study of metapopulation processes, in *Ecology, Genetics, and Evolution in Metapopulations*, ed. by I. Hanski and O.E.

Gaggiotti. Elsevier Academic Press, Amsterdam, pp. 367-386.

S. Gandon and F. Rousset (1999), Evolution of stepping-stone dispersal rates, *Proceedings of the Royal Society of London Series B - Biological Sciences* **266**: 2507-2513.

D.E. Goldberg and A.M. Barton (1992), Patterns and consequences of interspecific competition in natural communities - A review of field experiments with plants, *American Naturalist* **139**: 771-801.

E.E. Goldberg and R. Lande (2006), Ecological and reproductive character displacement on an environmental gradient, *Evolution* **60**: 1344-1357.

S. Gripenberg and T. Roslin (2005), Host plants as islands: Resource quality and spatial setting as determinants of insect distribution, *Annales Zoologici Fennici* **42**: 335-345.

C. Haag, M. Saastamoinen, J. Marden and I. Hanski (2005), A candidate locus for variation in dispersal rate in a butterfly metapopulation, *Proceedings of the Royal Society of London Series B - Biological Sciences* **272**: 2449-2456.

N. Haddad (1999), Corridor and distance effects on interpatch movements: A landscape experiment with butterflies, *Ecological Applications* **9**: 612-622.

W.D. Hamilton and R.M. May (1977), Dispersal in stable habitats, *Nature* **269**: 578-581.

I. Hanski (1983), Coexistence of competitors in patchy environment, *Ecology* **64**: 493-500.

I. Hanski (1994), A practical model of metapopulation dynamics, *Journal of Animal Ecology* **63**: 151-162.

I. Hanski (1999), *Metapopulation Ecology*, Oxford University Press, New York.

I. Hanski (2005), *The Shrinking World: Ecological consequences of Habitat Loss*, Oldendorf/Luhe, International Ecology Institute.

I. Hanski (2008), Spatial patterns of coexistence of competing species in patchy habitats, *Theoretical Ecology* **1**: 29-43.

I. Hanski (2009), The theories of island biogeography and metapopulation dynamics. Science marches forward, but the legacy of good ideas lasts for a long time, in *The Theory of Island Biogeography at 40*, ed. by R. Ricklefs and J. Losos, Princeton University Press, Princeton.

I. Hanski, J. Alho and A. Moilanen (2000), Estimating the parameters of survival and migration of individuals in metapopulations, *Ecology* **81**: 239-251.

I. Hanski, C. Erälahti, M. Kankare, O. Ovaskainen and H. Sirén (2004), Variation in migration rate among individuals maintained by landscape structure, *Ecology Letters* **7**: 958-966.

I. Hanski and O.E. Gaggiotti, Eds. (2004), *Ecology, Genetics, and Evolution of Metapopulations*, Elsevier Academic Press, Amsterdam.

I. Hanski and M. Gilpin (1991), Metapopulation dynamics: Brief history and conceptual domain, in *Metapopulation Dynamics: Empirical and Theoretical Investigations,* ed. by M. Gilpin and I. Hanski, Academic Press, London, pp. 3-16.

I. Hanski and M. Gyllenberg (1997), Uniting two general patterns in the distribution of species, *Science* **275**: 397-400.

I. Hanski and O. Ovaskainen (2000), The metapopulation capacity of a fragmented landscape, *Nature* **404**: 755-758.

I. Hanski and O. Ovaskainen (2003), Metapopulation theory for fragmented landscapes, *Theoretical Population Biology* **64**: 119-127.

I. Hanski and J. Pöyry (2007), Insect populations in fragmented habitats, in *Insect Conservation Biology,* The Royal Entomological Society Symposium Volume, CABI, UK.

I. Hanski and E. Ranta (1983), Coexistence in a patchy environment: Three species of *Daphnia* in rock pools, *Journal of Animal Ecology* **52**: 263-280.

I. Hanski and I. Saccheri (2006), Molecular-level variation affects population growth in a butterfly metapopulation, *Plos Biology* **4**: 719-726.

I. Hanski and D.-Y. Zhang (1993), Migration, metapopulation dynamics and fugitive co-existence, *Journal of Theoretical Biology* **163**: 491-504.

A. Hastings (1987), Can competition be detected using species co-occurrence data, *Ecology* **68**: 117-123.

P.W. Hedrick (1986), Genetic-polymorphism in heterogeneous environments - A decade later, *Annual Review of Ecology and Systematics* **17**: 535-566.

M. Heino and I. Hanski (2001), Evolution of migration rate in a spatially realistic metapopulation model, *American Naturalist* **157**: 495-511.

J.B. Hestbeck, J.D. Nichols, and R.A. Malecki (1991), Estimates of movement and site fidelity using mark-resight data of wintering Canada geese, *Ecology* **72**: 523-533.

R. Hilborn (1991), Modeling the stability of fish schools: Exchange of individual fish between schools of skipjack tuna (*Katsuwonus pelamis*), *Canadian Journal of Fisheries and Aquatic Sciences* **48**: 1081-1091.

J.K. Hill, C.D. Thomas and D.S. Blakeley (1999), Evolution of flight morphology in a butterfly that has recently expanded its geographic range, *Oecologia* **121**: 165-170.

M.F. Hill and H. Caswell (1999), Habitat fragmentation and extinction thresholds on fractal landscapes, *Ecology Letters* **2**: 21-127.

M. Holyoak, M.A. Leibold and R.D. Holt, Eds. (2005), *Metacommunities: Spatial Dynamics and Ecological Communities*, University of Chicago Press, Chicago.

C.B. Huffaker (1958), Experimental studies on predation: dispersion factors and predator-prey oscillations, *Hilgardia* **27**: 343-383.

J.A. Jacquez and C.P. Simon (1993), The stochastic SI model with recruitment and deaths I. Comparison with the closed SIS model, *Mathematical Biosciences* **117**: 77-125.

I. Jonsen and P.D. Taylor (2000), *Calopteryx* damselfly dispersions arising from multi-scale responses to landscape structure, *Conservation Ecology* **4**(2): 4. [online] URL: http://www.consecol.org/vol4/iss2/art4/.

L.A. Katz and R.G. Harrison (1997), Balancing selection on electrophoretic variation of phosphoglucose isomerase in two species of field cricket: *Gryllus veletis* and *G. pennsylvanicus*, *Genetics* **147**: 609-621.

J. Kolasa and T.N. Romanuk (2005), Assembly of unequals in the unequal world of a rock pool metacommunity, in *Metacommunities: Spatial Dynamics and Ecological Communities,* ed. by M. Holyoak, M. A. Leibold and R. D. Holt, Chicago University Press, Chicago, pp. 212-232.

A. Kruess (2003), Effects of landscape structure and habitat type on a plant-herbivore-parasitoid community, *Ecography* **26**: 283-290.

R.J. Kryscio and C. Lefèvre (1989), On the extinction of the S-I-S stochastic logistic epidemic, *Journal of Applied Probability* **27**: 685-694.

R. Lande (1987), Extinction thresholds in demographic models of territorial populations, *American Naturalist* **130**: 624-635.

R. Lande (1993), Risks of population extinction from demographic and environmental stochasticity and random catastrophes, *American Naturalist* **142**: 911-927.

O. Leimar and U. Nordberg (1997), Metapopulation extinction and genetic variation in dispersal-related traits, *Oikos* **80**: 448-458.

H. Levene (1953), Genetic equilibrium when more than one ecological niche is available, *American Naturalist* **87**: 331-333.

R. Levins (1969), Some demographic and genetic consequences of environmental heterogeneity for biological control, *Bulletin of the Entomological Society of America* **15**: 237-240.

R. Levins (1970), Extinction, *Lecture Notes in Mathematics* **2**: 75-107.

R. Levins and D. Culver (1971), Regional coexistence of species and competition between rare species. *Proceedings of the National Academy of Sciences of the USA* **68**: 1246-1248.

R. H. MacArthur and E.O. Wilson (1963), An equilibrium theory of insular zoogeography, *Evolution* **17**: 373-387.

R.H. MacArthur and E.O. Wilson (1967), *The Theory of Island Biogeography*, Princeton University Press, Princeton.

T.R. Malthus (1798), *An Essay on the Principle of Population as it Affects the Future Improvement of Society, with Remarks on the Speculations of Mr. Godwin, M. Condorcet, and other Writers,* J. Johnson, London.

S. Manel, M.K. Schwartz, G. Luikart and P. Taberlet (2003), Landscape genetics: Combining landscape ecology and population genetics, *Trends in Ecology and Evolution* **18**: 189-197.

M.A. McCarthy, D.B. Lindenmayer and M. Dreschler (1997), Extinction debts and risks faced by abundant species, *Conservation Biology* **11**: 221-226.

R.P. McIntosh (1985), *The Background to Ecology: Concept and Theory*, Cambridge University Press, Cambridge.

A. Moilanen (1999), Patch occupancy models of metapopulation dynamics: efficient parameter estimation using implicit statistical inference, *Ecology* **80**: 1031-1043.

A. Moilanen and I. Hanski (2001), On the use of connectivity measures in spatial ecology, *Oikos* **95**: 147-151.

A. Moilanen and M. Nieminen (2002), Simple connectivity measures for spatial ecology, *Ecology* **83**: 1131-1145.

S. Nee and R.M. May (1992), Dynamics of metapopulations: Habitat destruction and competitive coexistence, *Journal of Animal Ecology* **61**: 37-40.

A.J. Nicholson (1933), The balance of animal populations, *Journal of Animal Ecology* **2**: 132-178.

M. Nieminen, M. Siljander and I. Hanski (2004), Structure and dynamics of *Melitaea cinxia* metapopulations, in *On the Wings of Checkerspots: A Model System for Population Biology,* ed. by P. R. Ehrlich and I. Hanski, Oxford University Press, New York, pp. 63-91.

K. Niitepõld, A. D. Smith, J.L. Osborne, D.R. Reynolds, N.L. Carreck, A.P. Martin, J.H. Marden, O. Ovaskainen, and I. Hanski (2009), Flight metabolic rate and *Pgi* genotype influence butterfly dispersal rate in the field, *Ecology*, in press.

R.H. Norden (1982), On the distribution of the time to extinction in the stochastic logistic population model, *Advances in Applied Probability* **14**: 687-708.

I. Nåsell (1996), The quasi-stationary distribution of the closed endemic SIS model, *Advances in Applied Probability* **28**: 895-932.

I. Olivieri and P.-H. Gouyon (1997), Evolution of migration rate and other traits: The metapopulation effect, in *Metapopulation Biology,* ed. by I.A. Hanski and M.E. Gilpin. Academic Press, San Diego, pp. 293-324.

I. Oppenheim, K.E. Shuler and G.H. Weiss (1977), Stochastic theory of nonlinear rate processes with multiple stationary states, *Physica A* **88**: 191-214.

O. Ovaskainen (2001), The quasi-stationary distribution of the stochastic logistic model, *Journal of Applied Probability* **38**: 898-907.

O. Ovaskainen (2002), The effective size of a metapopulation living in a heterogeneous patch network, *American Naturalist* **160**: 612-628.

O. Ovaskainen (2004), Habitat-specific movement parameters estimated using mark-recapture data and a diffusion model, *Ecology* **85**: 242-257.

O. Ovaskainen and S.J. Cornell (2006), Space and stochasticity in population dynamics, *Proceedings of the National Academy of Sciences of the United States of America* **103**: 12781-12786.

O. Ovaskainen and I. Hanski (2001), Spatially structured metapopulation models: global and local assessment of metapopulation capacity, *Theoretical Population Biology* **60**: 281-304.

O. Ovaskainen and I. Hanski (2002), Transient dynamics in metapopulation response to perturbation, *Theoretical Population Biology* **61**: 285-295.

O. Ovaskainen and I. Hanski (2003), Extinction threshold in metapopulation models, *Annales Zoologici Fennici* **40**: 81-97.

O. Ovaskainen and I. Hanski (2003), The species-area relation derived from species-specific incidence functions, *Ecology Letters* **6**: 903-909.

O. Ovaskainen and I. Hanski (2004), Metapopulation dynamics in highly fragmented landscapes, in *Ecology, Genetics, and Evolution in Metapopulations,* ed. by I. Hanski and O.E. Gaggiotti, Elsevier Academic Press, Amsterdam, pp. 73-104.

O. Ovaskainen, H. Rekola, E. Meyke and E. Arjas (2008), Bayesian methods for analyzing movements in heterogeneous landscapes from mark-recapture data, *Ecology* **89**: 542-554.

O. Ovaskainen, K. Sato, J. Bascompte and I. Hanski (2002), Metapopulation models for extinction threshold in spatially correlated landscapes, *Journal of Theoretical Biology* **215**: 95-108.

O. Ovaskainen, A.D. Smith, J.L. Osborne, R.D. Reynolds, N.L. Carreck, A.P. Martin, K. Niitepõld and I. Hanski (2008), Tracking butterfly movements with harmonic radar reveals an effect of population age on movement distance, *Proceedings of the National Academy of Sciences of the United States of America 105*: 19090-19095.

T. Pakkala, I. Hanski and E. Tomppo (2002), Spatial ecology of the three-toed woodpecker in managed forest landscapes, *Silva Fennica* **36**: 279-288.

J. Pellet, E. Fleishman, D.S. Dobkin, A. Gander and D.D. Murphy (2007), An empirical evaluation of the area and isolation paradigm of metapopulation dynamics, *Biological Conservation* **136**: 483-495.

S. Petit, A. Moilanen, I. Hanski and M. Baguette (2001), Metapopulation dynamics of the bog fritillary butterfly: movements between habitat patches, *Oikos* **92**: 491-500.

J. Pither and P.D. Taylor (1998), An experimental assessment of landscape connectivity, *Oikos* **83**: 166-174.

D.A. Roff (1975), Population stability and the evolution of dispersal in a heterogeneous environment, *Oecologia* **19**: 217-237.

D.A. Roff and D.J. Fairbairn (1991), Wing dimorphisms and the evolution of migratory polymorphisms among the Insecta, *American Zoologist* **31**: 243-251.

O. Ronce (2007), How does it feel to be like a rolling stone? Ten questions about dispersal evolution, *Annual Review of Ecology, Evolution, and Systematics* **38**: 231-253.

O. Ronce and I. Olivieri (2004), Life history evolution in metapopulations, in *Ecology, Genetics, and Evolution of Metapopulations,* ed. by I. Hanski and O.E. Gaggiotti, Elsevier Academic Press, Amsterdam, pp. 227-257.

O. Ronce, I. Olivieri, J. Clobert and E. Danchin (2001), Perspectives on the study of dispersal evolution, in *Dispersal,* ed. by J. Clobert, E. Danchin, A.A. Dhondt and J.D. Nichols, Oxford University Press, Oxford, pp. 341-357.

M.L. Rosenzweig (1995), *Species Diversity in Space and Time*, Cambridge University Press, Cambridge.

M. Saastamoinen (2007), Life-history, genotypic, and environmental correlates of clutch size in the Glanville fritillary butterfly, *Ecological Entomology* **32**: 235-242.

M. Saastamoinen (2008), Heritability of dispersal rate and other life history traits in the Glanville fritillary butterfly, *Heredity* **100**: 39-46.

M. Saastamoinen and I. Hanski (2008), Genotypic and environmental effects on flight activity and oviposition in the Glanville fritillary butterfly, *American Naturalist* **171**: E701-E712.

M. Saastamoinen, S. Ikonen, and I. Hanski (2009), Significant effects of *Pgi* genotype and body reserves on lifespan in the Glanville fritillary butterfly, *Proceedings of the Royal Society of London Series B - Biological Sciences*, in press.

I.J. Saccheri, M. Kuussaari, M. Kankare, P. Vikman, W. Fortelius and I. Hanski (1998), Inbreeding and extinction in a butterfly metapopulation, *Nature* **392**: 491-494.

P. Schippers, J. Verboom, P. Knaapen and R.C. Van Apeldoorn (1996), Dispersal and habitat connectivity in complex heterogeneous landscapes: an analysis with a GIS based random walk model, *Ecography* **19**: 97-106.

T.W. Schoener (1983), Field experiments on interspecific competition, *American Naturalist* **122**: 240-285.

A. Storfer, M.A. Murphy, J.S. Evans, C.S. Goldberg, S. Robinson, S.F. Spear, R. Dezzani, E. Delmelle, L. Vierling and L.P. Waits (2007), Putting the 'landscape' in landscape genetics, *Heredity* **98**: 128-142.

C.D. Thomas, J.K. Hill and O.T. Lewis (1998), Evolutionary consequences of habitat fragmentation in a localised butterfly, *Journal of Animal Ecology* **67**: 485-497.

D. Tilman and P. Kareiva (1997), *Spatial Ecology*, Princeton University Press, Princeton.

L. Tischendorf and L. Fahrig (2000), On the usage of landscape connectivity, *Oikos* **90**: 7-19.

T. Tscharntke, T.A. Rand and F.J.J.A. Bianchi (2005), The landscape context of trophic interactions: insect spillover across the crop-noncrop interface, *Annales Zoologici Fennici* **42**: 421-432.

M.G. Turner, R.H. Gardner and R.V. O'Neill (2001), *Landscape Ecology in Theory and Practice,* Springer, New York.

M.C. Urban, M.A. Leibold, P. Amarasekare, L. De Meester, R. Gomulkiewics, M.E. Hochberg, C.A. Klausmeier, N. Loeuille, C. De Mazancourt, J. Norberg, J.H. Pantel, S.Y. Strauss, M. Vellend and M.J. Wade (2008), The evolutionary ecology of metacommunities, *Trends in Ecology and Evolution* **23**: 311-317.

N. Wahlberg, T. Klemetti, V. Selonen and I. Hanski (2002), Metapopulation structure and movements in five species of checkerspot butterflies, *Oecologia* **130**: 33-43.

S. van Nouhuys (2005), Effects of habitat fragmentation at different trophic levels in insect communities, *Annales Zoologici Fennici* **42**: 433-447.

L. van Valen (1971), Group selection and the evolution of dispersal, *Evolution* **25**: 591-598.

W.B. Watt (1983), Adaptation at specific loci. II. Demographic and biochemical-elements in the maintenance of the *Colias Pgi* polymorphism, *Genetics* **103**: 691-724.

W.B. Watt, F.S. Chew, L.R.G. Snyder, A.G. Watt and D.E. Rothschild (1977), Population structure of pierid butterflies. 1. Numbers and movements of some montane *Colias* species, *Oecologia* **27**: 1-22.

G.H. Weiss and M. Dishon (1971), On the asymptotic behavior of the stochastic and deterministic models of an epidemic, *Mathematical Biosciences* **11**: 261-265.

M.C. Whitlock (2004), Selection and drift in metapopulations, in *Ecology, Genetics, and Evolution of Metapopulations,* ed. by I. Hanski and O. Gaggiotti, Elsevier Academic Press, Amsterdam, pp. 153-173.

R.J. Whittaker (1998), *Island Biogeography: Ecology, Evolution, and Conservation*, Oxford University Press, Cambridge.

M. Williamson (1981), *Island Populations*, Oxford University Press, Oxford.

K.A. With and A.W. King (1999), Extinction thresholds for species in fractal landscapes, *Conservation Biology* **13**: 314-326.

S. Wright (1931), Evolution in Mendelian populations, *Genetics* **16**: 97-159.

CHAPTER 10

Metapopulation perspectives on the evolution of species' niches

Robert D. Holt
University of Florida

Michael Barfield
University of Florida

Abstract. The tapestry of the history of life reveals striking examples of both niche conservatism, and rapid niche evolution, where "niche" is used in the Grinnellian sense as that set of conditions, resources, etc. which permit populations of a species to persist in a locality without recurrent immigration. Recent years have seen the development of a rich body of theoretical studies aimed at understanding when one might expect niche conservatism vs. evolution in spatially and temporally heterogeneous environments. This literature has illuminated the role of many factors, such as genetic architecture, density dependence, and asymmetries in dispersal, in determining the likelihood of niche conservatism. However, most studies have assumed very simple spatial scenarios, such as a single source population (with conditions within a species' niche) supplying immigrants into a sink population (where conditions are outside the niche). In this contribution, after summarizing key insights from this prior literature, we will present the results of theoretical studies which examine how the spatial structure of the landscape can modulate the direction and pattern of niche evolution.

10.1 Introduction

The term "niche" refers to the range of conditions, resources – and indeed all biotic and abiotic factors – that permit populations of a species to persist (deterministically) in a given habitat without immigration. In effect, the niche is a mapping of population dynamics onto an abstract environment space (e.g., with axes of temperature, pH, food availability, predator density, etc.; Hutchinson 1958, Maguire 1973, Holt and Gaines 1992), emphasizing in particular the limits outside of which a species faces extinction. Formally, if environmental conditions in a given habitat are such that the low-density intrinsic rate of growth r (instantaneous per capita birth rate – per capita death rate) is negative, then conditions by definition are outside the niche, and introductions of the species should fail. By contrast, introductions into a habitat with $r > 0$ should tend to increase. So the niche of a species in effect partitions the world into areas where it can persist, and areas where it faces extinction. (For a species with discrete generations, sources and sinks can be defined in terms of the average fitness at low density, with unity being the threshold.)

To a first approximation, the geographical distribution of a species should be determined by its niche (Pulliam 2000), as should its habitat distribution at a more local, landscape scale. Understanding niches is of great practical importance, for instance in predicting how changes in climate might lead to shifts in distribution, and changes in land use can lead to altered patterns of abundance on a landscape. But all such predictions – and the scientific literature is replete with them – rest on the assumption that species' niches remain unchanged, even as the world changes. Such evolutionary conservatism, or the lack of change in the niche in a heterogeneous world, is called "*niche conservatism*."

The literature of evolutionary biology contains many examples that suggest niche conservatism, from short to long time scales (Bradshaw 1991, Wiens and Graham 2005). There are also many instances of rapid niche evolution, such as the evolution of antibiotic resistance in microbes, and the evolution of tolerance to heavy metal toxins. Understanding the factors that lead to niche conservatism, on the one hand, and rapid niche evolution, on the other, has been the focus of considerable theoretical attention and an increasing amount of empirical study (Holt 1996, Kawecki 2008). There are two circumstances in which one might look for niche evolution, or try to understand what leads to niche conservatism. First, in a spatially closed population (e.g., on an oceanic island), a temporal change in the environment can force a species to experience conditions outside its niche. Alternatively, in a spatially open population existing in a heterogeneous landscape, dispersal can take individuals out of habitats within the niche – source habitats – and place them into habitats outside the niche, This is a likely scenario at the edge of a species' range, for instance. Given genetic variation, evolution can potentially occur in both circumstances, so that sink populations can be transformed into source populations. Alternatively, even though genetic variation is present, sinks may remain sinks, and niche conservatism will be observed. The goal of theory is to provide insights into conditions under which each of these outcomes will occur.

Prior theory has largely focused on very simple landscapes, comprised either of species with random dispersal distributed over smooth gradients or a single source patch coupled by dispersal to a single sink patch. In this paper, we take steps towards examining niche evolution in more complex landscapes. We first review highlights (including previously unpublished results) from studies of models of niche evolution for sources and sinks coupled by dispersal, and then use these to motivate models for evolution in metapopulations comprised of two kinds of patches linked by dispersal.

We consider two limiting cases of a metapopulation. In both, the models track presence and absence of a species. The first is a "mainland-island" scenario of asymmetrical colonization. A species is established on a mainland, where it is adapted to one habitat type. The mainland population provides colonists onto islands made of the second habitat type, where the colonists are initially maladapted, and sufficiently so that the islands are sink habitats. In the absence of evolution, successful colonization is impossible. The question is how island area and distance influence colonization and extinction rates, taking into account the effects of selection and gene flow on adaptive colonization outside the niche. The second limiting case is that of classic metapopulation theory, which assumes we can ignore the details of spatial arrangements of the patches, and focus instead on the aggregate rates of colonization determined by average occupancy across the entire landscape.

10.2 Models for adaptive colonization into sink habitats

Theoreticians often assume that rates of dispersal are fixed parameters (e.g., a constant diffusion parameter). In reality, dispersal rates can often be highly variable. For instance, physical transport processes (e.g., the wind) can fluctuate greatly in strength, and source populations for dispersal propagules (or dispersal vectors) may vary greatly in density. Boreal forest bird species such as pine siskins and crossbills may be absent from the southern United States for many years, and then experience a large pulse of movement southward after failure of their food supply. The bottom line is that dispersal onto distant islands or habitat patches can be episodic, so that there is a substantial time lag between successive colonizing attempts. This assumption is implicit in classic island biogeography theory and much of metapopulation theory. We start with an island biogeographic perspective, which assumes that species persist and are at evolutionary equilibrium on a mainland, but colonize onto islands where persistence is enhanced by adaptive evolution to conditions on the islands. We then will move to a heterogeneous metapopulation, where colonization in effect is among islands in an archipelago or patches in a landscape.

We consider first a single episode of attempted colonization onto an island, where the colonists find themselves "outside the niche," hence declining in numbers. The fate of this population depends on the outcome of a race between demography and evolution. Without genetic variation, extinction is inevitable. If genetic variation is present in the dispersal propagule, or generated *in situ* via mutation, natural selection may increase the growth rate sufficiently to make it positive in the novel environment. However, before this can occur, the population might reach low levels at which it risks extinction. Fig. 10.1 schematically shows the expected pattern of population growth.

A quantitative genetics model for adaptation to a sink habitat

Gomulkiewicz and Holt (1995) (see also Holt and Gomulkiewicz 1997) provided a first step towards examining this process. They assumed that a single quantitative trait is undergoing selection. At each time step, the population declines (or grows) multiplicatively. The rate of growth itself changes over time, due to a single quantitative trait that is under selection in the novel environment. They assumed that evo-

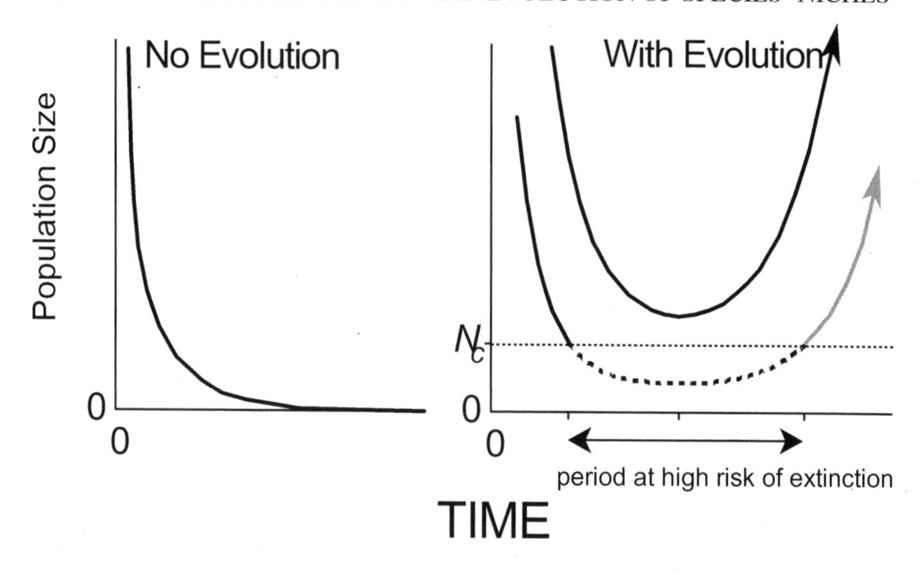

Figure 10.1 Population dynamics for introduction into a sink habitat. Left panel: Without evolution, extinction is ensured. Right panel: With evolution, the population may persist. However, if a population starts at low densities, and/or adapts slowly, it might spend time at very low densities, where it risks extinction. Adapted from Gomulkiewicz and Holt (1995).

lution fits the standard assumptions of quantitative genetics (Falconer 1989). The model is deterministic in both its demography and genetics; to heuristically address extinction, they assumed that there is a critical population size, N_c, below which a population is quickly vulnerable to extinction (e.g., due to Allee effects, or because of demographic stochasticity). Here we describe the assumptions of the model and some conclusions, and refer the reader to the original papers for derivations.

The basic scenario is depicted in Fig. 10.2. There is a single phenotypic trait z. On the mainland, stabilizing selection occurs, and genetic variation is maintained at a constant level (presumably by mutation, though this is implicit, not explicit). The colonizing propagule thus should have a distribution (assumed to be normal, which is typical for a quantitative trait) around the optimum on the mainland, d_0. P is the phenotypic variance of this distribution, which includes nongenetic sources of variation among individuals, such as developmental noise, as well as heritable variation. On the island, there is also potentially stabilizing selection on the trait, but around a new optimum (scaled to 0 in the figure; fitness is given by the dashed line). The fitness of an individual with phenotype z in the sink is given by a Gaussian function

$$W(z) = W_{\max} \exp[-z^2/2\omega], \tag{10.1}$$

where $W_{\max}$ is the fitness an individual enjoys when it has the optimal phenotype on the island, and ω is an inverse measure of the strength of selection. When ω is high, a small deviation of an individual's phenotype from the local optimum is not very costly; when small, selection severely acts against such individuals. The initial mean

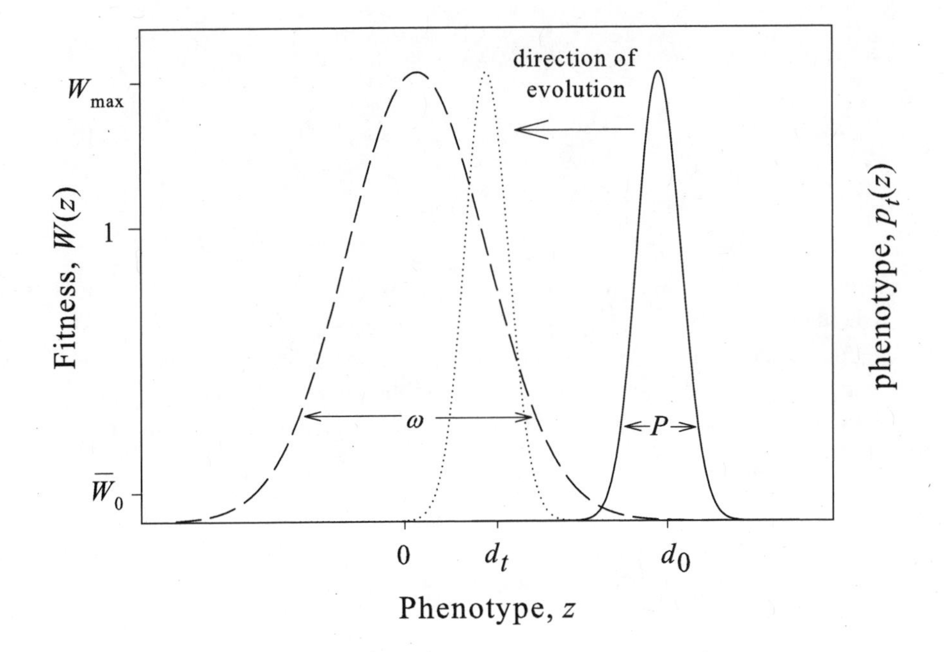

Figure 10.2 Evolution of Gaussian character z in a population with discrete generations. The solid line is the initial phenotypic distribution, the dashed line is the fitness function, and the dotted line is the phenotypic distribution after a period of evolution. The initial fitness is low, so the population size will decline initially, but could rebound once evolution has occurred, if the population avoids extinction after the initial decline. Adapted from Gomulkiewicz and Holt (1995).

trait value of a group of colonists introduced onto the island is d_0, which means they are initially maladapted; the larger is d_0, the lower is their initial fitness. Directional selection acts on the colonists, pushing their mean phenotypic value in the direction of the local optimum, and so reducing their degree of maladaptation (measured by z). The rate at which this happens is determined by the character's heritability, h^2 (which we assume fixed; this is one of many assumptions relaxed in the individual-based models discussed below).

Because the average trait value of individuals in the colonizing propagule is well displaced from the island optimum, the initial fitness of the colonizers is assumed to be well below one (the criterion for a sink with discrete generations), and so the population initially declines towards extinction. Propagules that potentially could persist after a period of adaptation may nonetheless initially decline so much that they risk extinction. Gomulkiewicz and Holt (1995) developed a discrete-time, deterministic quantitative genetic model for a population initially declining, but adapting to a sink environment, which based on the above assumptions led to the following equations for coupled demographic and evolutionary change:

$$N_{t+1} = \bar{W}_t N_t,$$

$$d_{t+1} = k d_t, \tag{10.2}$$

$$\bar{W}_t = \hat{W} \exp\left[\frac{-d_t^2}{2(P+\omega)}\right].$$

Here,

$$\hat{W} = W_{\max}\sqrt{\omega/(P+\omega)} \tag{10.3}$$

is the population growth rate when the mean phenotype has reached the local optimum; this is less than the maximal possible growth rate because it reflects an average over the distribution of trait values, and this distribution at evolutionary equilibrium includes individuals with suboptimal phenotypes. The rate of evolution is determined by the quantity

$$k = \frac{\omega + (1-h^2)P}{P+\omega}, \tag{10.4}$$

which can be viewed as a measure of evolutionary inertia. If heritability is very low, k is near unity, so the character changes very slowly; if ω is large, selection is weak, and again evolution is slow.

This pair of coupled difference equations can be solved in closed form, leading to

$$N_t = N_0 \hat{W}^t \exp\left[\frac{-d_0^2(1-k^{2t})}{2(P+\omega)(1-k^2)}\right]. \tag{10.5}$$

One can then calculate a number of quantities, such as the combination of initial conditions and parameter values that lead an introduced population to experience times when its abundance is below N_c, and for those populations that do dip below this value, how long they will stay there. If a population is strongly maladapted to start with, its numbers will plummet, and even though it has the genetic potential to persist in the new environment, the model suggests it is highly likely to go extinct first. Populations that evolve slowly (high k) are also likely to go extinct, as are populations which are initially low in numbers (even if they are evolving rapidly). In effect, this exercise provides qualitative insight into the likelihood of adaptive colonization, as a function of the degree of maladaptation in the novel environment, and the number of immigrants, among other ecological and genetic factors.

Stochastic models

The above paragraph used the word "likely," which is strictly speaking inaccurate. The model is deterministic and treats N as a continuous variable, and so numbers will not actually reach zero. Ergo, no extinction. A rigorous analysis of extinction (i.e., $N = 0$) requires one to grapple with the fact that organisms are discrete, and births and deaths are probabilistic. This is a large and challenging problem. Holt and

Gomulkiewicz (1997) used a branching process approach to examine this problem, assuming genetic variation at a single haploid genetic locus. They developed a probability generating function, and found that the qualitative conclusions drawn from the deterministic model are upheld. Recently, Orr and Unckless (2007) have developed stochastic models that also include novel mutations, and reached similar conclusions. But for stochastic models to be analytically tractable, they have to simplify many of the complex phenomena that occur in declining populations. When a population is declining towards extinction, while simultaneously evolving, many stochastic processes are at play at the same time. Genetic variation itself can be changing due to selection, and as numbers get small the vicissitudes of demographic stochasticity loom large. Gene frequencies and genetic variation change due to drift, and when multiple genetic loci are considered (as is appropriate for quantitative traits such as body size and thermal tolerance), linkage disequilibrium can shift stochastically. If populations decline slowly, mutational input can provide a significant source of genetic variation.

To develop an understanding of coupled evolutionary and demographic dynamics when all these processes are occurring at once, in previous papers we have reported the results of simulation studies based on individual-based models in which we track each individual and its genotype in source and sink environments (e.g., Holt et al. 2005). These models include all the above sources of stochastic variability. Here we just briefly sketch the assumptions of the models, and present a few results, that help motivate the metapopulation model presented below.

The basic life-history framework of these models is shown in Fig. 10.3. Individuals move synchronously through a series of life history stages. Selection occurs on a trait that influences juvenile survival, and density dependence is imposed as a ceiling number of breeding adults (K). In our genetic assumptions, we follow those used by Burger and Lynch (1995) in exploring evolution in a constantly changing environment. There are n loci that contribute additively to a single quantitative trait z, with free recombination. In the source, mutational input maintains variation (according to a continuum-of-alleles model), with a Gaussian distribution of mutational effects, and an environmental noise term (a zero-mean, unit-variance Gaussian random variable). Therefore, heritability emerges as an output of the model, rather than being a fixed quantity [as in the above model (10.1)-(10.4)]. Mutation can also occur (and at the same rate) in the sink. Juvenile survival is a Gaussian function of an individual's phenotype (z), with different habitats having different phenotypes at which survival reaches its maximum (so an individual adapted to the source generally has low survival in the sink). We allow the source population to reach an evolutionary equilibrium, with an emergent heritability of the trait reflecting the balance between mutation, selection, and drift, and then we pluck a propagule of adults at random and place them as immigrants in the sink habitat. After doing this a large number of times, and across a wide range of parameter values, patterns emerge that characterize when one might observe colonization outside the niche.

Persistence in the sink requires adaptation, and because colonization is occurring outside the niche, and adaptation is not instantaneous, colonization attempts can readily

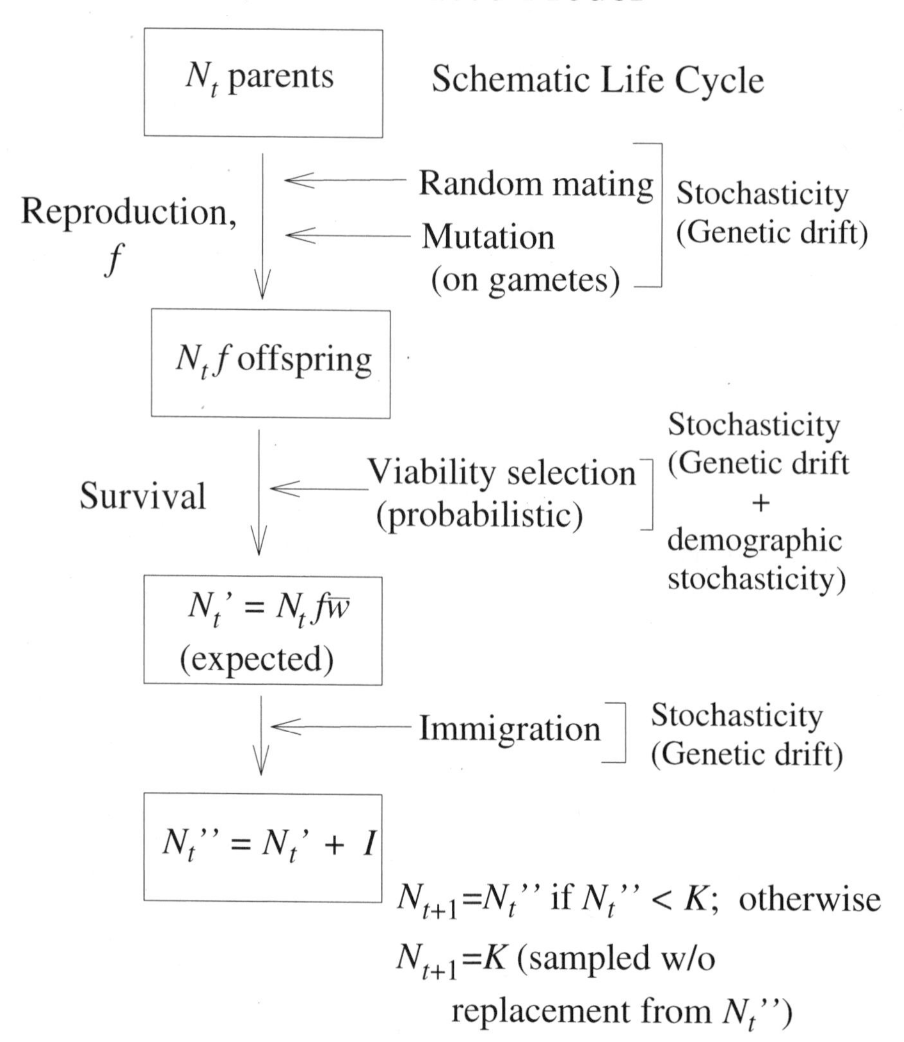

Figure 10.3 Schematic diagram of the life cycle in each habitat of the individual-based model, indicating the sources of stochasticity included. Note that migration from the source to the sink occurs before density regulation, and immigrants and residents have equal chances entering the mating pool.

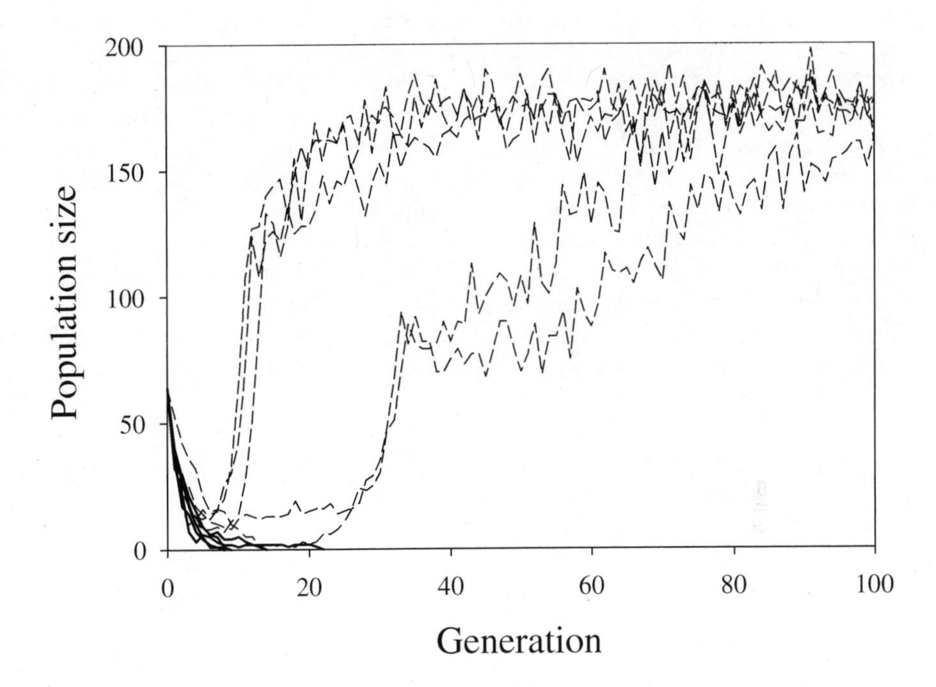

Figure 10.4 Sample trajectories for adult population size for populations introduced into a sink habitat. Initially, all populations decline in abundance, some going extinct (solid), but some rebounding (dashed). $K = 64$, mutational rate per haplotype = 0.01, mutational variance $\alpha^2 = 0.05$, strength of selection $\omega^2 = 1$, propagule size = 64; 4 births per pair. The difference between source and sink phenotypic optima is 2.5.

fail. Fig. 10.4 shows examples of time-series of population size against time. In these examples, 64 individuals are introduced into the sink. Some attempted colonizations (the solid lines) fail, but others succeed (dashed lines), after an initial period of decline. Even though all colonizing propagules are drawn from the same type of source population, there is considerable heterogeneity among successful replicate colonizing episodes (see Discussion).

With such simulations in hand, we can quantify adaptive colonization as a function of the degree of initial maladaptation and the number of colonists in the initial propagule. The maladaptation is the difference between the phenotypic optima of source and sink, a larger value indicating a lower expected fitness of source individuals introduced into the sink. Fig. 10.5 shows two patterns, emphasizing the relationship between adaptive colonization and on one hand the harshness of the sink environment, and on the other the number of individuals in the colonizing propagule. In Fig. 10.5a (adapted from Holt et al. 2005), we depict the probability of adaptive colonization as a function of the degree of maladaptation experienced in the sink by immigrants drawn from the source, for three different propagule sizes (numbers of introduced individuals). In the figure the top axis translates maladaptation (the bottom horizontal axis) into fitness. Even in favorable environments inside the niche, where fitness exceeds unity at low densities, demographic stochasticity can doom small propagules, but large propagules should be able to establish with a probability near one. However, in unfavorable environments, where fitness is initially less than unity, in the absence of genetic variation extinction is ensured regardless of initial population size. Given that genetic variation is present (as in the examples of Fig. 10.4), adaptive colonization becomes possible. The harsher the sink environment, however, the less likely this will occur. Basically, there is a footrace between demography (pushing a population towards extinction), and evolution by natural selection (increasing fitness). When initial fitness is low, and propagule size is small to modest, demography will overwhelm evolution, and colonization will fail.

The larger the number of individuals, the greater the chance of adaptive colonization. Fig. 10.5b shows that the likelihood of persistence over a thousand generations (which essentially always requires adaptation to the sink environment) has a sigmoidal dependence upon the logarithm of the number of individuals introduced into the sink. Recent experiments using yeast introduced into experimental sink habitats (created by increasing the salt concentration of the medium to be outside the initial niche of the species) by Andy Gonzalez and Graham Bell at McGill University (pers. comm.) have demonstrated a sigmoidal dependence of population survival on the logarithm of initial numbers in a sink, consistent with the prediction of this individual-based model. A variety of different assumptions about the genetic architecture underlying trait variation can also generate this relationship between initial population size and persistence (R. Gomulkiewicz, pers. comm.). A function that gives a good phenomenological fit to the output of these individual-based simulations is a logistic function of $\ln N_0$ and d_0:

$$\mathrm{Prob}(\text{adaptive colonization}|N_0, d_0) = \frac{N_0^a}{N_0^a + a' \exp\{a'' d_0\}} \tag{10.6}$$

where a, a', and a'' are all positive constants.

Of course, if there are repeated attempts at colonization, as long as there is a nonzero probability of adaptive colonization, eventually adaptation to the sink will occur. If the probability of adaptive colonization per colonizing bout is p, the probability of successful colonization after n colonization attempts is $1 - (1-p)^n$. In a mild sink, where initial fitness is not much below unity, p is not far below one, and adaptive colonization is likely over reasonably short time-horizons. But in a severe sink, where p is very low, there can be a very long lag before successful colonization occurs. Niche conservatism thus may not be absolute, but reflect quasi-equilibrial, long-term transients.

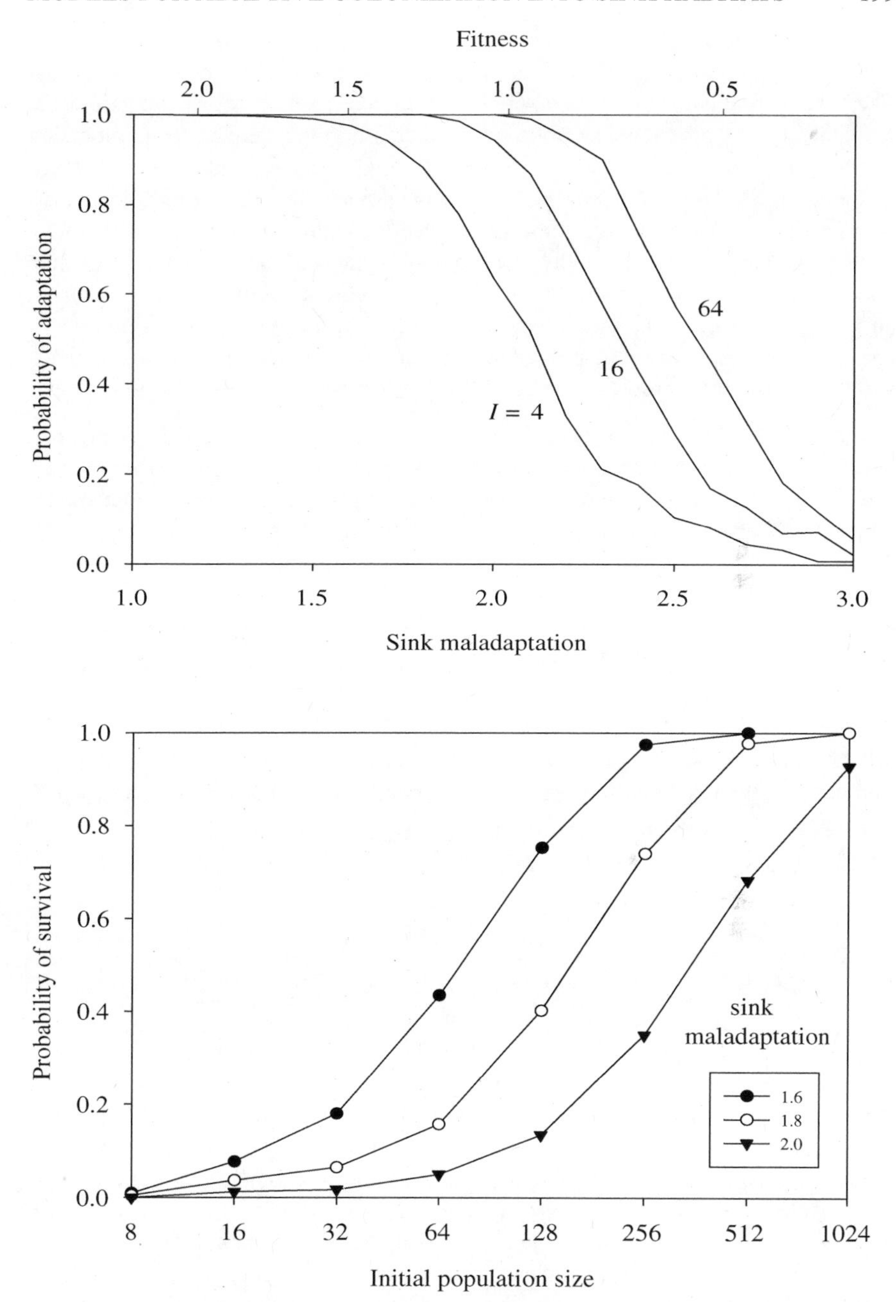

Figure 10.5 The probability of persistence and adaptation, as a function of (a)(top panel) degree of initial maladaptation in the sink habitat, for three different sizes of initial colonizing propagule, and (b)(bottom panel), initial population size. Other parameters as in Figure 10.4, except panel (b) has a fecundity of 2 rather than 4.

10.3 An island-mainland model with infrequent adaptive colonization

The bottom line is that in a metapopulation, in colonizing empty habitats outside the niche, higher propagule numbers, or an increase in the frequency of colonizing attempts, should facilitate adaptive colonization. This could lead to both distance and area effects on the rates of adaptive colonization. The number of colonization attempts into an island per unit time should decline with increasing distance from a source. The number of viable individuals in a colonizing propagule could also decline with distance (e.g., due to mortality in transit). The number of propagules landing on an island might increase with island size. Productive sources, or sources large in area, are more likely to be the progenitors of adaptive colonization into sink habitats, simply because more colonization attempts should emerge from such sources.

We can modify the familiar equilibrial model of island biogeography (MacArthur and Wilson 1967) to include adaptive colonization as follows (Holt and Gomulkiewicz 1997). Each island can be in one of three states: empty, recently colonized and maladapted, and adapted. The fraction of islands in each state are respectively P_0, P_m, and P_a. A simple dynamical model describing transitions among these states is:

$$\frac{dP_m}{dt} = c_m(1 - P_a - P_m) - EP_m - e_m P_m,$$
$$\frac{dP_a}{dt} = EP_m - e_a P_a, \tag{10.7}$$

where c_m is the rate of colonization, e_m is the rate of extinction of maladapted populations, e_a is the rate of extinction of adapted populations, and E is the rate at which maladapted populations become adapted. (The sum of the three fractions is 1, so $P_0 = 1 - P_m - P_a$.)

At equilibrium,

$$P_a^* = \frac{E}{e_a} P_m^*,$$
$$P_m^* = \frac{c_m e_a}{c_m(E + e_a) + (E + e_m)e_a}. \tag{10.8}$$

The total occupancy is $P^* = P_a^* + P_m^*$. The fraction of occupied islands that are adapted is $E/(E + e_a)$. Adaptation means that there will be genetic differentiation between the island and mainland populations, and so this quantity is the fraction of occupied islands that have endemic species. A little manipulation of (10.8) shows that adaptation increases occupancy if $e_a < e_m$, which makes intuitive sense. It is interesting that the degree of endemism on occupied islands is not affected by either the colonization rate, or the rate of extinction of maladapted populations, but only the rate of evolution and the rate of extinction of adapted populations. This conclusion is altered if there is heterogeneity among islands or species in extinction rates (R.D. Holt, unpublished results).

At this point we could use expression (10.6) to craft some more quantitative predictions about how island area and distance might affect the likelihood of niche evolution. Rather than pursue that route, we instead note that there is an important evolutionary process that we have not yet considered which complicates predictions about the relationship between distance (between the island and mainland) and the likelihood of observing niche evolution – gene flow.

10.4 Gene flow and population extinction

The expected relationship between island distance and the likelihood of adaptive colonization could break down if dispersal is sufficiently frequent that there are immigrants entering the population each generation, because recurrent gene flow can hamper local adaptation. The classic view of the evolutionary impact of dispersal is that it leads to gene flow that can force local populations away from their local adaptive optima. The genetic reason is that in a sexual species with random mating, if selection in the local environment leads towards local adaptation, on average immigrants should carry genes that lower fitness, compared to the genes carried by residents. The offspring of crosses between a resident and an immigrant should thus have lower expected fitness than do the offspring of crosses between two residents. This reproductive cost is what drives the classic scenario of gene flow "swamping" selection, potentially permanently preventing local adaptation. On top of this, a high rate of immigration can lead to ecological effects such as competition which depress the fitness of residents, and thus hamper selection improving local adaptation.

Fig. 10.6 shows an example of this effect for the individual-based model described above, for two habitats coupled by equal per capita rates of movement. Initially, we allow a population in each habitat to reach evolutionary equilibrium. There is ceiling density dependence, with 64 breeding adults in each habitat. The two habitats differ from each other very sharply in phenotypic optima, however (a difference of 6 on the scale shown in Fig. 10.5a). Each generation, there is a probability of 0.1 that an individual will move from its natal habitat (here we are allowing two-way dispersal, and not just a flow of individuals from the source to the sink). The figure shows the trajectory of population size in each habitat (censused after selection, but before density dependence is imposed). Because of demographic stochasticity, there is fluctuation in population size around its equilibrium. Initially, in some generations, one habitat has more individuals; in others, the other habitat does (the thin line; the dashed line indicates equal population sizes), so the two habitats remain roughly demographic equals. But eventually the system drifts to a state in which there are consistently more individuals in one habitat than the other (heavy line), and the system then collapses to a state in which the species is completely adapted to one habitat, and no individuals survive selection in the other habitat. The reason is that asymmetries in abundance between habitats lead to more individuals leaving the high-abundance habitat than returning to it. This implies that relatively more matings in the low-abundance habitat are between residents and immigrants, which on average degrades local adaptation in this habitat, which in turn further decreases population size. Thus, relatively modest asymmetries in abundance are quickly magnified by a positive feedback process, enhancing the role of gene flow suppressing local selection. Therefore, once the lo-

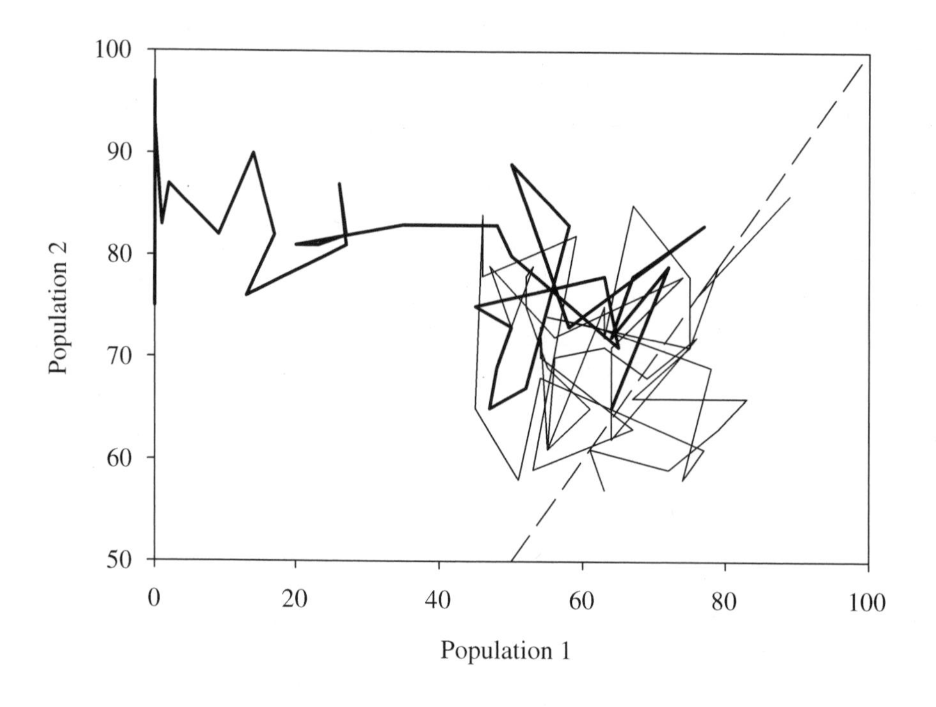

Figure 10.6 Population size phase plot for the individual-based model for two populations differing in phenotypic optima by 6 (a large amount) and with migration in each generation of 10% of each population to the other habitat. Each habitat is limited to 64 mating individuals; other parameters as in Figure 10.4. Initially (thin line), the habitat with the larger population size varied with time. Eventually, however, the population in habitat 1 starts to decline, and due to positive feedback this leads to its maladaptation and extinction (heavy line), i.e., no individuals survive the phase of the life cycle where selection occurs.

cal population is moderately maladapted, it quickly loses its ability to replace itself, and so relies entirely upon immigration. If we now were to cut off migration, the individuals found in the "wrong" habitat would be so strongly maladapted there, that extinction would be inevitable.

Ronce and Kirkpatrick (2001) called this phenomenon "migrational meltdown." Harding and McNamara (2002) suggest that this perverse effect of recurrent dispersal on persistence might be called an "anti-rescue" effect. The basic idea is that asymmetrical dispersal can lead to a kind of suppression of natural selection. The example shown in the figure is for a single pair of patches. But much the same phenomenon should emerge in metapopulations comprised of a mixture of distinct kinds of habitats, where selection operates in different directions in different habitats (e.g., optimal body size might vary with temperature or food availability). Too much dispersal from one habitat type to another could lead to enhanced extinction rates.

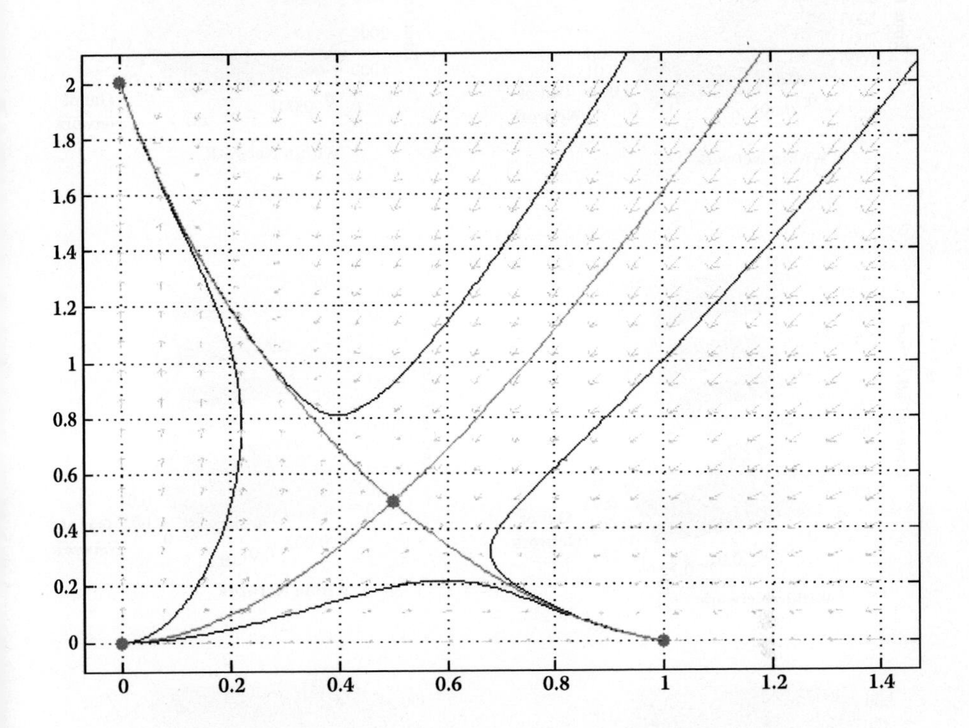

Figure 3.9 Phase portrait of the competition model (3.24). The stable manifold of $(u*, v*)$ (connecting orbit from the origin) is the threshold manifold which separates the basins of attraction of two stable equilibria; and the unstable manifold of $(u*, v*)$ (connecting orbits from stable equilibria) is the carrying simplex.

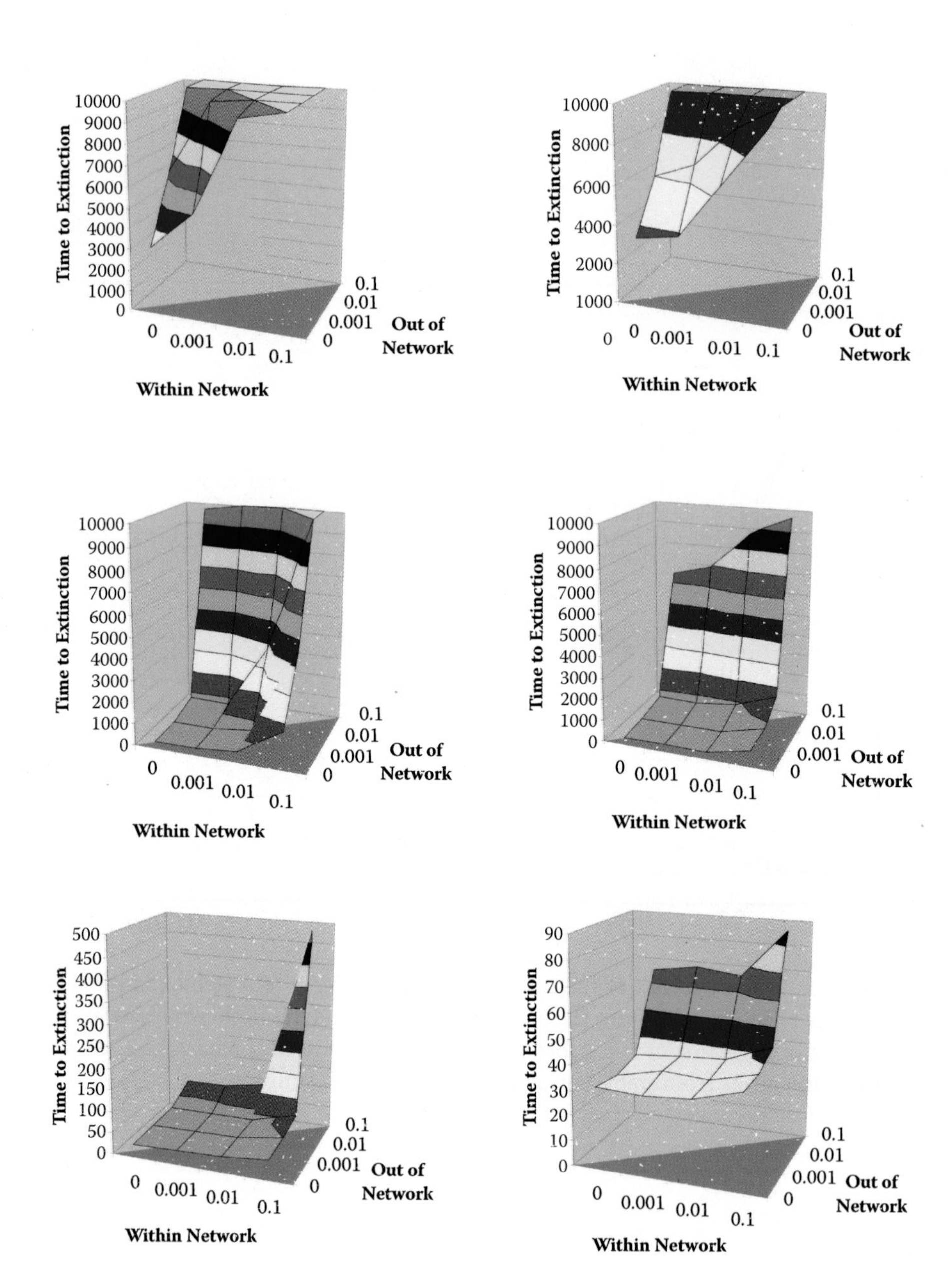

Figure 5.3 Effects of river network 'branchiness' on extinction risk in 15-patch dendritic metapopulations. Panels on the left are from a Full dendritic network, and on the right are from the Pruned network. Three extinction probabilities were modeled (0.001, top row; 0.01, middle; 0.1 bottom row), under combinations of within- and out-of-network dispersal probabilities (0, 0.001, 0.01, 0.1).

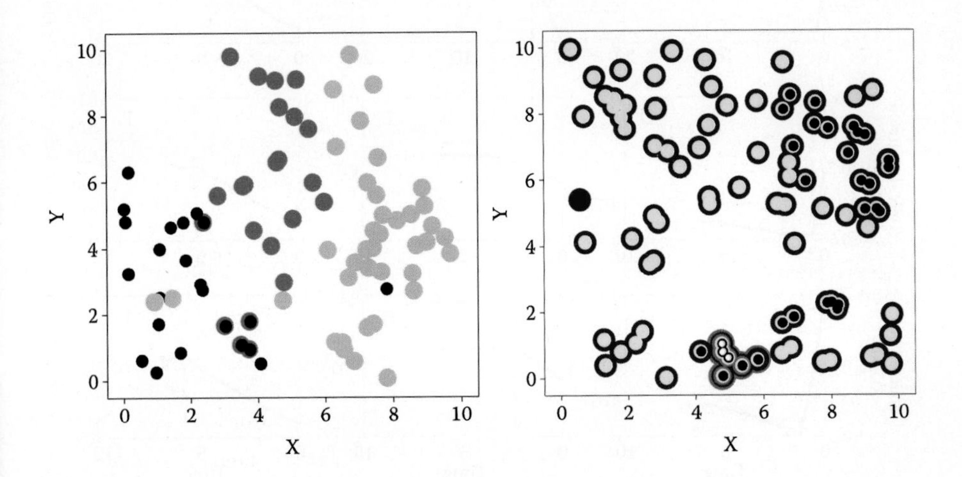

Figure 9.1 Two examples of limited spatial distributions of many similar competing species with short range of migration. The colours indicate the presence of different species in different habitat patches with probability of occupancy (p) greater than 0.2. (a) Very strong competitors exhibit practically exclusive spatial distributions, while (b) moderately strong competitors may end up with partly nested distributions.

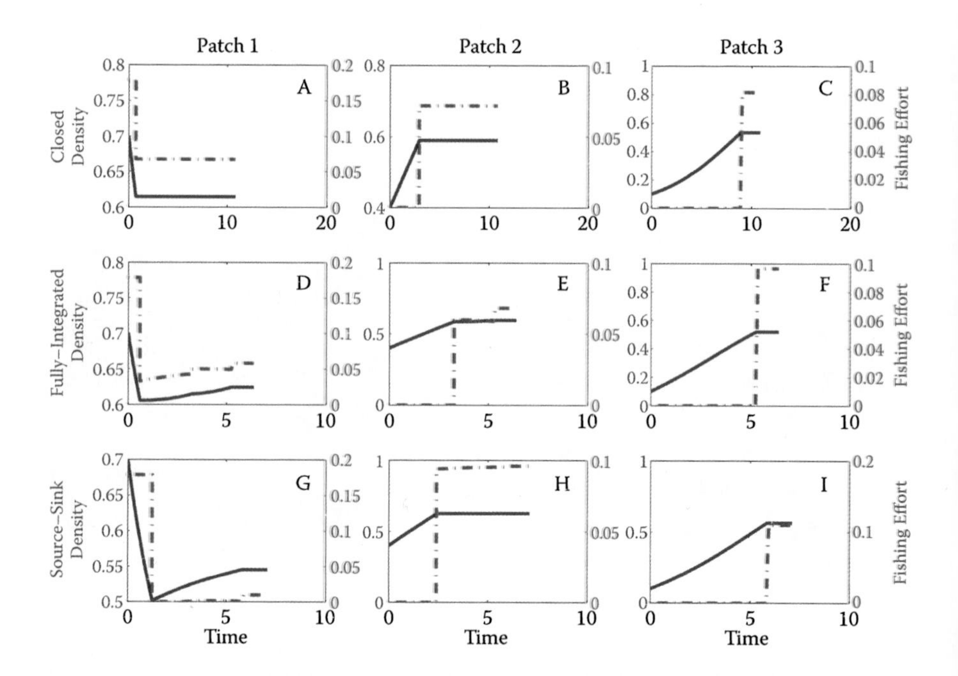

Figure 16.2 Optimal management of a metapopulation for the closed (panels A-C), fully integrated system (panels D-F) and source-sink system with patch 1 as the source patch (panels G-I). The left y-axis measures the density of the population in the patch (solid lines), and the right y-axis measures the amount of fishing effort (dashed lines). The x-axis is in the time units of the simulation and should not be interpreted in calendar units, such as years. The parameters used in the numerical analysis are: $(c_1, c_2, c_3) = (.48, .42, .3)$, $(p_1, p_2, p_3) = (1, 1, 1.05)$, $(q_1, q_2, q_3) = (1.5, 1.5, 1.5)$, $(r_1, r_2, r_3) = (.26, .26, .26)$, $d_{ij} = b = .0525$ for $i \neq j$ and $d_{ij} = 2b$ for $i = j$, and $\delta = .05$.

Broadly, we can imagine three avenues through which gene flow between habitats could elevate extinction rates in a metapopulation. First, there could be direct extinction, as in the example of migrational meltdown shown in Fig. 10.6. Second, gene flow could lead to depressed average population size (an example is in Holt 1983), and thus increase the risk of local extinction due to demographic stochasticity. Finally, a population which is displaced from its local adaptive optimum is likely to suffer a reduced growth rate when rare, which means that it is harder for it to rebound following a disturbance.

10.5 A metapopulation model with maladaptive gene flow

We now develop a metapopulation model that captures the flavor of these microevolutionary processes, and show that the enhancement of local extinction rates by gene flow can lead to alternative evolutionary states in a heterogeneous landscape. In this model, space is implicit, rather than explicit. A species occupies two distinct habitat types ($i = 1, 2$), each of which occupy a fraction h_i of the patches on a landscape. The fraction of the total patches that are of type i and occupied is p_i. The colonization rate from patch type i to patch type j is c_{ji}. Because adaptive colonization should be more difficult than colonization that does not require adaptation, we assume that cross-habitat colonization, though it may occur, happens at a lower rate than does colonization within a given habitat type.

If dispersal is at random, there should be an increasing rain of propagules across the two habitats, as the occupancy in either habitat increases. This means that the opportunity for migrational meltdown (or the other mechanisms by which gene flow can increase extinction listed above) in a patch of type i should increase with the occupancy of patch type j. This is modeled by making the extinction rate for each patch type an increasing function of the occupancy of the other patch type, with baseline extinction rates e_i; the extinction rates then increase with p_j at proportional rates γ_{ij}. A metapopulation model that permits both adaptive colonization, and anti-rescue due to migrational meltdown, is as follows:

$$\begin{aligned} \frac{dp_1}{dt} &= (h_1 - p_1)(c_{11}p_1 + c_{12}p_2) - e_1(1 + \gamma_{12}p_2)p_1, \\ \frac{dp_2}{dt} &= (h_2 - p_2)(c_{22}p_2 + c_{21}p_1) - e_2(1 + \gamma_{21}p_1)p_2. \end{aligned} \tag{10.9}$$

The first terms on the right-hand side describe colonization of empty habitats of each habitat type, due to dispersers moving both within- and among-habitat types, in a metapopulation that is a mixture of two habitats (Holt 1997).

As a limiting case of the above model, we assume that there is no cross-colonization into empty habitats, $c_{12} = c_{21} = 0$, so the equations reduce to:

$$\begin{aligned} \frac{dp_1}{dt} &= (h_1 - p_1)c_{11}p_1 - e_1(1 + \gamma_{12}p_2)p_1, \\ \frac{dp_2}{dt} &= (h_2 - p_2)c_{22}p_2 - e_2(1 + \gamma_{21}p_1)p_2. \end{aligned} \tag{10.10}$$

For Eq. (10.10), an equilibrium with neither species present is stable if and only if

$$e_i > c_{ii}h_i \tag{10.11}$$

for each habitat type. If this is true for habitat type i but not for habitat j, then the species can increase when rare in the latter habitat, and will go to the stable equilibrium density $p_j = (c_{jj}h_j - e_j)/c_{jj}$ (while fixing $p_i = 0$). This equilibrium can also be stable if inequality (10.11) is violated for both habitat types, because the presence of the species in one habitat type increases the extinction rate in the other, and therefore makes it harder for the species to persist there (or increase when rare).

The condition for p_i to increase when rare at the above (p_j only) equilibrium is

$$e_i[1 + \gamma_{ij}(c_{jj}h_j - e_j)/c_{jj}] < c_{ii}h_i. \tag{10.12}$$

Assuming $\gamma_{ij} > 0$, this condition requires a lower basic extinction rate e_i (or higher $c_{ii}h_i$) than would be required if $\gamma_{ij} = 0$ [or $p_j = 0$, either of which give the condition $e_i < c_{ii}h_i$, which is the reverse of condition (10.11)]. Similarly, if the species is established in habitat i, it can prevent invasion of habitat j in some cases for which habitat j could otherwise be invaded. Therefore, there is the possibility of two stable alternative equilibrial landscapes, in each of which adaptation to one habitat suppresses presence and adaptation to the other. These alternative landscape states arise when inequality (10.11) is violated for each habitat type in turn (i.e., each habitat type could be invaded if the other one was not already occupied), and inequality (10.12) is also violated for each habitat type (i.e., neither can be invaded if the other is at its equilibrium). In the symmetrical case, this reduces to $c < \gamma e$ (where $c_{11} = c_{22} = c$, $\gamma_{12} = \gamma_{21} = \gamma$ and $e_1 = e_2 = e$). In this symmetrical case, there is an equilibrium with both habitats occupied, but it can be shown that this equilibrium is unstable, if the two single-habitat equilibria are both stable.

In the case above, the presence of the species in one habitat type has only a negative effect on the species in the other habitat type, through increased extinction rate, because we assumed there was no cross-colonization. If there is cross-colonization, then the presence of the species in one habitat type can increase its occupancy in the other through colonization. However, it is still possible for there to be alternative stable equilibria, if the negative effect on extinction is greater than the positive effect of cross-colonization. But it is reasonable to expect that alternative stable equilibria will be less likely with cross-colonization.

Without cross-colonization, we showed above that the species in one habitat type can completely exclude it in the other (the alternative stable equilibria have 0 occupancy for one habitat type). If there is cross-colonization, then the presence of the species in one habitat type guarantees its persistence in the other through colonization from

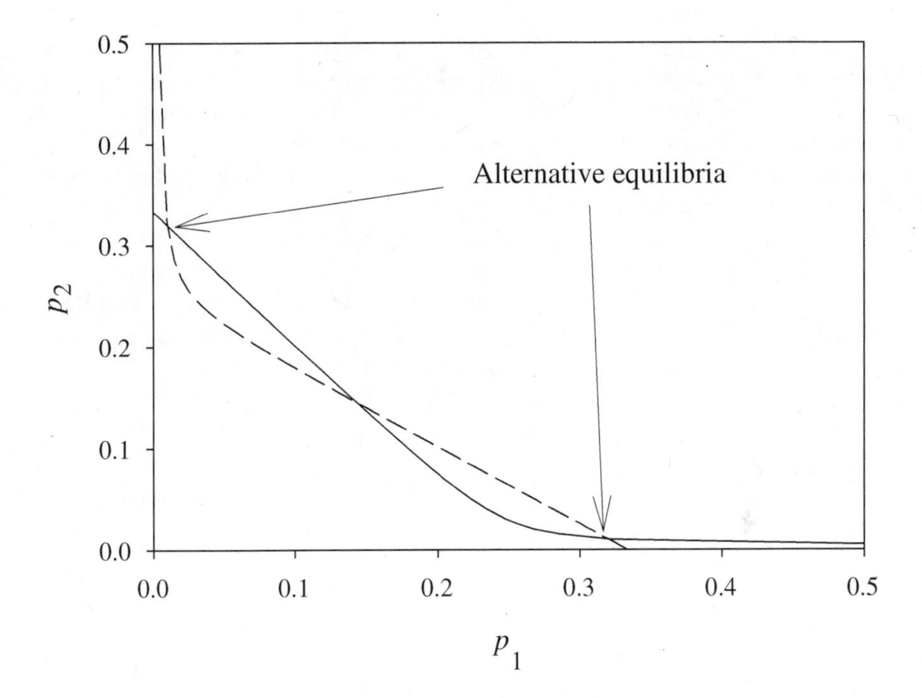

Figure 10.7 Isoclines for a symmetric metapopulation model with two alternative equilibrial states. Dashed line is isocline for habitat type 1. Parameters are $h = 0.5$, $c = 0.3$, $c_x = 0.001$, $e = 0.1$, and $\gamma = 4$. The species, if adapted to one habitat type, by gene flow sufficiently elevates extinction in the other habitat type that it remains maladapted there and hence sparsely occupies the available habitat patches..

one habitat type to the other. Therefore, if there are alternative stable equilibria, both habitat types will have a positive occupancy in both equilibria (assuming both cross-colonization terms are positive). The system [Eq. (10.9)] is now more difficult to analyze, because all equilibria (other than $p_1 = p_2 = 0$) have both habitats occupied, and must be solved by setting the derivatives in (10.9) to 0 and solving for p_1 and p_2. Unfortunately, there are no simple closed-form expressions for these equilibria in general.

One case that can be solved with cross-colonization is to assume symmetry. So again let $c_{11} = c_{22} = c$, $\gamma_{12} = \gamma_{21} = \gamma$, and $e_1 = e_2 = e$, and in addition let $c_{12} = c_{21} = c_x$ ("x" for cross). In this case, there is a symmetric equilibrium, which can be solved by setting the derivative in (10.9) to 0, setting $p_1 = p_2 = p$, and solving for p. This gives the symmetric equilibrium

$$p = [h(c + c_x) - e]/(c + c_x + e\gamma). \tag{10.13}$$

It is instructive to examine the isoclines for the model (Fig. 10.7). For example, the isocline for p_1 is found by setting the derivative in the first equation of (10.9) to 0, giving an equation relating p_1 and p_2. This isocline is hyperbolic. It has a vertical asymptote at $p_1 = h_1c_{12}/(e_1\gamma_{12} + c_{12})$ and intersects the positive p_1 axis at $(h_1c_{11} - e_1)/c_{11}$. The isoclines always cross in the first quadrant (assuming both cross-colonization terms are positive). For some parameters, the isoclines cross only once, but for others they can cross three times (Fig. 10.7). In the symmetric case, if the magnitude of the slope of the p_2-isocline is higher at the symmetric equilibrium (as in Fig. 10.7), then this isocline is higher than the p_1-isocline for p_1 values just below the equilibrium. However, the p_1-isocline has a vertical asymptote at a positive p_1, while the p_2-isocline is approaching an oblique asymptote. Therefore, the isoclines must cross again at a lower p_1, and by a similar logic they must also cross at a higher p_1. In this case, the symmetric equilibrium is unstable, and there are alternative stable equilibria. The condition for this is

$$(e\gamma - c)(hc - e) > hc_x(2c + c_x + e\gamma) + ec_x. \tag{10.14}$$

If $c_x = 0$, this reduces to the symmetric result above (the species cannot persist unless $hc > e$, so the second term on the left must be positive). The presence of cross colonization makes alternative stable states more difficult, since not only must $\gamma e > c$, but it must be higher by a greater amount, for greater c_x. The parameters used in Fig. 10.7 satisfy inequality (10.14), and therefore alternative equilibria exist, as shown.

Thus, migrational meltdown can lead to alternative stable states in a metapopulation, assuming cross-colonization between habitat types is not too common. It can also lead to other effects, which we note below.

10.6 Discussion

We have presented several complementary models that provide building blocks for examining niche evolution in heterogeneous landscapes. We started with models that look closely at evolutionary processes in particular habitats that have conditions outside a species' niche requirements, where with rare dispersal, extinction is inevitable unless there is adaptive evolution, and with frequent dispersal, recurrent gene flow can hamper adaptation.

The first deterministic model [Eqs. (10.2)-(10.4)] leads to heuristic insights about how initial population size and the degree of maladaptation influence the likelihood of extinction rather than adaptive changes sufficient to permit persistence in a sink habitat. These results motivate studies of individual-based models (IBMs) that incorporate stochasticity in both demography and genetics. These IBMs confirm the suggestions drawn from the deterministic models and help highlight issues that warrant closer theoretical scrutiny.

One of these issues is distinguishing among distinct sources of variation in adaptation to sink environments. Consider again the populations of Fig. 10.4. Although all colonizing propagules are drawn from the same type of source population, the surviving populations show considerable heterogeneity in their patterns of evolutionary rescue. Some populations start to evolve higher fitness permitting persistence quite quickly, and then rapidly reach their maximum population, at which they are fully adapted. Others barely hang on, and then even after they evolve sufficiently to persist, take longer to increase fitness and eventually reach full adaptation (and maximum population size). To understand this heterogeneity in responses, it is useful to reflect on the sources of genetic variation in these novel populations and how this variation is altered by drift, recombination, and mutation.

There are only two possible sources of genetic variation in the sink. First, colonizing propagules can sample preexisting variation in the source. Second, there can be mutational input. Without novel mutations arising in the sink, evolutionary rescue entirely depends upon genotypes with expected fitness greater than unity being potentially present in this initial sample from the source (the genotypes may only be "potentially" present because they are generated by mating and recombination among the immigrants and their descendents, rather than literally present in the initial generation). At low population sizes, genetic variation is lost by drift. The longer a population spends at low numbers, the greater the amount of variation brought in by sampling from the source that will be lost by drift. If a population persists in a genetically depleted state after going through such a long bottleneck, further evolution may largely depend upon the input of novel mutations, which will typically play out over a longer time scale than the reassortment of variation present in the initial propagule. In Fig. 10.4, the populations that spend the greatest time at low densities also seem to have the most sluggish rate of evolution, once they have adapted sufficiently to survive.

Models of demographic stochasticity show that initial population size has a large effect on population persistence, even in favorable environments. If mean fitness is less than one, and there is no evolution, the probability of extinction is unity. With genetic variation permitting adaptive colonization, we have shown that initial population size again has a strong influence on population persistence. There are several distinct reasons that initial population size matters in adaptive colonization into a sink. First, a larger colonizing propagule means more variation from the source is sampled. Second, for a given rate of decline in the sink, a larger initial population provides a larger demographic window for novel mutations to arise and potentially rescue the declining population. In a homogeneous population declining at a constant rate, a classic result in branching process theory is that the number of replication events that occur before extinction for a population initially at size N_0 and declining at average rate R is $N_0/(1-R)$ (Feller 1968, p. 299). Since mutation happens during replication, the potential input of novel mutations should be governed by the number of replication events. All else being equal, larger initial populations have greater scope to experience novel mutations permitting adaptation and persistence, before extinction, than do small populations. In like manner, the less harsh the sink environment (i.e., the closer initial fitness is to unity), the larger the number of replication events that will be observed before extinction, and so the greater the opportunity for the input of novel mutations. An interesting challenge for future theoretical work is to tease apart the relative roles of sampling from established populations and *in situ* mutation as sources of genetic variation for selection to act upon in sink populations. (A similar partitioning pertains to recurrent immigration; variation can be sampled from the source, or generated by mutation in the sink.)

Environmental heterogeneity provides an opportunity for local adaptation, but gene flow can prevent this from occurring. When adaptation is required for persistence, gene flow can enhance extinction risks for some local populations. Our model for a metapopulation in a landscape comprised of two distinct habitat types shows that alternative landscape states are possible, in which a species by being initially adapted to one habitat prevents itself from becoming adapted to the other. The model suggests that evolutionary "dominance" in a metapopulation is more likely if 1) cross-habitat, adaptive colonization is difficult (i.e., in our quantitative genetics model, there is a large difference in adaptive optima in the two habitats); 2) recurrent gene flow across habitats substantially increases extinction risks in the recipient habitat; and 3) one habitat is sparse in the landscape, or high in intrinsic extinction rate, or low in intrinsic colonization rate, relative to the other habitat. Given these conditions, "success breeds success," and the habitat that a species becomes adapted to can indirectly suppress adaptation in the other habitat, and thus constrain the fraction of the landscape occupied by the species.

The model helps point out the importance of historical contingencies for determining the ultimate habitat range of a species. A species that colonizes this landscape may evolve in a number of different directions, leading to different ultimate patterns of habitat specialization. If it is difficult to colonize across habitats, but the anti-rescue effect is unimportant, a species initially adapted to just one habitat type may invade and rapidly fill up those habitats to which it is initially well-adapted, and then begin to colonize the other habitat (Fig. 10.8a). If adaptive colonization is difficult, then this may be a slow process. If dispersal is sufficient in magnitude to lead to anti-rescue effects (migrational meltdown), then a variety of additional phenomena may occur. A species may initially be a generalist, adapted to both habitats. But if one habitat is sparse, and the other widespread, generalization may be lost, because adaptation is biased towards the more common habitat. Or a species may actually be adapted initially to the sparser habitat, but then switch in its habitat specialization over to the other, more widespread, habitat, and lose its ability to persist in its ancestral habitat (Fig. 10.8b). In this case, niche evolution is actually a niche switch between habitats. Note that there is only one stable equilibrium for the parameters of both panels of Fig. 10.8. The equilibrium is symmetric for the parameter choices leading to Fig. 10.8a, but very asymmetric for the parameter choices used in Fig. 10.8b, with habitat 1 having a very low occupancy.

One limitation in the above model is that when the species occupies a substantial fraction of both habitat types in a landscape, the immigrants showing up in any given occupied patch are likely to be a mixture of emigrants from each of the habitat types. This observation does not affect our conclusions about the existence of alternative

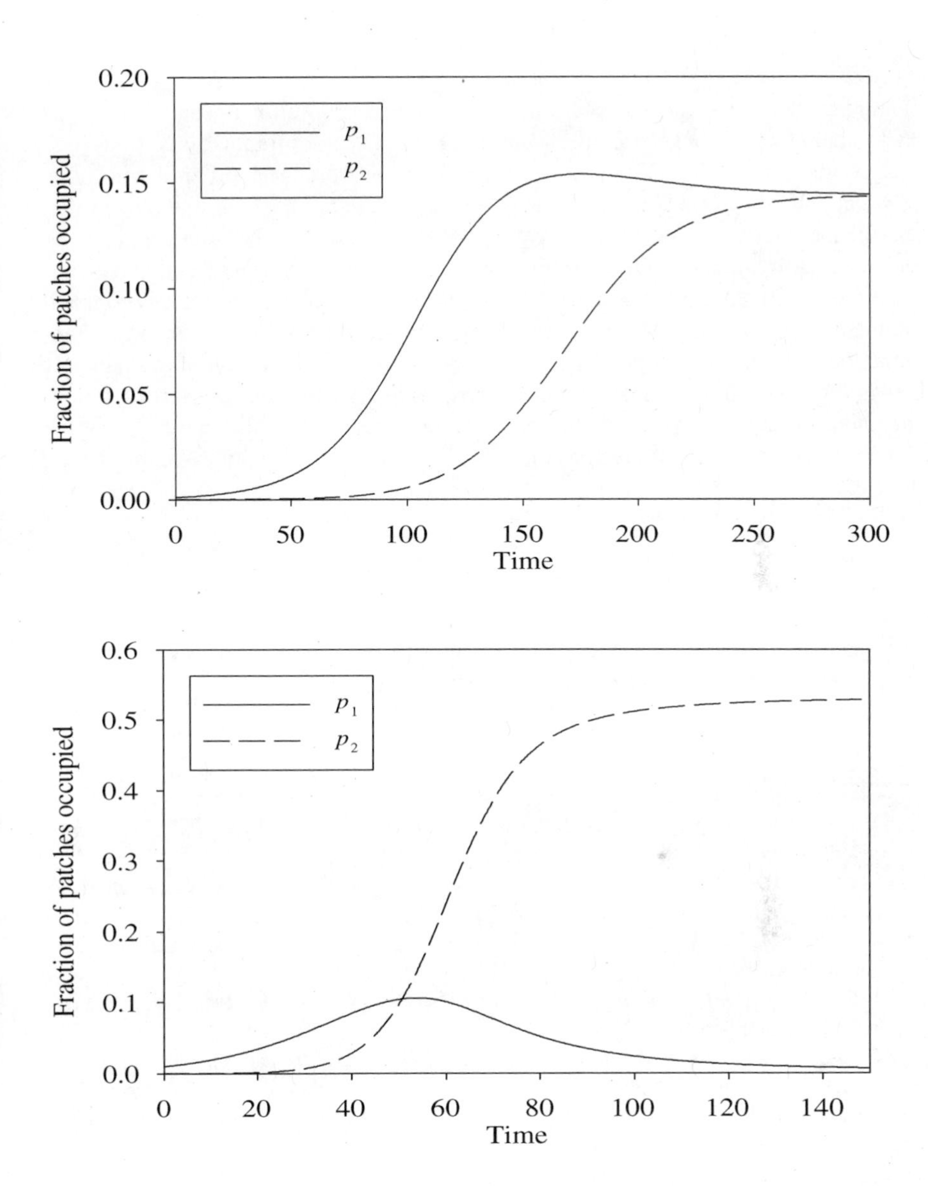

Figure 10.8 (a) Time plots for symmetric metapopulation model with $h = 0.5$, $c = 0.3$, $c_x = 0.001$, $e = 0.1$, and $\gamma = 0.5$. Initially, habitat type 2 is empty, while habitat type 1 has an occupancy of 0.001. Because within-habitat-type colonization is much higher than cross-colonization, habitat type 2 is occupied only after a lag. (b) Time plots for the metapopulation model that is symmetric except for abundance of the two habitat types and colonization rates. Parameters are $h_1 = 0.3$, $h_2 = 0.7$, $c_1 = 0.4$, $c_2 = 0.3$, $c_x = 0.001$, $e = 0.05$, and $\gamma = 4$. Initially, habitat type 2 is empty, while habitat type 1 has an occupancy of 0.01. Because habitat type 2 is more abundant on the landscape, the species there is able to suppress the species in habitat type 1.

stable equilibria, but could shift the range of parameter values where one observes this outcome.

Future extensions of this work will include examining evolution in spatially explicit landscapes, and a consideration of multiple habitat types, arranged in various spatial configurations. Studies with individual-based models in landscapes with three distinct habitats reveal some unexpected effects, reflecting how the interplay of dispersal and selection affects the entire distribution of allelic values, within and among habitats (Holt and Barfield, in prep.). Understanding niche conservatism and evolution requires a simultaneous consideration of how the structure of the environment influences the pattern and strength of natural selection, and how selection in conjunction with other evolutionary forces modifies the pool of variation available for evolution. Grappling with this issue is central to many basic questions in evolutionary biology, and is also of urgent practical importance, given the rapidly changing environments we humans are currently forcing the biota of the globe to experience.

10.7 Acknowledgments

We thank the editors for their invitation to contribute to this volume, and the University of Florida Foundation for financial support.

10.8 References

A.D. Bradshaw (1991), Genostasis and the limits to evolution, *Phil. Trans. Roy. Soc. B, Biological Sciences* **333**:289-305.

R. Burger and M. Lynch (1995), Evolution and extinction in a changing environment: A quantitative-genetic analysis, *Evolution* **49**:151-163.

D.S. Falconer (1989), *An Introduction to Quantitative Genetics,* 3rd ed., Longman Scientific, Harlow, UK.

W. Feller (1968), *An Introduction to Probability Theory and Its Applications,* Vol. 1. 3rd ed., John Wiley and Sons, New York.

R. Gomulkiewicz and R.D. Holt (1995), When does evolution by natural selection prevent extinction?, *Evolution* **49**:201-207.

K.C. Harding and J.M. McNamara (2002), A unifying framework for metapopulation dynamics, *American Naturalist* **160**:173-185.

R.D. Holt (1983), Immigration and the dynamics of peripheral populations, in *"Advances in Herpetology and Evolutionary Biology,"* ed. by K. Miyata and A. Rhodin, Harvard University, Cambridge, pp. 680-694.

R.D. Holt (1996), Demographic constraints in evolution: Towards unifying the evolutionary theories of senescence and niche conservatism, *Evolutionary Ecology* **10**:1-11.

R. D. Holt (1997), From metapopulation dynamics to community structure: Some consequences of spatial heterogeneity, in *"Metapopulation Biology,"* ed. by I. Hanski and M. Gilpin, Academic Press, New York, pp. 149-164.

R.D. Holt and M.S. Gaines (1992), The analysis of adaptation in heterogeneous landscapes: Implications for the evolution of fundamental niches, *Evolutionary Ecology* **6**:433-447.

R.D. Holt and R. Gomulkiewicz (1997), The evolution of species' niches: a population dynamic perspective, in *"Case Studies in Mathematical Modelling: Ecology, Physiology, and Cell Biology,"* ed. by H. Othmer, F. Adler, M. Lewis and J. Dallon, Prentice-Hall, pp. 25-50.

R.D. Holt, M. Barfield and R. Gomulkiewicz (2005), Theories of niche conservatism and evolution: Could exotic species be potential tests? in *"Species Invasions: Insights into Ecology, Evolution, and Biogeography,"* ed. by D. Sax, J. Stachowicz and S.D. Gaines, Sinauer Associates, Sunderland, MA, pp. 259-290.

G.E. Hutchinson (1958), Concluding remarks, *Cold Spring Harbor Symp. Quant. Biol.* **22**:415-427.

T. Kawecki (2008), Adaptation to marginal habitats, *Annu. Rev. Ecol. Evol. Syst.* **39**:321-342.

R.H. MacArthur and E.O. Wilson (1967), *The Theory of Island Biogeography,* Princeton University Press, Princeton.

B. Maguire Jr. (1973), Niche response structure and the analytical potentials of its relationship to the habitat, *The American Naturalist* **107**:213-246.

H.A. Orr and R.L. Unckless (2007), Population extinction and the genetics of adaptation, *The American Naturalist* **172**: 160-169.

H.R. Pulliam (2000), On the relationship between niche and distribution, *Ecology Letters* **3**:349-361.

O. Ronce and M. Kirkpatrick (2001), When sources become sinks: Migrational meltdown in heterogeneous habitats, *Evolution* **55**:1520-1531.

J.J. Wiens and C.H. Graham (2005), Niche conservatism: Integrating evolution, ecology, and conservation biology, *Annu. Rev. Ecol. Evol. Syst.* **36**:519-39

CHAPTER 11

Evolution of dispersal in heterogeneous landscapes

Robert Stephen Cantrell
University of Miami

Chris Cosner
University of Miami

Yuan Lou
Ohio State University

Abstract. Dispersal is the mechanism by which populations distribute themselves across landscapes. As such, its study is an essential aspect of spatial ecology. Habitats themselves are heterogeneous across space and time. Dispersal can reflect purely random movement or may be conditioned on properties of the environment or the presence of other organisms. Understanding what forms of dispersal confer selective advantage in what types of habitats is an issue that has recently come to the forefront of spatial ecology and its interface with evolutionary theory. The connection between ecology and evolutionary theory is usually expressed through the concepts of evolutionarily stable strategy and invasibility. There is some dichotomy in theoretical predictions of selective advantage. Dispersal of some sort is favored in a metapopulation framework. Unconditional dispersal is generally not favored in temporally constant environments in a discrete diffusion setting. Unconditional dispersal may, however, be favored in this framework if there is temporal variability in the habitat. Conditional dispersal may be favored when there is spatial variation. Such results have been extended to both reaction-advection-diffusion and integrodifference modeling frameworks. This essay will review the development of the theory of evolution of dispersal, describe the current state of understanding in the subject, and highlight important open questions and issues.

11.1 Introduction

The dispersal of organisms is clearly an important aspect of many ecological processes. It drives biological invasions, allows populations to colonize empty habitats, and allows individuals to track resources and avoid predators or competitors. It plays a significant role of the life histories of many organisms. Yet, despite the fact that dispersal is ubiquitous, our understanding of its evolutionary causes and ecological effects is still quite limited. In their introduction to the book "Dispersal" (Clobert et al., 2001), the editors remark that "dispersal is probably the most important life history trait involved in both species persistence and evolution" and that "One of the most studied yet least understood concepts in ecology and evolutionary biology is the movement of individuals, propagules, and genes." There are a number of factors that can influence the evolution of dispersal, and correspondingly there are a number of different modeling approaches that have been used to study it. Factors that are commonly invoked to explain the evolution of dispersal can be either genetic or ecological (Gandon and Michalakis, 2001). Genetic factors include kin selection, i.e., reduction of competition between related individuals (Hamilton and May, 1977), and avoidance of inbreeding (Gandon, 1999). The main ecological factors involve environmental heterogeneity in time and/or space (McPeek and Holt, 1992). In the present article we will focus our attention on ecological factors, especially spatial heterogeneity. Most of the analysis of the ecological aspects of the evolution of dispersal has been based on ecological models rather than explicitly evolutionary models. Evolutionary conclusions typically have been drawn from ecological models by means of the notion of evolutionarily stable strategies. A strategy is said to be evolutionarily stable if a population using it cannot be invaded by a small population using any other strategy. The idea is that the strategies observed in natural systems are those that are evolutionarily stable, because they can resist invasion. If two strategies are compared and the first is found to be evolutionarily stable relative to invasion by the second while the second is not evolutionarily stable with respect to the first then the interpretation is that the first should be able to invade and displace the second. On the other hand, if neither strategy is evolutionarily stable with respect to the other then each can invade the system when rare and hence they may be expected to coexist in some sort of stable polymorphism. (The theory of uniform persistence or permanence gives a rigorous mathematical formulation for this idea; see Hutson and Schmitt (1992).) Most of the analysis we will describe in this article is motivated by the idea of evolutionary stability.

It is clear that in some sorts of temporally varying environments there should be selection for some amount of dispersal. In particular, for populations inhabiting patchy environments where they are subject to local extinctions, persistence is possible only if the population can recolonize empty patches. A collection of local populations distributed across a network of patches is called a metapopulation. The idea that local populations may be subject to extinction but that empty patches can be recolonized by individuals dispersing from other patches is the basis for patch occupancy models for metapopulations. Those models do not include explicit population dynamics; they only track the probabilities that patches are occupied. In that modeling framework dispersal is viewed as a factor in the rate of colonization so some amount of dispersal is essential to prevent extinction of the entire metapopulation. Patch occupancy models should be distinguished from discrete diffusion models which keep track of population densities but do not necessarily incorporate local extinctions or other forms of temporal variability (see Hanski (1999, 2001)). Even in the context of patch occupancy models or stochastic individual based models that allow local extinctions there are interesting questions about the evolution of dispersal, but we will not pursue those here. We refer the interested reader to Heino and Hanski (2001). For many types of plants, only seeds can disperse under normal conditions, so again the process of dispersal is tightly connected to the process of recruitment. Indeed, patch occupancy models where each patch represents a location where a single plant can grow have been widely used to study dispersal and competition in plants; see for example Tilman (1994). There are various modeling approaches that can be used to study the evolution of dispersal; see Levin et al. (2003) and Clobert et al. (2001). We will discuss the evolution of dispersal, including the effects of temporal variation, in the context of reaction-diffusion models, their generalizations, and their discrete analogues. Even in the context of reaction-diffusion or discrete-diffusion models it turns out that temporal variation can cause selection for dispersal. This phenomenon was observed by McPeek and Holt (1992) in numerical experiments on discrete diffusion, studied further in that context from the viewpoint of adaptive dynamics by Parvinen (1999), and studied analytically and numerically by Hutson et al. (2001) in the reaction-diffusion context.

The effects of spatial heterogeneity on the evolution of dispersal in systems where the environment is uniform in time are rather subtle. Hastings (1983) obtained analytic results on reaction-diffusion models and their spatially discrete analogues that suggested there would be selection for slow dispersal in spatially varying but temporally constant environments. However, Hastings' results were based on assumptions about the process of dispersal and the patterns of spatial distribution of populations that it would produce that are not universally satisfied; in particular they do not hold in some models incorporating dispersal behavior that depends on environmental conditions. McPeek and Holt (1992) made a number of observations on the basis of numerical experiments on two-patch discrete-time models. They found that there was selection for slow dispersal in the spatially varying but temporally constant case if the dispersal process was independent of environmental conditions, but there was not when the dispersal process depended on environmental conditions in the right way. They also found that there could be selection for fast dispersal in environments with both spatial and temporal variation even if the dispersal process was independent of environmental conditions. (In later work, Holt and McPeek (1996) found that chaotic population dynamics can induce selection for dispersal in a manner similar to the effects of extrinsic spatiotemporal variation.) McPeek and Holt (1992) introduced the terms "conditional" and "unconditional" respectively to describe dispersal processes that do or do not depend on environmental conditions. The particular form of conditional dispersal that McPeek and Holt found to be evolutionarily stable in spatially varying but temporally constant environments has the feature that it results in an equilibrium distribution of the population where all individuals have the same fitness (as measured by reproduction rate), independent of their location, and there is no net movement of individuals at equilibrium. Such a distribution is consistent with a descriptive theory of how organisms should distribute themselves developed by Fretwell and Lucas (1970) called the ideal free distribution. Conditional dispersal that leads to an ideal free distribution of population is sometimes called "balanced dispersal." The population dynamics arising from the movement of individuals from regions of greater fitness to regions of lower fitness by unconditional dispersal are sometimes called "source-sink" dynamics. There has been some empirical study of whether natural populations display balanced dispersal or source-sink dynamics, or perhaps neither. The empirical study in Doncaster et al. (1997) supports the view that some populations display a form of balanced dispersal; see also Cantrell et al. (2007a), Holt and Barfield (2001), and Morris et al. (2004) for additional discussion and references related to the ideal free distribution, balanced dispersal, source-sink dynamics, and the evolution of dispersal.

11.2 Random dispersal: Evolution of slow dispersal

Hastings (1983) asked whether spatial variation alone can lead to selection for increased dispersal in a spatially inhomogeneous but temporal constant environment. To that end, he envisioned a scenario where an environment was inhabited by a resident species at a stable equilibrium density, and some mutation occurred, thus introducing a small mutant population into the environment. He considered both reaction-diffusion and discrete diffusion models in continuous time as models for such a scenario. Specifically, in the reaction-diffusion case, the model for the resident population took the form

$$\begin{aligned} u_t &= D\nabla \cdot [\mu(x)\nabla u] + F(x,u)u \qquad && \text{in } \Omega \times (0,\infty), \\ \frac{\partial u}{\partial n} &= 0 && \text{on } \partial\Omega \times (0,\infty), \end{aligned} \tag{11.1}$$

where $u(x,t)$ is a population density, the habitat Ω is a bounded region in $\mathcal{R}^N$ with smooth boundary $\partial\Omega$, $\nabla\cdot$ is the divergence operator, ∇ denotes the gradient operator, $\mu(x) > 0$ describes how the rate of diffusion varies spatially, $D > 0$ describes the overall rate of diffusion, n is the outward unit normal vector on $\partial\Omega$, and the boundary condition means that no individuals cross the boundary of the habitat. We will refer to such boundary conditions as "zero-flux." Note that the specific form taken by zero-flux boundary conditions depends on the flux, so that zero-flux boundary conditions may involve additional terms, e.g., in cases where the dispersal terms involve advection. In (11.1) and in most of the models described in this article we interpret the local population growth rate $F(x,u)$ as being determined by the level of resources available at location x to a population living at density u. We will also use the local population growth rate as a measure of the fitness of an individual at the point x when the population density is u. Hastings assumed that the model (11.1) had a stable positive equilibrium u^* with $F(x,u^*)$ not identically zero, modeled a small invading mutant population v as satisfying

$$v_t = d\nabla \cdot [\mu(x)\nabla v] + F(x,u^*+v)v \quad \text{in} \quad \Omega \times (0,\infty), \tag{11.2}$$

also with zero-flux boundary conditions, and determined when the model predicted that the mutant population could successfully invade the resident population. The model in (11.2) was based on the assumption that the mutant population is so small that it has a negligible effect on the resident population. The main finding in Hastings

(1983) was that if the mutant differs from the resident species only by having a different dispersal rate, then it can invade when rare if and only if its dispersal rate is less than that of the resident species. Hastings obtained a similar result for a spatially discrete analogue of (11.1); we will return to that model later in our discussion of the ideal free distribution. Analogous results for the discrete-time case were obtained for the case where dispersal is unconditional (so that a hypothesis analogous to having $F(x, u^*)$ not identically zero is satisfied) by numerical experiments in McPeek and Holt (1992) and proved analytically in Parvinen (1999).

The criterion for whether or not a mutant could invade the system described by (11.1) is the instability or stability of the equilibrium $v = 0$ in (11.2). In this case and many others, the stability of such an equilibrium can be determined by a linear stability analysis. Linear second order elliptic operators on bounded domains typically have a principal eigenvalue which has a larger real part than any other eigenvalue and is characterized by having a positive eigenfunction. This eigenvalue is analogous to the principal eigenvalue of a primitive matrix. Its existence follows from the Krein-Rutman theorem, which is an extension of the Perron-Frobenius theorem on matrices to the infinite dimensional case. It turns out that second order parabolic equations with periodic coefficients also have a principal eigenvalue. See Cantrell and Cosner (2003), Section 2.5, for a discussion of principal eigenvalues. The stability or instability of equilibria in most of the models we will discuss can thus be determined by the sign of the principal eigenvalue of the linearized problem. In some cases the principal eigenvalue may be zero, so that a nonlinear stability analysis is needed. In the analysis of (11.2), Hastings showed that if $F(x, u^*)$ is not identically zero then the principal eigenvalue of the linearization of (11.2) around $v = 0$ is positive if and only if $d < D$ in (11.2). The conclusion about invasibility follows immediately.

A possible biological reason for the evolution of slow dispersal is that passive diffusion takes individuals from more favorable locations to less favorable locations more often than it does the reverse (Hastings, 1983), since it typically moves individuals from regions of high density to regions of lower density. In terms of resource matching, one consequence of random diffusion is to cause the resident species to undermatch the best resources at equilibrium. In fact, the zero-flux boundary condition in (11.1) and the divergence theorem imply that at the equilibrium u^* the integral of $F(x, u^*)u^*$ over Ω is zero, so that if $F(x, u^*)$ is nonzero at equilibrium it must change sign so that the population overmatches the resources in some places but undermatches them in others. When a slower diffusing mutant population is introduced, it can grow at locations where the resident undermatches the resources in the habitat (which would typically be the locations with the best resources), and is more likely to remain in those locations, so it can thus invade successfully. It is interesting to note that the analysis in Hastings (1983) breaks down if the assumption that $F(x, u^*)$ is nonzero at equilibrium is removed. If $F(x, u^*) = 0$ on Ω then the resident matches the resources perfectly. Furthermore, if certain technical conditions are satisfied, it can be shown that if u^* is unique for each D and there is a unique positive solution $u = K(x)$ to the equation $F(x, u) = 0$ then $u^* \to K(x)$ on the interior of Ω as $D \to 0$; see Cantrell and Cosner (2003), Proposition 3.16. In a logistic model $K(x)$

would represent the local carrying capacity of the environment. Thus, a population that diffuses sufficiently slowly will come closer to matching the available resources than one that diffuses more rapidly.

Hastings' result is a local one in the sense that it concerns only the invasion of invading species when it is rare. After the invasion of the mutant, can it drive the resident species to extinction or will it coexist with the resident species? This led Dockery et al. (1998) to consider the following continuous-time continuous-space model for two randomly diffusing competing species:

$$\begin{cases} u_t = \mu\Delta u + u[m(x) - u - v] & \text{in } \Omega \times (0,\infty), \\ v_t = \nu\Delta v + v[m(x) - u - v] & \text{in } \Omega \times (0,\infty), \\ \dfrac{\partial u}{\partial n} = \dfrac{\partial v}{\partial n} = 0 & \text{on } \partial\Omega \times (0,\infty), \end{cases} \tag{11.3}$$

where $u(x,t)$ and $v(x,t)$ represent the population densities of competing species with respective dispersal rates μ and ν. The symbol Δ stands for the Laplace operator ($\Delta = \nabla^2$), which is the composition of the divergence and gradient operators and models the random dispersal of the species. The scalar function $m(x)$ represents their common intrinsic growth rates and it reflects the quality and quantity of resources available at the location x. The habitat Ω is as in (11.1). The zero-flux boundary condition in (11.3) means that no individuals cross the boundary of the habitat. The most notable feature of (11.3) is that these two species are identical except their dispersal rates.

Dockery et al. (1998) showed that if the dispersal rate of the mutant is smaller than that of the resident species, then the mutant not only can invade but also can drive the resident species to extinction, i.e., a slower diffusing species always emerges as the winner of the competition. For nonlocal dispersions, some similar results hold (see Hutson et al. (2003)). However, when the intrinsic growth rate varies periodically in time, it is shown in McPeek and Holt (1992) for patch models and in Hutson et al. (2001) for diffusion models that the slower diffuser may not always be the winner, and faster dispersal can be selected in some situations. A challenging open problem is whether the slowest diffuser always wins the competition in the context of k competing species with $k \geq 3$ (Dockery et al., 1998).

11.3 Random dispersal vs. conditional dispersal

In reality, species do not always move randomly. As resources are often distributed heterogeneously across the habitat, a species can often sense local environment change and its movement may be affected by environmental factors such as resource distributions and population density. One of the simplest modeling approaches is to assume that organisms display taxis and can move up along the gradient of a local population growth rate. Such biased movement upward along resource gradients is an example of conditional dispersal and has been considered in Belgacem and Cosner (1995) and

Cosner and Lou (2003) for a single species. Among other things, Belgacem and Cosner (1995) and Cosner and Lou (2003) showed that conditional dispersal involving both random diffusion and directed movement up resource gradients can sometimes (but not always) make persistence of a single species more likely. For two-patch models, McPeek and Holt (1992) showed that in spatially varying but temporally constant environments certain types of conditional dispersal can be advantageous.

Hence, it is of interest to compare a random dispersal strategy with a conditional dispersal strategy such as biased movement along a resource gradient, and determine which dispersal strategy will evolve. This led Cantrell et al. (2006, 2007b) to introduce the model

$$\begin{cases} u_t = \nabla \cdot [\mu \nabla u - \alpha u \nabla m] + [m(x) - u - v]u & \text{in } \Omega \times (0, \infty), \\ v_t = \nu \Delta v + v[m(x) - u - v] & \text{in } \Omega \times (0, \infty), \\ \mu \frac{\partial u}{\partial n} - \alpha u \frac{\partial m}{\partial n} = \frac{\partial v}{\partial n} = 0 & \text{on } \partial\Omega \times (0, \infty), \end{cases} \tag{11.4}$$

where the two species have different dispersal strategies: the species with density v disperses only by random diffusion, the other species disperses by a combination of random diffusion and a directed movement towards more favorable habitats, where α is a positive parameter which measures the tendency of biased movement along the resource gradient. Both still satisfy zero-flux boundary conditions.

When $\alpha = 0$, from the previous section we know that the slower diffusing species always wins the competition. What happens if $\alpha > 0$? It turns out that the answer is rather delicate and depends on both the magnitude of α and the geometry of the habitat Ω.

It is shown in Cantrell et al. (2006, 2007b) that for convex habitats, the competitor that moves upward along the resource gradient may have a competitive advantage even if it diffuses more rapidly than the other competitor, i.e., a faster diffuser with some (weak) advection along the resource gradient can win the competition. It means that the advantage gained from the directed movement upward along resource gradients can compensate for the disadvantage created by faster diffusion, at least for convex habitats.

The case $\mu = \nu$ also depends on the geometry of the habitat. For convex habitats, we show in Cantrell et al. (2007b) that for small positive α, the species with density u always wins. Hence, at least for convex habitats, species with a small amount of biased movement have the advantage. That is, the dispersal strategy with some biased movement can evolve there. On the other hand, there are some nonconvex habitats, as constructed in Cantrell et al. (2007b), such that the species u always loses. It is interesting that the geometry of the habitat can play an important role in the evolution of dispersal, and this may have potential applications to the conservation of species. For example, it may be helpful in understanding how habitat fragmentation affects the loss of species.

If we further increase α, it seems that the species with density u becomes "smarter" and hence will continue to win the competition. Surprisingly, for sufficiently large α, one often can expect that the two competing species can coexist (Cantrell et al., 2007b). In other words, strong advection upward along environmental gradients can induce the coexistence of species and provide a mechanism for the coexistence of competing species. If we interpret the competitors as different genotypes of the same species, this situation would correspond to a stable polymorphism. (In at least some species there appears to be a genetic basis for some aspects of dispersal ability; see Roff (1994).)

From the biological point of view, such coexistence results are surprising, at least at the first look. Given any pair of $\mu < \nu$, when α is positive and small, the species u always wins the competition, i.e., the slower diffuser still wins. As α increases, the species with density u has the tendency to move toward more favorable regions, so it seems to have more competitive advantage than the species with density v and should still win the competition. However, the results in Cantrell et al. (2007b) show that the "smarter" species may coexist with the other species, which is randomly diffusing with a larger random diffusion rate. A possible explanation for such coexistence is that as α becomes large, the "smarter" competitor moves toward and concentrates at places with the locally most favorable environments, leaving enough resources elsewhere for the other species to survive. Thus, there is a type of spatial segregation of the competitors which leads to coexistence. These biological intuitions are justified by some rigorous analytical results from Chen and Lou (2008) in the case when there is only one local maximum of resource density.

In terms of resource matching, a big difference between random diffusion and biased movement along the resource gradient is that random diffusion leads the species to undermatch the best resources, while the biased movement along the resource gradient can lead the species to better match the resources if the advection rate is suitable, or overmatch the best resources if the advection rate is too large. Whether a dispersal strategy is evolutionarily stable or not seems to rely crucially on how well the species can apply the dispersal strategy to match the resources.

11.4 Evolution of conditional dispersal

What happens if both competing species disperse by random diffusion and advection along environmental gradients? Intuitively, one possible consequence of biased movement up a resource gradient is to cause a certain degree of crowding in the favorable regions of the habitat which might change the outcome of the competition. To understand the evolution of conditional dispersal, Chen et al. (2008) considered the model:

$$\begin{cases} u_t = \nabla \cdot [\mu \nabla u - \alpha u \nabla m] + [m(x) - u - v]u & \text{in } \Omega \times (0, \infty), \\ v_t = \nabla \cdot [\nu \nabla v - \beta v \nabla m] + [m(x) - u - v]v & \text{in } \Omega \times (0, \infty), \\ \mu \dfrac{\partial u}{\partial n} - \alpha u \dfrac{\partial m}{\partial n} = \nu \dfrac{\partial v}{\partial n} - \beta v \dfrac{\partial m}{\partial n} = 0 & \text{on } \partial\Omega \times (0, \infty). \end{cases} \tag{11.5}$$

When $\beta = 0$ and α is large, from the previous section we know that the two species can often coexist with each other. Hence, neither of the two dispersal strategies is the winning one. What happens if $\beta > 0$? It turns out that at least two scenarios can occur (Chen et al., 2008):

(i) If only one species has a strong tendency to move upward the environmental gradients, e.g., β is small and α is large, the two species can coexist since one species mainly pursues resources at places of locally most favorable environments while the other relies on resources from other parts of the habitat. This is the same as the case when $\beta = 0$.

(ii) If both species have a strong tendency to move upward the environmental gradients, e.g., β is large and α is even larger, it can lead to overcrowding of the whole population at places of locally most favorable environments, which causes the extinction of the species with stronger biased movement. From the biological point of view, strong biased movement along the resource gradient of both species can induce overmatching of resources for both species at places of locally most favorable environments. This is particularly disadvantageous to the species with stronger biased movement as it puts all of its bets on such places.

These results seem to imply that selection is against excessive advection along environmental gradients due to overmatching of the best resources, and they also suggest that an intermediate biased movement rate may evolve in the model.

To further understand the evolution of conditional dispersal, Hambrock and Lou (2008) recently considered the situation when the advection rates α and β are close to each other (different from the case when one is much larger than the other as in previous case), and their findings also support the conjecture that an intermediate biased movement rate may evolve in the model. More precisely, suppose that $\mu = \nu$ and if both advection rates are small, then the species with the larger advection rate always wins; if $\mu = \nu$ and both advection rates are suitably large, then the species with the smaller advection rate always wins.

Another interesting finding in Hambrock and Lou (2008) is that the evolution of random diffusion rates also depends on the magnitude of the advection rates and will change direction if the advection rates vary from small to large. More precisely, suppose that $\alpha = \beta > 0$. Then for small advection rates, the slower diffuser always wins (this is the same as the case when $\alpha = \beta = 0$). However, when the advection rates are large, the faster diffuser is always the winner in the competition.

11.5 Dispersal and the ideal free distribution

Ideal free distribution (IFD) theory describes how organisms should distribute themselves in space if they could move freely to optimize their fitness (Fretwell and Lucas, 1970). It says that individuals should locate themselves so that no individual can increase its fitness by moving to another location. Thus, it predicts that at equilibrium the fitness of individuals should be the same in all locations, and there should be no net movement at equilibrium. (This is in contrast to the dynamics of many source-sink models where the fitness in the source is larger than that in the sink, which is typically negative, and the sink population is sustained by net movement from the source to the sink; see Pullian (1988).) McPeek and Holt (1992) observed in discrete-time discrete diffusion models that there could be selection for dispersal in spatially varying but temporally constant environments if the dispersal rates had the feature that the equilibria of the system were the same with and without dispersal. If we interpret the fitness of an individual on a given patch with a given population density as being given by the population growth rate on that patch at that density, this feature means that at equilibrium every individual would have fitness zero, which is consistent with the ideal free distribution. It turns out that such a form of conditional dispersal is evolutionarily stable in many situations, see Cantrell et al. (2007a) and Holt and Barfield (2001). To make these ideas more precise, let us consider a discrete diffusion model of the type studied by Hastings (1983):

$$\frac{du_i}{dt} = F_i(u_i)u_i + \sum_{\substack{j=1 \\ j\neq i}}^{n} [d_{ij}u_j - d_{ji}u_i] \quad \text{for } i = 1, \ldots, n. \tag{11.6}$$

Suppose that for each $i = 1, \cdots, n$, $u_i^* > 0$ is a stable equilibrium of $du/dt = F_i(u)$, so that $F_i(u_i^*) = 0$ for $i = 1, \cdots, n$, with $dF/du < 0$ for $u = u_i^*$. Suppose further that for some dispersal strategy determined by nonzero dispersal coefficients $\{d_{ij}\}$, u^* is also a positive equilibrium of (11.6). That implies

$$\sum_{\substack{j=1 \\ j\neq i}}^{n} \left[d_{ij}u_j^* - d_{ji}u_i^*\right] = 0 \quad \text{for } i = 1, \ldots, n. \tag{11.7}$$

It turns out that under these conditions the strategy defined by $\{d_{ij}\}$ is evolutionarily stable relative to strategies which do not satisfy (11.7). Furthermore, any dispersal strategy leading to an equilibrium u^{**} that does not have $F_i(u_i^{**}) = 0$ for $i = 1, \cdots, n$ cannot be evolutionarily stable; see Cantrell et al. (2007a). This result extends to some models for competition and predator-prey interactions; related results are obtained in Cressman and Krivan (2006), Kirkland et al. (2006), and Padrón and Trevisan (2006). If the model for invasibilty by a small invading population (that is, the model corresponding to a discrete version of (11.2)) is linearized around zero, the resulting linear model is neutrally stable, so asymptotic stability arises from higher order effects. For a full model for two populations with competing strategies at arbitrary densities, analogous to a spatially discrete version of (11.3), (11.4), and (11.5), different strategies satisfying (11.7) have a type of neutral stability with respect to each other. This is consistent with the findings of McPeek and Holt (1992).) Since $F_i(u_i^*) = 0$ for $i = 1, \cdots, n$, all patches have the same fitness at equilibrium. Also, by (11.7), there is no net movement at equilibrium. Thus, the evolutionarily stable strategies represent forms of balanced dispersal in that they lead to a population distribution that is ideal free. Note that the condition $F_i(u_i^*) = 0$ for $i = 1, \cdots, n$, is exactly the negation of the condition that $F_i(u_i^*)$ is not identically zero relative to i imposed by Hastings (1983) and by Parvinen (1999) in results showing selection for slow dispersal in the spatially discrete case. Furthermore, the case of condition (11.7) with $n = 2$ is equivalent to the condition for evolutionary stability found by McPeek and Holt (1992). The analysis in Cantrell et al. (2007a) depends on the fact that the models are finite dimensional. The problem of extending the results of Cantrell et al. (2007a), Cressman and Krivan (2006), Kirkland et al. (2006), and Padrón and Trevisan (2006) to the infinite dimensional case is interesting and largely open.

A novel variation on these ideas was introduced by Wilson (2001) who developed a habitat occupancy model for a source-sink situation. The model has a form similar to a coupled pair of patch occupancy models, but with one model describing a source habitat and the other a sink habitat. As usual in habitat occupancy models, there must be at least some dispersal within the source patch for persistence to be possible, but the question is whether or not dispersal into the sink habitat can evolve. The source patch is assumed to have a stable equilibrium proportion p_1^* of occupied habitat in isolation, so that without dispersal there is no positive equilibrium, and the equilibrium $(p_1^*, 0)$ is stable. However, in some cases there is an evolutionarily stable dispersal strategy with nonzero dispersal that results in positive proportions of both the source and sink habitats. It turns out that under this strategy the fitness in both source and sink habitats can be seen to be zero, and "surprisingly" (Wilson, 2001, p. 30) the equilibrium proportion of occupied habitat in the source is still p_1^*. Perhaps in view of the results described previously this last feature is not really so surprising.

It is natural to ask whether an ideal free distribution of population can arise from dispersal that is conditional on local information but does not require global knowledge of the environment, as in reaction-diffusion-advection models. A version of the ideal free distribution in continuous space was introduced in Kshatriya and Cosner (2001). A dynamic model whose equilibria can be expected to fit such a distribution recently has been developed via advection-diffusion equations in Cosner (2005), under the assumptions that organisms move upward along the local gradient of fitness and that fitness varies spatially and is reduced by crowding. The model in Cosner (2005) has the form

$$u_t = -\alpha \nabla \cdot [u \nabla f(x, u)] \quad \text{on} \quad \Omega \times (0, \infty),$$

with the no-flux boundary condition

$$u \frac{\partial f(x, u)}{\partial n} = 0 \quad \text{on} \quad \partial\Omega \times (0, \infty),$$

where $f(x, u) = m(x) - u(x)$ represents the local effective growth rate of the species, $m(x)$ is the intrinsic per capita growth rate, and $u(x)$ is the population density.

Cantrell et al. (2008) considered a variation on that model which also includes random diffusion as part of the dispersal process, and it has the form

$$u_t = \nabla \cdot [\mu \nabla u - \alpha u \nabla f(x, u)] + u f(x, u) \quad \text{in } \Omega \times (0, \infty), \tag{11.8}$$

with no-flux boundary conditions

$$\mu \frac{\partial u}{\partial n} - \alpha u \frac{\partial f(x,u)}{\partial n} = 0 \quad \text{on } \partial\Omega \times (0,\infty). \tag{11.9}$$

See Grindrod (1988) for a similar model which addresses different questions. One of the main findings in Cantrell et al. (2008) is that as the rate of movement up fitness gradients becomes large and/or the rate of random diffusion becomes small, the density of organisms approximately matches the availability of resources everywhere in the habitat. This differs significantly from both unconditional dispersal by random diffusion and conditional dispersal where organisms tend to move up gradients of resource density without reference to crowding effects. Both of those dispersal strategies lead to population distributions where the density overmatches resource in some locations but undermatches it in others. This fact is the essential reason why there is selection for slow dispersal in models with purely diffusive dispersal, because for such models the only way for the equilibrium population density to approximately match the distribution of resources is for the diffusion rate to go to zero. It is also the reason why too strong a tendency to move up resource gradients without regard to crowding effects can sometimes make a population subject to invasion by another population using a different strategy.

11.6 Dispersal in temporally varying environments

In contrast to spatial heterogeneity, temporal variation in environments can sometimes select for unconditional dispersal. It can also lead to coexistence of different strategies in a stable polymorphism. Much of the work on the evolution of dispersal in time varying environments involves at least some numerical computation because analytic results are harder to obtain than in the temporally constant case. Some analytic results are derived in Hutson et al. (2001) for a reaction-diffusion model of the general form shown in (11.3) but with $m(x)$ replaced by $m(x,t)$ where $m(x,t)$ is periodic in t. In spatially homogeneous but temporally varying environments, the results of McPeek and Holt (1992) (based on numerical experiments on two-patch discrete-time discrete diffusion models) and those of Hutson et al. (2001) (obtained analytically for reaction-diffusion models) indicate that there is no selection for or against unconditional dispersal. In both of those studies the models had stable equilibria; in the case of models that support periodic or chaotic solutions the situation can be different. We will return to that case later. McPeek and Holt (1992) observed that when there is variation in time but not in space then as in the spatially and temporally constant case, there can be selection against forms of dispersal that cause the population to undermatch resources in one patch and overmatch them in the other, but there is no selection for or against uniform unconditional dispersal. In the case of environments with both spatial and temporal variability, McPeek and Holt (1992) found that if only unconditional strategies are considered then except in certain special cases, the system would evolve to a polymorphism consisting of a slow dispersal strategy and a relatively fast dispersal strategy. Hutson et al. (2001) obtained similar analytic results provided that the time average of the coefficient $m(x,t)$ over a period is positive and some additional technical conditions are satisfied. Hutson et al. (2001) did not consider conditional dispersal. McPeek and Holt (1992) did; they found that there was selection for a specific conditional strategy that satisfied an "ideal free" or "balanced dispersal" condition analogous to (11.7). (In this situation the heterogeneity was obtained by drawing carrying capacities for discrete-logistic within-patch models at random from some distribution, so the equilibria u_i^* in (11.7) would be replaced by the means of those carrying capacities.) This is in contrast with the temporally constant case, where within the class of strategies satisfying (11.7) any number of strategies were seen in McPeek and Holt (1992) to be able to coexist in a state of neutral stability. It would be of interest to consider the evolution of conditional dispersal in the reaction-diffusion setting used in Hutson et al. (2001).

In discrete-time models variability in time does not require extrinsic variation in the environment. Such models can have periodic or chaotic dynamics without it. In Holt and McPeek (1996), it was observed that in a two-patch discrete-time model with equal growth rates on the two patches, chaotic dynamics generally favor the evolution of some amount of unconditional dispersal; if the carrying capacities of the patches are different, chaotic dynamics can support a polymorphism of slower and faster dispersal strategies. Those results were refined and extended in Doebeli and Ruxton (1997) and Parvinen (1999), where it was observed that if growth rates as well as carrying capacities differ between patches then evolutionary branching leading to a polymorphism can occur even if the population dynamics are cyclic. (In situations where patches are ecologically identical, the dispersal rate tends to evolve until the dynamics on the patches are synchronized, after which there is no more selection, so that case is special.)

11.7 Future directions

It would be of interest to study the evolutionary stability of ideal free dispersal relative to other conditional dispersal strategies in spatially varying but temporally constant environments. Using the modeling approach of Cantrell et al. (2006, 2007b), Chen and Lou (2008), and Dockery et al. (1998) in that context would lead to a system of the form of

$$\begin{cases} u_t = \nabla\cdot[\mu\nabla u - \alpha u\nabla f(x,u+v)] + uf(x,u+v) & \text{in } \Omega\times(0,\infty), \\ v_t = \nabla\cdot[\nu\nabla v - \beta v\nabla g(x,u+v)] + vf(x,u+v) & \text{in } \Omega\times(0,\infty), \end{cases} \tag{11.10}$$

with no-flux boundary conditions

$$\mu\frac{\partial u}{\partial n} - \alpha u\frac{\partial f(x,u+v)}{\partial n} = \nu\frac{\partial v}{\partial n} - \beta v\frac{\partial g(x,u+v)}{\partial n} = 0 \quad \text{on } \partial\Omega\times(0,\infty), \tag{11.11}$$

where $f(x,w) = m(x) - b(x)w$ (or perhaps some other or more general form of population growth term with crowding effects) and g represents part of an alternate dispersal strategy. For example, $g = 0$ would correspond to unconditional dispersal by simple diffusion, $g = m$ would correspond to advection up resource gradient without consideration of crowding, $g = -(u+v)$ would correspond to avoidance of crowding without reference to resource distribution, and $g = m - \delta(u+v)$ or $g = m - \delta b(x)(u+v)$ would correspond to a combination of advection up resource gradient and avoidance of crowding.

Many of the results on dispersal in spatially and temporally varying environments or for populations with chaotic dynamics have been obtained through numerical simulation. It would be of interest to extend the range and scope of rigorous analytic results in that area. As noted previously, McPeek and Holt (1992) found that in spatially and temporally varying environments, selection typically favors a certain specific fixed conditional dispersal strategy. It would be of interest to try to see if something similar is true in other types of models. It would also be of interest to examine dispersal strategies which themselves could include variation in time, such as movement along the gradient of a temporally varying resource. Ultimately it might be possible to connect ideas about the evolution of local dispersal in temporally and spatially variable environments to the evolution of migration.

All of the models we have described so far operate on a single trophic level and treat the resource upon which the focal species depends as being extrinsically determined. It is natural to ask how including explicit trophic interactions where the resource itself is dynamic and may even coevolve with the consumer might influence the predictions of models for the evolution of dispersal. In Schreiber et al. (2000) it was shown that in a discrete-time patch model for a host-parasitoid system with coevolution of patch selection, a version of the ideal free distribution is evolutionarily stable. In Cantrell et al. (2007a), balanced dispersal leading to an ideal free distribution was shown to be evolutionarily stable in discrete-diffusion models for predator-prey systems provided that the model incorporates some type of self limitation or intraspecific competition by the predators. It would be of interest to examine extensions of models along the lines of (11.3), (11.4), (11.5), or (11.10) where the resource was explicitly modeled as a dynamic variable and the dispersal strategies of the consumers might include various forms of preytaxis. Two sorts of dispersal that organisms may use to track resources are movement upward along resource gradients and area-restricted search or kinesis, where organisms slow down their movements in regions where resources are dense but speed them up where resources are rare; see Farnsworth and Beecham (1999) and Kareiva and Odell (1987). To compare dispersal strategies for the consumers in such a setting one would use models similar to the following:

$$\begin{cases} u_t = \nabla \cdot [\mu(w)\nabla u - \alpha u \nabla f(u+v,w)] + u(eh(u+v,w) - d), \\ v_t = \nabla \cdot [\nu(w)\nabla v - \beta v \nabla g(u+v,w)] + v(eh(u+v,w) - d), \\ w_t = \nabla \cdot [\rho \nabla w] + (m(x) - w)w - (u+v)h(u+v,w) \end{cases} \tag{11.12}$$

in $\Omega \times (0,\infty)$ with no-flux boundary conditions

$$\mu(w)\frac{\partial u}{\partial n} - \alpha u \frac{\partial f(u+v,w)}{\partial n} = \nu(w)\frac{\partial v}{\partial n} - \beta v \frac{\partial g(u+v,w)}{\partial n} = \frac{\partial w}{\partial n} = 0 \tag{11.13}$$

on $\partial\Omega \times (0,\infty)$. In (11.12) and (11.13) u and v are consumers that are ecologically identical except for their dispersal strategies, w is a resource, and h is a functional response. The diffusion rates for u and v are allowed to depend on w to model area-restricted search. The dispersal terms f and g could incorporate advection up the gradient of w, or of h, or down the gradient of $u + v$. Clearly there are many reasonable variations on the general form shown in (11.12). It would also be possible to model coevolution of dispersal by the consumer and the resource, but that would require a model with four equations. Incorporating trophic interactions more widely into models for the evolution of dispersal would be an interesting but challenging direction for future research.

11.8 Acknowledgments

This research was partially supported by the NSF grants DMS-0816068 (RSC, CC) and DMS-0615845 (YL).

11.9 References

F. Belgacem and C. Cosner (1995), The effects of dispersal along environmental gradients on the dynamics of populations in heterogeneous environment, *Can. Appl. Math. Quart.* **3**:379-397.

R.S. Cantrell and C. Cosner (2003), *Spatial Ecology via Reaction-Diffusion Equations*, Series in Mathematical and Computational Biology, John Wiley and Sons, Chichester, UK.

R.S. Cantrell, C. Cosner, D.L. DeAngelis, and V. Padrón (2007a), The ideal free distribution as an evolutionarily stable strategy, *J. Biol. Dyn.* **1**:249-271.

R.S. Cantrell, C. Cosner, and Y. Lou (2006), Movement towards better environments and the evolution of rapid diffusion, *Math Biosci.* **204**:199-214.

R.S. Cantrell, C. Cosner, and Y. Lou (2007b), Advection mediated coexistence of competing species, *Proc. Roy. Soc. Edinb.* **137A**:497-518.

R.S. Cantrell, C. Cosner, and Y. Lou (2008), Approximating the ideal free distribution via reaction-diffusion-advection equations, *J. Diff. Eqs.* **245**:3687-3703.

X.F. Chen and Y. Lou (2008), Principal eigenvalue and eigenfunction of elliptic operator with large convection and its application to a competition model, *Indiana Univ. Math. J.* **57**:627-658.

X.F. Chen, R. Hambrock, and Y. Lou (2008), Evolution of conditional dispersal: a reaction-diffusion-advection mode, *J. Math Biol.* **57**:361-386.

J. Clobert, E. Danchin, A. Dhondt, and J. Nichols eds., *Dispersal*, Oxford University Press, Oxford, 2001.

C. Cosner (2005), A dynamic model for the ideal free distribution as a partial differential equation, *Theor. Pop. Biol.* **67**:101-108.

C. Cosner and Y. Lou (2003), Does movement toward better environments always benefit a population? *J. Math. Anal. Appl.* **277**:489-503.

R. Cressman and V. Krivan (2006), Migration dynamics for the ideal free distribution, *Am. Nat.* **168**:384-397.

J. Dockery, V. Hutson, K. Mischaikow, and M. Pernarowski (1998), The evolution of slow dispersal rates: A reaction-diffusion model, *J. Math. Biol.* **37**:61-83.

M. Doebeli and G. D. Ruxton (1997), Evolution of dispersal rates in metapopulation models: Branching and cyclic dynamics in phenotype space, *Evolution* **51**:1730-1741.

C.P. Doncaster, J. Clobert, B. Doligez, L. Gustafsson, and E. Danchin (1997), Balanced dispersal between spatially varying local populations: An alternative to the source-sink model, *Am. Nat.* **150**:425-445.

K. Farnsworth and J. Beecham (1999), How do grazers achieve their distribution? A continuum of models from random diffusion to the ideal free distribution using biased random walks, *Am. Nat.* **153**:509-526.

S. Fretwell and H. Lucas Jr. (1970), On territorial behavior and other factors influencing habitat selection in birds: Theoretical development. *Acta Biotheoretica* **19**:16-36.

S. Gandon (1999), Kin competition, the cost of inbreeding, and the evolution of dispersal. *J. Theor. Biol.* **82**:345-364.

S. Gandon and Y. Michalakis, Multiple causes of the evolution of dispersal, in *Dispersal,* ed. by J. Clobert, E. Danchin, A. Dhondt, and J. Nichols, Oxford University Press, Oxford, 2001, pp. 155-167.

P. Grindrod (1988), Models of individual aggregation or clustering in single and multiple-species communities, *J. Math. Biol.* **26**:651-660.

R. Hambrock and Y. Lou, The evolution of mixed dispersal strategies in spatially heterogeneous habitats, *Bull. Math. Biol.*, in revision, 2008.

W.D. Hamilton and R. May (1977), Dispersal in stable habitats, *Nature* **269**:578-581.

I. Hanski, *Metapopulation Ecology*, Oxford Univ. Press, Oxford, 1999.

I. Hanski, Population dynamic consequences of dispersal in local populations and metapopulations, in *Dispersal,* ed. by J. Clobert, E. Danchin, A. Dhondt, and J. Nichols, Oxford University Press, Oxford, 2001, pp. 283-298.

A. Hastings (1983), Can spatial variation alone lead to selection for dispersal? *Theor. Pop. Biol.* **24**:244-251.

M. Heino and I. Hanski (2001), Evolution of migration rate in a spatially realistic metapopulation model, *Am. Nat.* **157**:495-511.

R. Holt and M. Barfield, On the relationship between the ideal free distribution and the evolution of dispersal, in *Dispersal,* ed. by J. Clobert, E. Danchin, A. Dhondt, and J. Nichols, Oxford University Press, Oxford, 2001, pp. 83-95.

R.D. Holt and M.A. McPeek (1996), Chaotic population dynamics favors the evolution of dispersal, *Am. Nat.* **148**:709-718.

V. Hutson and K. Schmitt (1992), Permanence and the dynamics of biological systems, *Math. Biosci.* **111**:1-71.

V. Hutson, S. Martinez, K. Mischaikow, and G.T. Vickers (2003), The evolution of dispersal, *J. Math. Biol.* **47**:483-517.

V. Hutson, K. Mischaikow, and P. Poláčik (2001), The evolution of dispersal rates in a heterogeneous time-periodic environment, *J. Math. Biol.* **43**:501-533.

P. Kareiva and G. Odell (1987), Swarms of predators exhibit "preytaxis" if individual predators use area restricted search, *Am. Nat.* **130**:233-270.

S. Kirkland, C.-K. Li, and S.J. Schreiber (2006), On the evolution of dispersal in patchy environments, *SIAM J. Appl. Math.* **66**:1366-1382.

M. Kshatriya and C. Cosner (2001), A continuum formulation of the ideal free distribution and its implications for population dynamics, *Theor. Pop. Biol.* **81**:277-284.

S.A. Levin, H.C. Muller-Landau, R. Nathan, and J. Chave (2003), The ecology and evolution of seed dispersal: A theoretical perspective, *Annu. Rev. Eco. Evol. Syst.* **34**:575-604.

M.A. McPeek and R.D. Holt (1992), The evolution of dispersal in spatially and temporally varying environments, *Am. Nat.* **140**:1010-1027.

D.W. Morris, J.E. Diffendorfer, and P. Lundberg (2004), Dispersal among habitats varying in fitness: reciprocating migration through ideal habitat selection, *Oikos* **107**:559-575.

V. Padrón and M.C. Trevisan (2006), Environmentally induced dispersal under heterogeneous logistic growth, *Math. Biosci.* **199**:160-174.

K. Parvinen (1999), Evolution of migration in a metapopulation, *Bull. Math. Biol.* **61**:531-550.

H.R. Pulliam (1988), Sources, sinks, and population regulation, *Am. Nat.* **132**:652-661.

D. Roff (1994), Habitat persistence and the evolution of wing dimorphism in insects, *Am. Nat.* **144**:772-798.

S.J. Schreiber, L.R. Fox, and W.M. Getz (2000), Coevolution of contrary choices in host-parasitoid systems, *Am. Nat.* **155**:637-648.

D. Tilman (1994), Competition and biodiversity in spatially structured habitats, *Ecology* **75**:2-16.

H.B. Wilson (2001), The evolution of dispersal from source to sink populations, *Evolutionary Ecology Research* **3**:27-35.

CHAPTER 12

Evolution of dispersal scale and shape in heterogeneous environments: A correlation equation approach

Benjamin M. Bolker
University of Florida

Abstract. Dispersal of offspring to new spatial locations is one of the fundamental ecological processes that determines both the outcome of interspecies competition and the spatial patterns formed in communities of sessile organisms. Classical studies in the evolution of dispersal, motivated by the observation of discrete polymorphisms in dispersal phenotype, have focused on the decision whether to disperse offspring out of the natal patch or not, balancing the risks of dispersal against increased competition in the natal patch. Ecologists have also quantified the distribution of dispersal distance, and on the consequences of different shapes of dispersal distributions (leptokurtic, fat-tailed, etc.); theoreticians have recently begun to study how these shapes could evolve in homogeneous landscapes. This chapter will present simulation and analytical results (using spatial moment equations) on the evolution of dispersal distributions in heterogeneous environments: in particular, I will show how the scale and shape of the spatial autocorrelation of environmental suitability affects the evolution of dispersal. Polymorphism (near vs. far) dispersal can occur either among the offspring of individuals, as shown by a leptokurtic dispersal curve, or among individuals within a population, as shown by an evolutionary branching point in dispersal scale.

12.1 Introduction

Organisms' dispersal ability, the spatial scale over which they relocate themselves or their offspring, is a fundamental life-history trait that determines a population's spatial pattern with respect to kin, conspecific and heterospecific competitors, and environmental gradients. Long-distance dispersal determines species' ability to recolonize new habitats after large-scale disturbance (Clark, 1998), while medium-distance dispersal affects their ability to coexist in relatively stable habitats (e.g., via competition-colonization tradeoffs (Holmes and Wilson, 1998)).

The evolution of dispersal is therefore a long-standing focus of evolutionary ecology

(Clobert et al., 2001; Levin et al., 2003). Studies of dispersal often focus on *natal dispersal*, the relocation of offspring relative to their parents that occurs once per generation (at the seed stage in plants, or often at maturity in animals). Evolutionary ecologists typically ask what natal dispersal strategies we should expect to evolve under different patterns of spatial and temporal heterogeneity in the environment, and under varying ecological constraints, such as costs of dispersal or correlations with other life-history traits (e.g., dispersal in plants is often correlated with seed size, which in turn trades off with seed number (Ezoe, 1998; Levin et al., 2000)). When do we expect less (or shorter-range) vs. more (or longer-range) dispersal (Hamilton and May, 1977; Harada and Iwasa, 1994)? When is there a single optimum dispersal strategy as opposed to a polymorphic evolutionary stable strategy (ESS): in ecological terms, when can multiple dispersal strategies coexist (Levin et al., 1984; Bolker et al., 2003)? How do the advantages of particular dispersal strategies depend on other life-history traits such as competitive ability or growth rate?

Such studies have typically focused on one of three characteristics of dispersal.

- *Propensity* or *rate* — the probability that an offspring will disperse a relatively long distance from its parent, or between suitable patches (Hamilton and May, 1977; Comins et al., 1980; Ludwig and Levin, 1991; Harada and Iwasa, 1994; Travis and Dytham, 1998).
- *Scale* — the typical distance that offspring disperse, quantified as mean or median dispersal distance (Ezoe, 1998; Murrell et al., 2002).
- *Shape* — the distribution of dispersal distances among offspring (i.e., the dispersal *kernel*), often quantified as leptokurtosis, tail shape, or frequency of long-distance dispersal events (Kot et al., 1996; Clark et al., 1999).

These studies provide a clear general picture of the benefits and costs of dispersal. All else equal, higher dispersal propensity (or longer dispersal scale) decreases crowding and therefore lessens kin or intraspecific competition. It minimizes the effects of temporal variation by substituting an average growth rate across spatial sites (an arithmetic average) for the temporal average growth rate that a lineage would experience at a single location (a geometric average); the geometric average is always lower in a variable environment and drops to zero if habitats ever experience lethal environmental conditions. Dispersal also decreases inbreeding depression. The costs of dispersal include the cost of constructing specialized structures such as parachutes or fruits to encourage dispersal, the risk of mortality during dispersal or when settling at a new site, and the risk of dispersing to unsuitable habitat (Cody and Overton, 1996; Bolker and Pacala, 1990; Cheptou et al., 2008).

While the evolutionary dynamics of the rate and scale of dispersal are relatively well understood, we know considerably less about the evolution of the shape of the dispersal kernel. Hovestadt et al. (2001) used an individual-based model to simulate the evolution of dispersal kernels in environments with varying degrees of autocorrelation (patchiness), generated by creating random landscapes with different fractal dimensions. They found that in contrast to homogeneous landscapes, where the dispersal kernel did not change from its initial uniform distribution, autocorrelated landscapes selected for power-law dispersal kernels which allowed a higher proportion of local dispersal while simultaneously spreading those seeds that were dispersed across a broad area. Rousset and Gandon (2002) complemented this study by developing analytical equations that predicted the evolution of dispersal shape under distance-dependent costs of dispersal in a homogeneous environment.

These two studies show the wide range of alternatives for considering the evolution of dispersal shape. While Rousset and Gandon (2002) used analytic tools to derive general conclusions, they did not explicitly consider the effects of environmental heterogeneity (although in a sense their distance-dependent costs of dispersal do include the risk of landing in unsuitable habitat). In contrast, Hovestadt et al. (2001) did consider different types of autocorrelated habitats, but their study was purely computational.

In this chapter I take an approach that complements these two studies. Like Hovestadt et al. (2001), it explicitly incorporates environmental heterogeneity; like Rousset and Gandon (2002), it uses a theoretical framework that allows for more general conclusions. I combine the general approach of adaptive dynamics, asking what strategies are invasible by others, with equations for the dynamics of populations based on the mean densities and spatial correlations of populations. I use this combination to analyze the invasion dynamics of competitors that differ only in the shape or scale of their dispersal kernels in heterogeneous environments and ask the typical questions: Under what conditions are longer vs. shorter or fatter vs. thinner tails advantageous? Why?

12.2 Methods

12.2.1 Competition model

I used a continuous-time stochastic point-process version of the logistic equation (Law et al., 2003). Individuals, which are all identical except for their dispersal kernels, are located at points in space. The model space is one-dimensional with periodic (i.e., wraparound) boundary conditions. While the dynamics of ecological models often depend on their dimensionality, past studies of this type of model have found quantitative, but not qualitative, differences between one and two dimensions (Bolker and Pacala, 1997). In an arbitrarily small time interval $(t, t + \Delta t)$ individuals reproduce with probability $f\Delta t$ and die with probability $(\mu(x) + \alpha C(x))\Delta t$, where $\mu(x)$ is the intrinsic death rate at x (see below for the model's description of environmental heterogeneity) and $C(x)$ is a measure of the local population density. A *competition kernel* $U(r)$ defines the local population density, determining the added mortality of a focal individual due to an individual located a distance r away. If $\{x_i\}$ is the set of locations of all competitors (i.e., all of the N individuals in the population) then the local population density at x is $\sum_{i=1}^{N} U(|x - x_i|)$. If we represent the population as a sum of Dirac delta functions ("spikes" integrating to 1) at these locations, then we can also represent the local density as $\int_{\Omega} U(|x-y|)\,dy$, where Ω is the entire space. I normalize the integral of U to 1, so that α represents the overall strength of competition. In the limit where U is uniform over the entire space (i.e., competition is independent of location) the model reduces to a stochastic logistic equation with population growth rate $f-\bar{\mu}$ and carrying capacity $(f-\bar{\mu})/\alpha$, where $\bar{\mu}$ is the spatially averaged mortality rate.

When individuals reproduce, their offspring disperse in a random direction θ, to a random distance r chosen according to a dispersal kernel $D(r,\theta)$.

I consider only interactions among monomorphic populations, making the usual adaptive dynamics assumptions that the mutation rate is low relative to the time scale of competitive exclusion, so that any new advantageous mutant will fix in the population before the occurrence of a new mutation (Abrams, 2001).

12.2.2 Dispersal curves

In contrast to the nonparametric dispersal kernels of Hovestadt et al. (2001) and Rousset and Gandon (2002), I used a parametric family of kernels, the *general error distribution*, also known as the exponential power distribution [Fig. 12.1] (Mineo and Ruggieri, 2005). I used the parameterization

$$D(r) = \frac{c_1 e^{-(c_1 r)^\theta}}{2\Gamma(1+1/\theta)}, \tag{12.1}$$

where Γ is the Gamma function and $c_1 = \Gamma(2/\theta)/(\theta s(\Gamma(1+1/\theta)))$, which sets the average dispersal distance to s independent of θ.

The general error distribution is well established as a model for dispersal kernels (Clark et al., 1999). Special cases of the general error distribution include the normal or Gaussian ($\theta = 2$) and the Laplacian or radial exponential ($\theta = 1$). Clark et al. (1999) point out that this distribution is limited because convexity close to the source is correlated with leptokurtosis (fat tails). When θ is small, the dispersal kernel is concave at the origin and is leptokurtic; when θ is large, it is convex at the origin and platykurtic.

12.2.3 Environmental heterogeneity

I assume static spatial environmental heterogeneity which is isotropic, homogeneous, and second-order stationary. These criteria essentially mean that the entire spatial pattern can be described by the *autocorrelation function*, which specifies the spatial dependence between two points a distance r apart. In the first part of the paper, I assume a homogeneous environment; in the second, I assume the environmental autocorrelation function is Laplacian ($e^{-c_1 r}$); and in the last, that it is power exponential, which is the spatial analogue of the general error distribution, with $0 < \theta \leq 2$ (Møller and Waagepetersen, 2004).

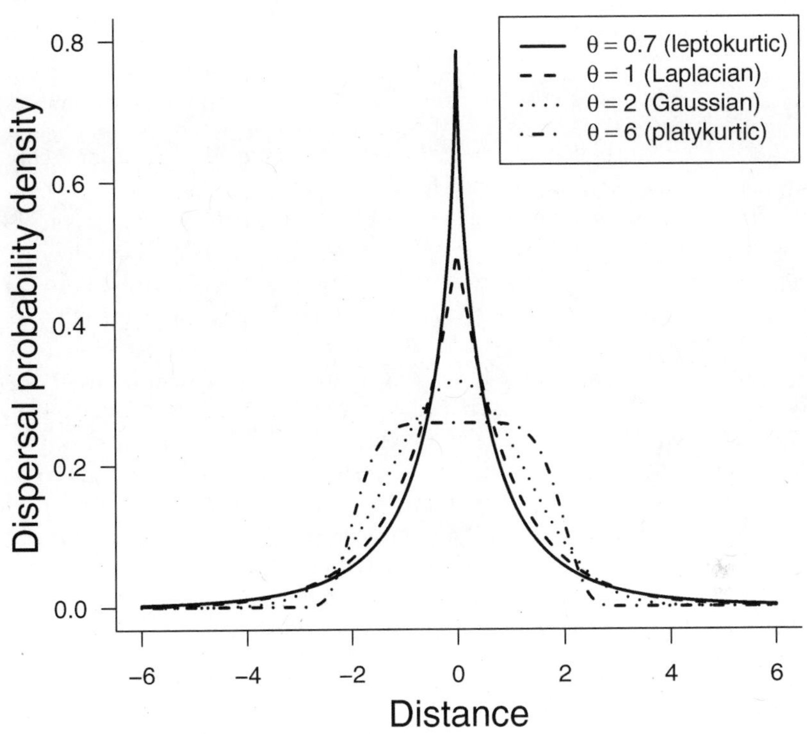

Figure 12.1 Examples of the generalized error distribution with average dispersal distance =1 and varying θ. Curves range from leptokurtic ($\theta < 1$) to platykurtic ($\theta > 2$).

12.2.4 Analysis

The resulting master equations for the number of dispersal type i — that is, the probabilities that the number of individuals at a location x will increase or decrease by 1 during a time interval Δt — are therefore:

$$\text{Prob}(N_i(x) \to N_i(x) + 1) : f \int D(|y - x|, \theta_i) N_i(y) dy, \tag{12.2}$$

$$\text{Prob}(N_i(x) \to N_i(x) - 1) : N_i(x)(\mu(x) + \alpha \int U(|y - x|) \sum_i N_i(y) dy). \tag{12.3}$$

From the master equations we can derive ordinary differential equations that describe the dynamics of n_i, the *average* density of a population with dispersal type i. Because we have assumed that the environment is stationary — that is, there are no gradients or other special points — we will assume that the average across stochastic realizations at any point is equal to that at any other point, and that this average will be equal to the spatially averaged density.

Taking these averages we get:

$$\frac{dn_i}{dt} = (f - \bar{\mu} - \alpha n_i)n_i - c_{\mu n}(0) - \alpha \int U(r) c_{nn}(r)\, dr. \qquad (12.4)$$

In this equation, $c_{\mu n}(0)$ is the covariance between habitat quality (local mortality rate) and population density, or the spatial cross-covariance at lag zero. This term represents *habitat association*, the tendency for populations to build up in favorable environments. Since $\mu(x)$ is the badness (rather than goodness) of the local environment, population density is small in areas where $\mu(x)$ is large, so $c_{\mu n}(0) < 0$, so the habitat association term has an overall positive effect on the population growth rate. Similarly, $\bar{c}_{nn} = \int U(r) c_{nn}(r)\, dr$ is a local crowding term, analogous to Lloyd's crowding index, which averages the local increase in population density of all dispersal types (if $\bar{c}_{nn} > 0$) around focal individuals, weighted by the strength of competition, to determine the negative effect of crowding. At equilibrium, the crowding term may be either positive or negative, corresponding to clustering ($\bar{c}_{nn} > 0$) or thinning ($\bar{c}_{nn} < 0$), depending on the relative strength of short dispersal (which increases crowding) and local competition (which decreases it) (Bolker and Pacala, 1999). (The same general framework could apply to conspecific facilitation — for example, through increased local densities of pollinators or other mutualists — if α were negative, but here I consider only competitive interactions.)

These equations can be solved if the spatial pattern of the population determining habitat association and crowding are known, but in general we have to extend the system of equations to include integrodifferential equations for the dynamics of the spatial covariances c_{nn} and $c_{\mu n}$ (the habitat pattern, described by the habitat autocovariance $c_{\mu\mu}$, is assumed to be static). These equations in turn involve triple correlations among the population density and habitat at different points. Some form of *moment closure* is required in order to close the system and get a set of equations we can analyze or (as in this case) solve numerically. The technical details of moment closure are discussed elsewhere: here we use a power-2 symmetric closure (Bolker, 2003), which like other moment closure schemes derives an approximation for the triple covariance among three points from the known covariances among each pair of points. I developed Mathematica code that derives the moment equations for a specified set of master equations and optionally uses the xtc numerical solver (by Bard Ermentrout) to compute population trajectories, equilibrium values, or invasion rates: all code is available at `www.zoo.ufl.edu/bolker/meqs/emonk/`.

In order to analyze the effects of different dispersal strategies I held the nonspatial parameters fixed at $\{f = 5, \bar{\mu} = 1, \alpha = 1\}$, corresponding to a carrying capacity of $K = 4$ and an intrinsic growth rate of $f - \bar{\mu} = 4$.

In spatial stochastic models, the intrinsic growth ratio $R = f/\bar{m}u$ determines the importance of stochasticity: the baseline value here, $R = 5$, represents a moderately large growth ratio. For example, for density-dependent establishment (as analyzed in Bolker and Pacala (1999)) a monoculture equilibrium is clustered if $R < 2$ and thinned (more even than random) if $R > 2$.

The baseline values of the spatial parameters in the model are the competition scale parameter (m_c) = 1; competition shape parameter (θ_c) = 1; dispersal scale parameter (m_d) = 1; dispersal shape parameter (θ_d) = 1; environment scale parameter (m_e) = 1; and environment shape parameter (θ_e) = 1. In other words, unless otherwise specified, all spatial kernels are Laplacian with scale 1. The competition kernel (U) was held constant throughout the analyses. While the scale m_c can be fixed at 1 without loss of generality, since it determines the overall scale of the system, changing the shape of the competition kernel θ_c does affect the dynamics (Birch and Young, 2006), but the possible interaction of dispersal, environment, and competition shape parameters is left for future studies.

For each set of parameters of interest, I numerically solved the equations for monoculture equilibrium, running the equations out until the population densities and correlations were stable to within a convergence tolerance of 10^{-4}. To determine invasibility, I ran the resident species out to equilibrium, then started the second species at a density of 10^{-4} with all covariances equal to zero (i.e., assuming a random initial distribution with respect to residents, conspecifics, and the environment), ran the equations out to $t = 1$ — potentially enough time for the population to increase by a factor of $e^r = e^4 \approx 55$ in the absence of competition — and calculated the proportional increase in density, (final-initial)/initial. To determine the evolutionary dynamics as a whole, I constructed pairwise invasion plots (PIP) that evaluate the invasion rate at each point on a grid representing combinations of resident and invader dispersal traits. In general, the geometries of PIPs determine which dispersal strategies or combinations of dispersal strategies will evolve in the long run, under the assumptions of adaptive dynamics that mutations are infrequent but not limiting.

12.3 Results

12.3.1 Dispersal scale in homogeneous landscapes

As expected from previous studies, longer dispersal is always better in a homogeneous environment if it does not bear a cost. Analyzing equilibrium population densities as a function of competition (m_c) and dispersal scale (m_d) shows that at any competition scale, increasing dispersal scale maximizes the population density (Fig. 12.2). Populations are generally evenly distributed ($\bar{c}_{nn} < 0$), with population densities greater than the nonspatial carrying capacity, when dispersal scale is large and competition scale is small, and clustered or aggregated ($\bar{c}_{nn} > 0$) under the opposite conditions. At large dispersal scales, the population density is highest for intermediate scales of competition: however, this analysis focuses on dispersal evolution and assumes that the scale of competition is fixed rather than evolving to an optimal value.

The first concern about this result, although it is completely expected, is to make sure that the moment closure approximation is reasonably consistent with the actual dynamics of the system. While the moment approximation overpredicts the equilibrium density for small dispersal scales and underpredicts it for large dispersal scales, the

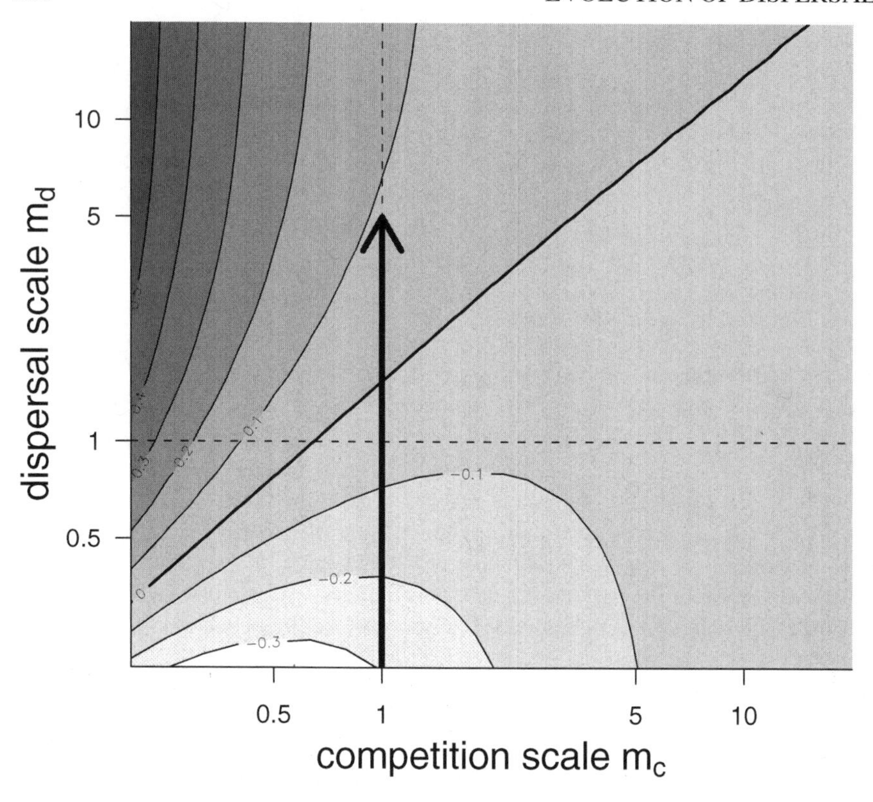

Figure 12.2 Equilibrium density (relative to nonspatial carrying capacity) as a function of dispersal and competition scale for a single species in a homogeneous environment. Thick line represents parameters where the effects of crowding and thinning cancel and the population density equals the nonspatial carrying capacity. Dashed lines show $m_d = 1$ and $m_c = 1$.

qualitative pattern is correct (Fig. 12.3). (The moment approximations fail badly for even smaller scales than those shown, where the simulated population goes extinct while the moment approximation predicts a gradual decrease in density.)

The second concern is whether we can really predict evolution trajectories from knowledge of equilibrium population sizes. Evolutionary studies often confound the naïve expectation that a population will evolve in the direction that maximizes population density; "luxury consumption" in order to reduce resources and deny them to competitors is one simple example (van Wijk et al., 2003). In this case, however (and in all cases explored) the PIP (not shown) agrees with the equilibrium density plot. At all values of the resident dispersal scale, the invader growth rate is positive if and only if it disperses farther than the resident. The expectation of adaptive dynamics is thus that mutants with progressively larger dispersal scales will invade and displace the shorter-dispersing resident population. (Because this concordance between traits

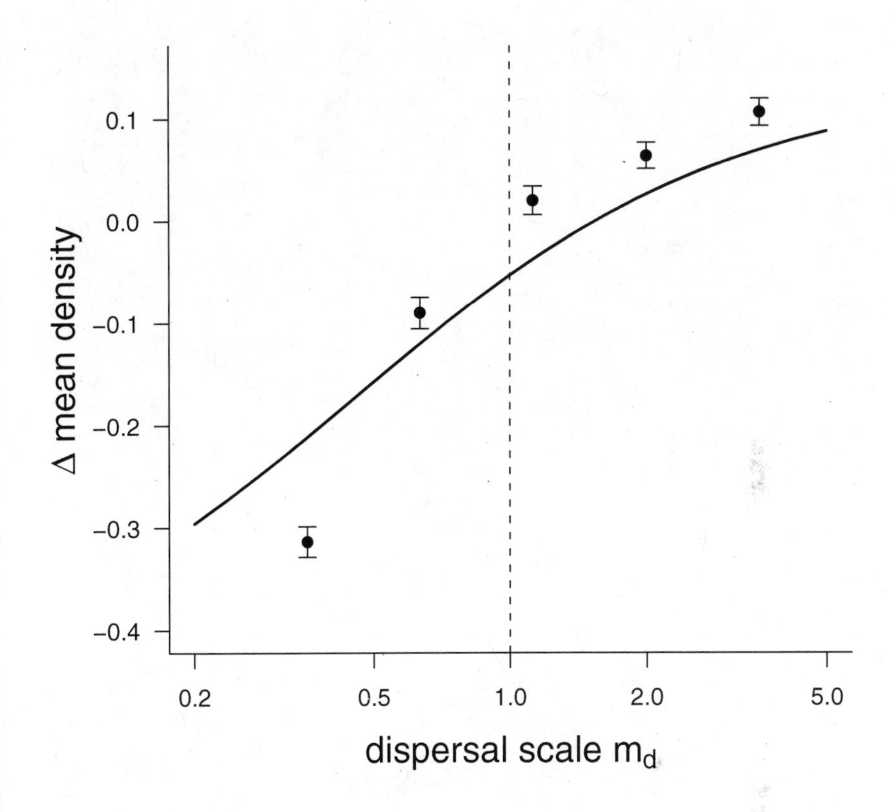

Figure 12.3 Comparison of predicted density (line: deviation from nonspatial carrying capacity $K = 4$ from moment equations) with simulation results (points with error bars: total time=100, transient=25, time step=0.01, total length=128). Parameters as in text, with $m_c = 1$ (corresponding to thick arrow in Fig. 12.2), $\theta_d = \theta_c = 1$. Error bars show ± 1 standard error.

leading to higher equilibrium densities and evolutionarily favored traits holds for all the cases I have explored in this study, I have not shown the PIPs.)

12.3.2 Dispersal shape in homogeneous environments

What about the evolution of dispersal shape? I analyzed the equilibrium density and invasion rate for dispersal shape parameters θ_d ranging from 0.25 (strongly leptokurtic) to 3 (moderately platykurtic), for competition scales (m_c) from 0.2 to 100. For any scale of competition, platykurtic dispersal ($\theta_d \to \infty$) gives the highest equilibrium density.

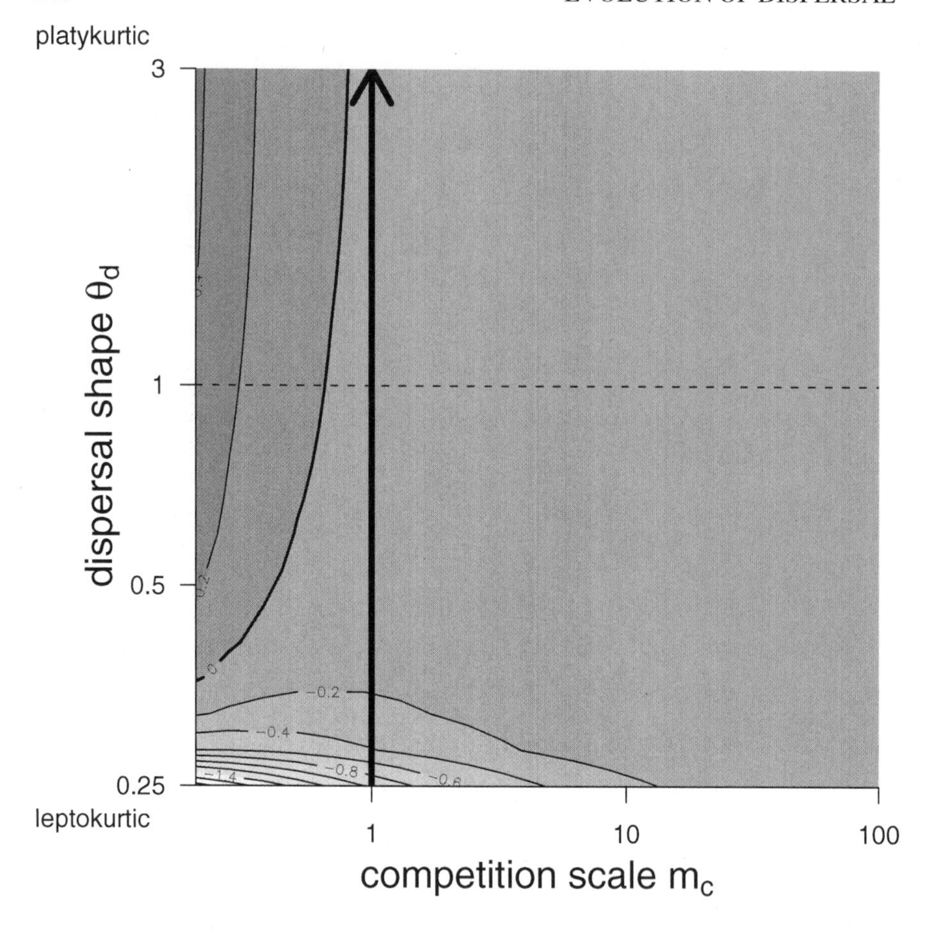

Figure 12.4 Equilibrium density (relative to nonspatial carrying capacity) as a function of dispersal shape and competition scale in a homogeneous environment. Thick line, density equal to nonspatial carrying capacity.

As with the evolution of dispersal scale, we need to check the accuracy of the moment closure approximation. Fig. 12.5 shows that it is reasonable, although the moment equations predict a sharp drop-off in equilibrium density for $\theta_d < 0.3$ while the simulations show a more gradual decline. Numerical analyses of the invasion criteria again agree with the equilibrium densities: for a resident with any dispersal kernel shape, a mutant invader with a more platykurtic dispersal kernel can invade.

12.3.3 Dispersal scale in heterogeneous environments

Moving to a heterogeneous environment (beginning with the baseline case of exponential spatial autocorrelation in mortality rate, $\theta_e = 1$), we see a more interesting

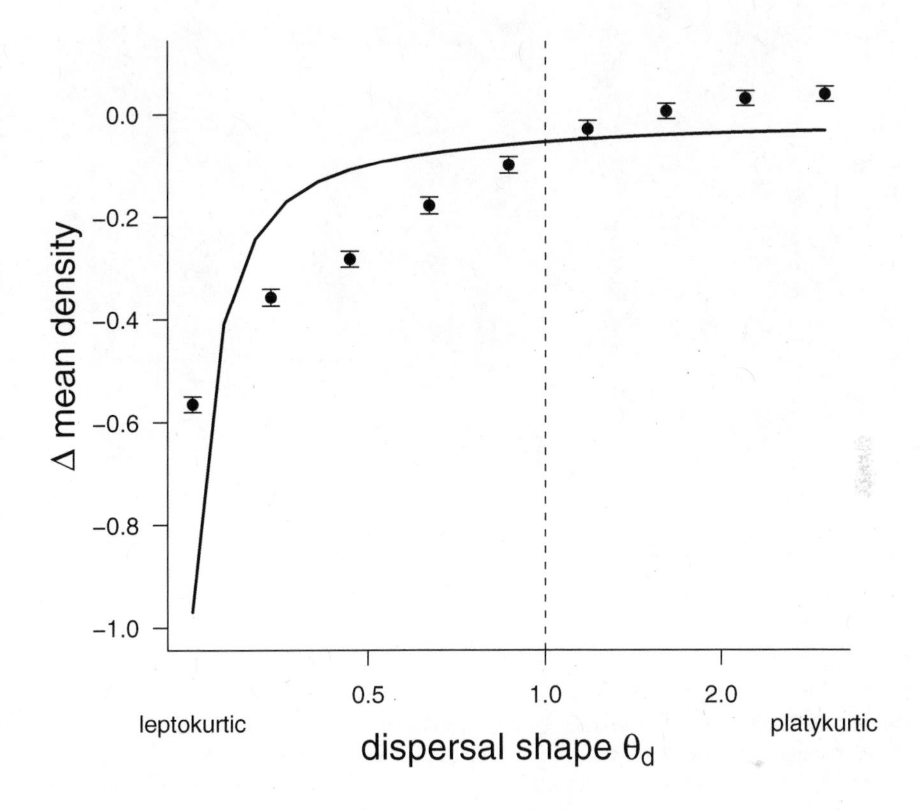

Figure 12.5 Comparison of predicted density from moment equations (lines) with simulation results (points): simulation parameters as in Fig. 12.3, except $m_d = 1$.

pattern where population density no longer changes monotonically with dispersal scale. For small competition scales (on the left side of Fig. 12.6), the system is *competition dominated*: that is, the effects of local competition are much stronger than the effects of environmental variation, and the effect of dispersal scale is similar to the homogeneous case, with long dispersal leading to higher equilibrium population densities. For large competition scales (on the right side of Fig. 12.6), the system is *environment dominated*: local competition is unimportant and the effects of environmental variation dominate, increasing equilibrium density through habitat association when the dispersal scale is short. For long competition scales ($m_c > 3$), Fig. 12.6 shows that a *minimum* population density occurs at intermediate dispersal scales.

Fig. 12.7 compares the moment equation predictions to simulation results, confirm-

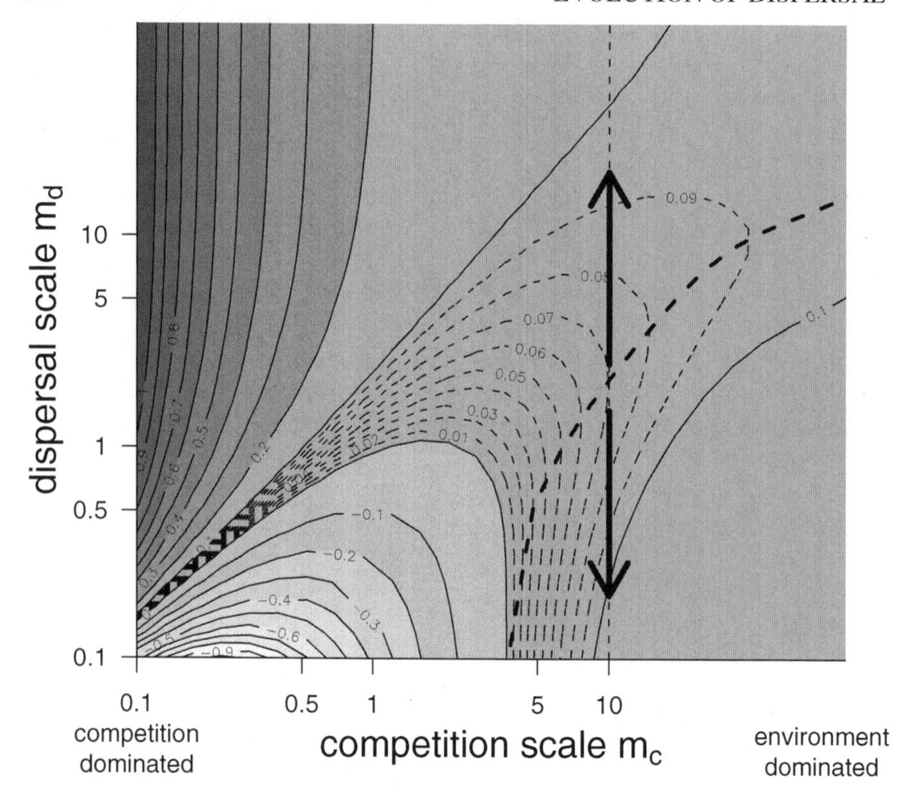

Figure 12.6 Equilibrium density (deviation from nonspatial carrying capacity) as a function of dispersal and competition scale for a single species in a heterogeneous environment ($m_e = 1$, $\theta_e = 1$). Solid lines show density contours at a spacing of 0.1, while (thin) dashed lines show density contours at a spacing of 0.01 for the region between 0 and 0.1. Heavy dashed line shows the location of the minimum density for each competition scale. Arrows show predicted direction of dispersal evolution.

ing the existence of a minimum in population density at intermediate dispersal scales. The scale of the effect is small (from a minimum of 0.05 density units greater than the nonspatial carrying capacity $K = 4$ up to a maximum of about 0.11). For very short dispersal scales ($m_d \leq 0.2$), the simulations show that clustering causes the population density to collapse in a way that is not captured by the moment equations.

Numerical analyses again confirm that the minimum in population density corresponds to an evolutionary branching point in the dispersal scale. Selection will drive populations starting above a dispersal scale of $m_d \approx 2$ to larger dispersal scales and those starting below to lower dispersal scales. Depending on the genetic and mating systems, such a divergence point may lead to an evolutionarily stable polymorphism in the population.

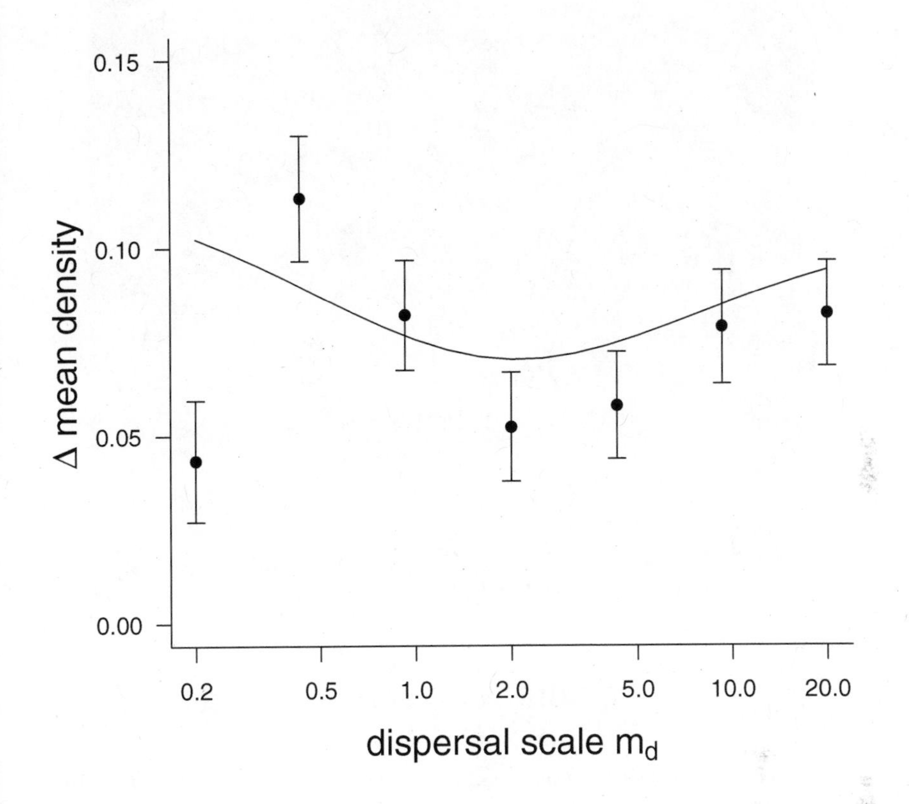

Figure 12.7 Comparison of predicted density from moment equations with simulation results: simulation parameters as in Fig. 12.3, except $m_d = m_e = 1$, $m_c = 10$, $\theta_d = \theta_e = \theta_c = 1$.

12.3.4 Dispersal shape in heterogeneous environments

The effects of dispersal shape on equilibrium density in a heterogeneous environment are similar to those of dispersal scale (Fig. 12.8). In competition-dominated systems ($m_c \ll m_e = 1$), the results are as for homogeneous environments, with platykurtic dispersal (high θ_d) giving the highest density. In environment-dominated systems ($m_c \gg m_e = 1$), the situation reverses and leptokurtic dispersal (low θ_d) maximizes density. In a transition range ($2 < m_c < 20$), an intermediate θ_d maximizes density (in contrast to the dispersal-scale case, where an intermediate value minimized density).

Simulation results are consistent with the moment equation predictions for $\theta_c = 10$ (Fig. 12.9). However, this cross-section also shows that the predicted maximum in density is closer to a threshold function of dispersal shape, with the predicted density

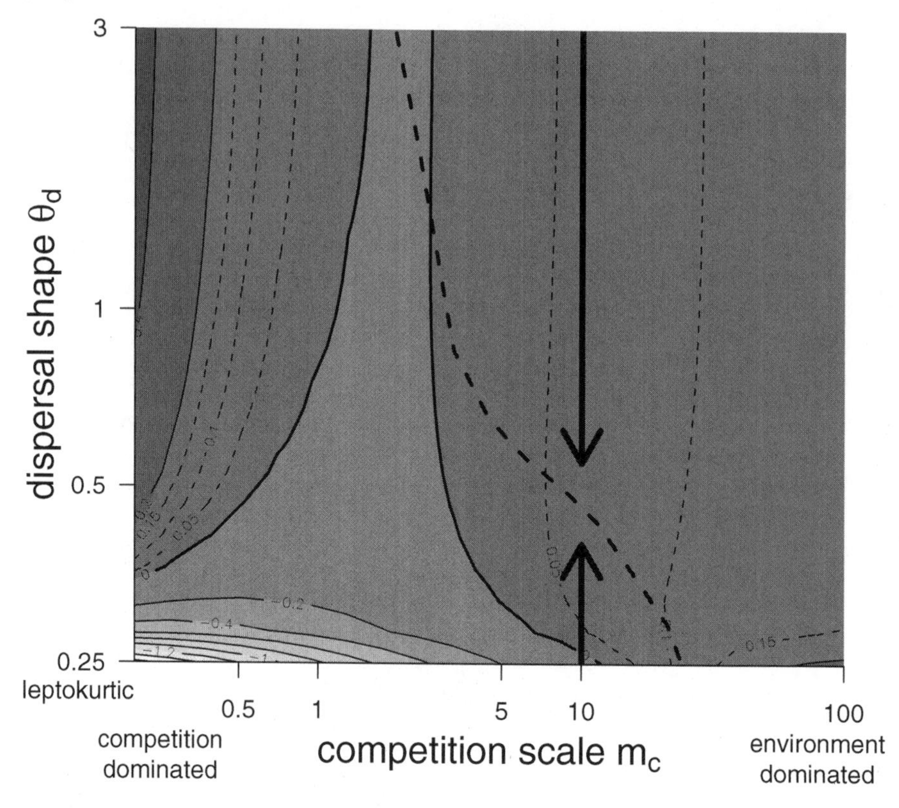

Figure 12.8 Equilibrium density as a function of competition scale and dispersal shape in a heterogeneous environment ($m_e = \theta_e = \theta_c = m_d = 1$).

barely dropping from its maximum value at ($\theta_d = 0.5$) as dispersal shape increases. The simulation results are noisy enough that the expected maximum is not detectable, if it exists.

12.4 Discussion and conclusions

How can we understand the above results in terms of the well-understood costs and benefits of dispersal?

Since these analyses neglect the costs of dispersal structures and the risk of dispersal mortality on the one hand and the benefits of avoiding inbreeding depression on the other, the only cost/benefit that applies in a homogeneous environment is avoidance of kin competition. The optimal dispersal scale, therefore, is to send one's offspring as far as possible — avoiding both intergenerational competition by putting offspring far away from the parent and sibling competition by spreading offspring far

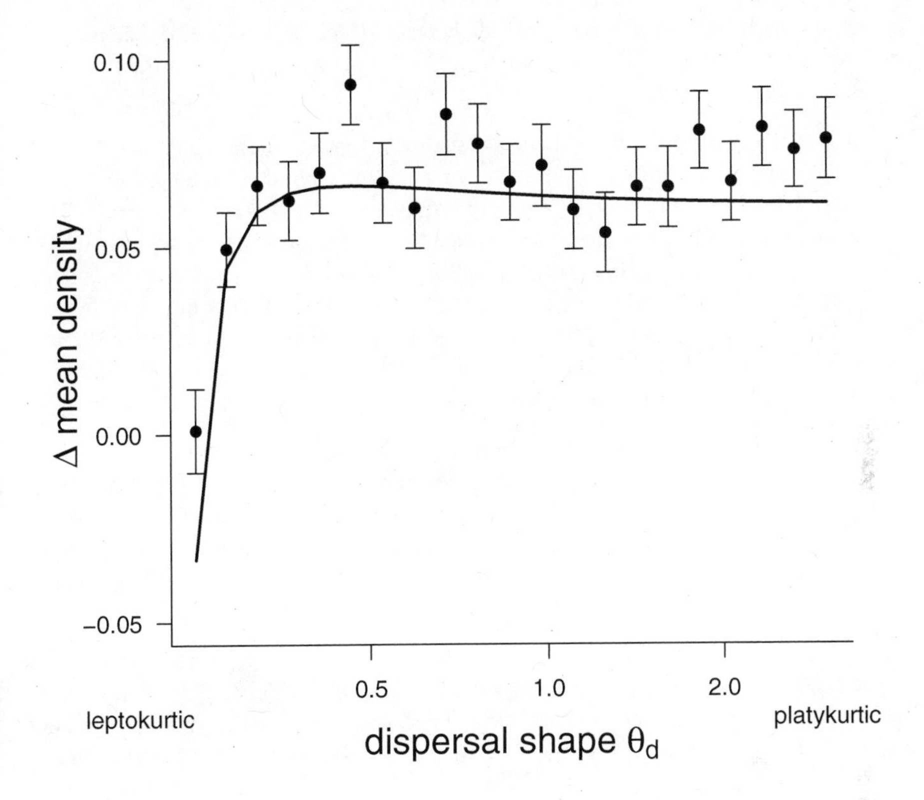

Figure 12.9 Comparison of predicted density from moment equations with simulation results: simulation parameters as in Fig. 12.3, except $m_d = m_e = 1$, $m_c = 10$, $\theta_e = \theta_c = 1$.

from each other. That the optimal dispersal shape in a homogeneous environment is platykurtic makes sense if the total dispersal scale is constrained: then the next best choice is to spread the offspring as evenly as possible.

In heterogeneous environments, the benefit of avoiding competition must now be balanced against the costs of dispersing one's offspring to a less favorable environment. An organism that is successfully surviving and reproducing is more likely to be located in favorable habitat, so sending its offspring away worsens their prospect. This benefit of short dispersal — habitat association (Bolker, 2003) or growth-density covariance (Snyder and Chesson, 2003) — is general, applying both to abiotic heterogeneity and to competition with slow-moving or short-dispersing but dominant competitors (Neuhauser, 1998; Bolker et al., 2003). How should organisms compromise between short and long dispersal to balance costs and benefits?

In the case of dispersal scale, when competition and environmental scales are similar (so that organisms must deal with both crowding and environmental effects), the correlation equations and simulations suggest a polymorphic strategy at the level of the population. Either longer or shorter dispersal can increase population densities and invasion rates, but an intermediate dispersal — trying to split the difference — minimizes fitness.

In contrast, the density plots for dispersal scale suggest an intermediate maximum for similar competition and environmental scales, a single optimal strategy. However, this intermediate maximum, at moderately high levels of leptokurtosis ($\theta_d \approx 0.5$), can be thought of as representing an *individual-level* polymorphism. As suggested by Snyder and Chesson (2003) in the context of temporal variation, leptokurtic dispersal allows an individual to send its offspring either short distances (offspring in the body of the dispersal kernel) or long (offspring in the tail). Because the density curve is nearly flat above a threshold value (12.9), we would expect to see a broad range of moderately to strongly leptokurtic dispersal kernels.

This evolutionary dynamic has consequences for the theory of dispersal itself: when dispersal kernels are strongly leptokurtic, then the extensive theory developed for inter-patch or "near vs. far" dispersal (e.g., Levin et al., 1984) should apply well because most offspring travel either a short or a long distance. Thus the evolutionary process itself may push organisms into a parameter regime where simpler theoretical frameworks can illuminate their interactions and further evolution.

The much-discussed tendency of leptokurtosis and "fat tails" to accelerate wave-like spread (Clark, 1998), on the other hand, may *not* be an important determinant of dispersal evolution. While a rapidly spreading population would be the first to colonize a newly opened habitat, this advantage would not persist long in a stable habitat mosaic of habitat patches. In Snyder and Chesson's (2003) analysis, the higher moments of dispersal kernels (skew, kurtosis, etc.) have extremely weak effects on coexistence. (The difference between their results and the present analysis, where kurtosis is favored, is that their analysis does not take intraspecific crowding into account.)

Being able to derive the equations for the dynamics of spatial correlations among species and between species and environmental factors has allowed an efficient survey of different dispersal kernels in a range of environmental conditions. Correlation equations also allow for analytical insights, in simple enough systems. In trying to tell a reasonably complete story about the effects of dispersal scale and shape in heterogeneous environments I have of course neglected several important factors, especially the cost of dispersal and the effects of temporal heterogeneity. Temporal heterogeneity in particular raises the problem that one cannot properly understand dispersal without considering the correlated life-history traits that make up a complete life-history strategy (Snyder, 2006): how do dormancy, growth rate, time to maturity, fecundity, and dormancy interact with dispersal strategies? What are the phenotypic and genotypic correlations among dispersal characters and other life-history traits (Rees, 1996)? I have also assumed that dispersal is neither density-dependent (Travis et al., 1999) nor directed (Wenny, 2001). While plants (the implicit focus of much of the literature on evolution of dispersal) may not appear to have much control over their dispersal, animal-dispersed seeds do have a great deal of control over their dispersers (and thus indirectly over the environment they land in), and even wind-dispersed seeds are affected by local density (Schurr et al., ms. in review).

Even after several decades the theory of spatial competition and evolution in heterogeneous environments remains incomplete, but new analytical tools open new avenues for understanding how organisms can balance the costs and benefits of dispersal at different scales. Many more challenges remain, including further integration of the theory with data and incorporation of other life history traits.

12.5 Acknowledgments

I am grateful to M. McCoy and M. Brooks for comments on the manuscript, and to B. Ermentrout (xtc), M. Schlather (RandomFields R package), and the R community (R Development Core Team, 2007) for developing useful software tools.

12.6 References

P. A. Abrams (2001), Modelling the adaptive dynamics of traits involved in inter- and intraspecific interactions: An assessment of three methods. *Ecology Letters* **4**:166–175.

D. A. Birch and W. R. Young (2006), A master equation for a spatial population model with pair interactions, *Theoretical Population Biology* **70**:26–42.

B. M. Bolker and S. W. Pacala (1997), Using moment equations to understand stochastically driven spatial pattern formation in ecological systems, *Theoretical Population Biology* **52**:179–197.

B. M. Bolker and S. W. Pacala (1999), Spatial moment equations for plant competition: Understanding spatial strategies and the advantages of short dispersal. *American Naturalist* **153**:575–602.

B. M. Bolker (2003), Combining endogenous and exogenous spatial variability in analytical population models, *Theoretical Population Biology* **64**:255–270.

B. M. Bolker (2004), Continuous-space models for population dynamics, in *"Ecology, Genetics, and Evolution of Metapopulations,"* ed. by I. Hanski and O. E. Gaggioti, Elsevier Science, San Diego, CA, pp. 45-69.

B. M. Bolker, S. W. Pacala, and C. Neuhauser (2003), Spatial dynamics in model plant communities: What do we really know? *American Naturalist* **162**(2):135–148.

P.-O. Cheptou, O. Carrue, S. Rouifed, and A. Cantarel (2008), Rapid evolution of seed dispersal in an urban environment in the weed *Crepis sancta*, *Proceedings of the National Academy of Sciences of the USA* **105**:3796–3799.

J. S. Clark (1998), Why trees migrate so fast: Confronting theory with dispersal biology and the paleorecord, *American Naturalist* **152**(2):21.

J. S. Clark, M. Silman, R. Kern, E. Macklin, and J. HilleRisLambers (1999), Seed dispersal near and far: Patterns across temperate and tropical forests, *Ecology* **80**(5):1475–1494.

J. Clobert, E. Danchin, A. A. Dhondt, and J. D. Nichols (2001), *Dispersal*. Oxford University Press, Oxford.

M. L. Cody and J. M. Overton (1996), Short-term evolution of reduced dispersal in island plant populations, *Journal of Ecology* **84**(1):53–61.

H. N. Comins, W. D. Hamilton, and R. M. May (1980), Evolutionarily stable dispersal strategies, *Journal of Theoretical Biology* **82**:205–230. Reprinted as Chapter 16 in *Narrow Roads of Gene Land: The Collected Papers of W.D. Hamilton*, vol. 1, "Evolution of Social Behavior" (1996), W.H. Freeman, New York.

H. Ezoe (1998), Optimal dispersal range and seed size in a stable environment, *Journal of Theoretical Biology* **190**(3):287–293.

W.D. Hamilton and R. M. May (1977), Dispersal in stable habitats, *Nature*, **269**:578–581. Reprinted as Chapter 12 in *Narrow Roads of Gene Land: The Collected Papers of W.D. Hamilton*, vol. 1, "Evolution of Social Behavior" (1996), W.H. Freeman, New York.

Y. Harada and Y. Iwasa (1994), Lattice population dynamics for plants with dispersing seeds and vegetative propagation, *Researches in Population Ecology* **36**(2):237–249.

E.E. Holmes and H.B. Wilson (1998), Running from trouble: Long-distance dispersal and the competitive coexistence of inferior species, *American Naturalist* **151**(6):578–586.

T. Hovestadt, S. Messner, and H. J. Poethke (2001), Evolution of reduced dispersal mortality and 'fat-tailed' dispersal kernels in autocorrelated landscapes, *Proceedings of the Royal Society of London B* **268**(1465):385–391.

M. Kot, M. A. Lewis, and P. van den Driessche (1996), Dispersal data and the spread of invading organisms, *Ecology* **77**:2027–2042.

R. Law, D. J. Murrell, and U. Dieckmann (2003), Population growth in space and time: Spatial logistic equations, *Ecology* **84**:252–262.

S. A. Levin, D. Cohen, and A. Hastings (1984), Dispersal strategies in patchy environments, *Theoretical Population Biology* **26**:165–191.

S. A. Levin and H. C. Muller-Landau (2000), The evolution of dispersal and seed size in plant communities, *Evolutionary Ecology Research* **2**:409–435.

S. A. Levin, H. C. Muller-Landau, R. Nathan, and J. Chave (2003), The ecology and evolution of seed dispersal: A theoretical perspective, *Annual Review of Ecology, Evolution and Systematics* **34**:575–604.

D. Ludwig and S. A. Levin (1991), Evolutionary stability of plant communities and the maintenance of multiple dispersal types, *Theoretical Population Biology* **40**(3):285–307.

A. M. Mineo and M. Ruggieri (2005), A software tool for the exponential power distribution: The normalp package, *Journal of Statistical Software* **12**:1–24.

J. Møller and R. P. Waagepetersen (2004), *Statistical Inference and Simulation for Spatial Point Processes*, CRC Press.

D. J. Murrell, J. M. J. Travis, and C. Dytham (2002), The evolution of dispersal distance in spatially-structured populations, *Oikos* **97**:229–236.

C. Neuhauser (1998), Habitat destruction and competitive coexistence in spatially explicit models with local interactions, *Journal of Theoretical Biology* **193**(3):445–463.

R Development Core Team (2007), *R: A Language and Environment for Statistical Computing*, R Foundation for Statistical Computing, Vienna, Austria. ISBN 3-900051-07-0.

M. Rees (1996), Evolutionary ecology of seed dormancy and seed size, *Philosophical Transactions of the Royal Society B* **351**:1299–1308.

F. Rousset and S. Gandon (2002), Evolution of the distribution of dispersal distance under distance-dependent cost of dispersal, *Journal of Evolutionary Biology* **15**(4):515–523.

R. E. Snyder (2006), Multiple risk reduction mechanisms: Can dormancy substitute for dispersal? *Ecology Letters* **9**:1106–1114.

R. E. Snyder and P. Chesson (2003), Local dispersal can facilitate coexistence in the presence of long-lasting spatial heterogeneity, *Ecology Letters* **6**:1–9.

J. M. J. Travis and C. Dytham (1998), The evolution of dispersal in a metapopulation: A spatially explicit, individual-based model, *Proceedings of the Royal Society B* **265**(1390):17–

23.

J. M. J. Travis, D. J. Murrell, and C. Dytham (1999), The evolution of density-dependent dispersal, *Proceedings of the Royal Society B* **266**(1431):1837–1842.

M. T. van Wijk, M. Williams, L. Gough, S. E. Hobbie, and G. R. Shaver (2003), Luxury consumption of soil nutrients: a possible competitive strategy in above-ground and below-ground biomass allocation and root morphology for slow-growing arctic vegetation? *Journal of Ecology* **91**:664–676.

D. G. Wenny (2001), Advantages of seed dispersal: A re-evaluation of directed dispersal, *Evolutionary Ecology Research* **3**:51–74.

CHAPTER 13

Spatiotemporal dynamics of measles: Synchrony and persistence in a disease metapopulation

Alun L. Lloyd
North Carolina State University

Lisa Sattenspiel
University of Missouri

Abstract. Measles incidence records provide some of the most detailed accounts of the spatiotemporal dynamics of a population. In a few situations they describe the ebb and flow of the disease at a large number of spatial locations at weekly or monthly intervals over a period of several decades. The interaction between a naturally damped nonlinear predator-prey oscillation and seasonal variation in transmission leads to the occurrence of a rich set of dynamical behavior in the incidence of the infection. Before the introduction of mass vaccination against the disease, large multi-annual oscillations in incidence (i.e., recurrent epidemics) were commonly seen in large cities of the developed world. Such oscillations were not characteristic of all locations, however; many smaller population centers exhibited irregular outbreaks interspersed by periods of absence of the infection. Work on identifying and characterizing the processes that govern the transmission dynamics of measles has been driven by attempts to understand the observed incidence patterns. In particular, three primary questions have emerged: 1) what governs the long-term dynamics of the disease within communities and allows oscillatory behavior to be maintained in some populations, but to be less apparent or absent in others, 2) what determines whether and how long measles can persist within a community, and 3) what determines the distribution and spatial spread of measles outbreaks among communities across a region. In this chapter we review the literature that has used statistical analyses and mathematical models to address these questions.

13.1 Introduction

The transmission dynamics of measles have long held a fascination for population modelers. The availability of high quality epidemiological data, sampled at large numbers of spatial locations and covering several decades, provides an almost unrivalled opportunity to study spatiotemporal dynamics in a population biology setting (Grenfell and Harwood, 1997).

The measles system exhibits a rich set of dynamical behavior, resulting from an interaction between a naturally damped nonlinear predator-prey oscillation and seasonal variations in transmission (Dietz, 1976). In large cities of the developed world, particularly before the introduction of mass vaccination against the infection, large multiannual oscillations in the incidence of disease—so-called recurrent epidemics—were commonly seen (Hamer, 1906; Soper, 1929). Notably, however, such oscillations were not uniform across all populations. Outside the large cities of Western Europe and the United States these patterns were less regular or even absent altogether (Bartlett, 1957; Cliff et al., 1981).

The development of mathematical models of the transmission of measles was stimulated by a desire to understand the underlying factors that generated these observed patterns. In particular, three primary questions have provided the motivation for most measles modeling work: 1) what governs the long-term dynamics of the disease within communities, and especially what causes regular measles cycles to be maintained in some populations, but to be less apparent or absent in others, 2) what determines the persistence of infection in communities, and 3) what determines the distribution and spatial spread of measles outbreaks among communities across a region.

Research on the epidemiology of measles as well as early mathematical modeling work identified several factors that are important in determining local persistence and regular cycling of the disease. For example, the widely differing timescales between the infection process, with infection lasting on the order of weeks, and the replacement of susceptibles by demographic turnover, means that stochastic effects have a major impact on the dynamics of measles. Thus, the patterns of incidence within a given community are strongly dependent on its size. This behavior was investigated by Bartlett (1956, 1957, 1960), who used both mathematical analysis and analysis of incidence data to determine that unless populations were of sufficient size (the critical population size), the measles transmission chain could not be maintained over time, leading to extinction of the disease rather than persistence in an endemic state. For cities that are large enough to maintain the infection, stochastic effects alone would lead to fluctuations in incidence about an endemic level, but cannot explain the tight periodicity of disease incidence seen in large Western cities. Further research (London and Yorke 1973; Yorke and London 1973) identified the important role that seasonality in disease transmission plays in shaping local patterns of measles incidence.

While simple mathematical models can separately depict either persistence patterns or dynamical patterns rather easily, the development of a model that can simultaneously perform these two tasks has proved to be more of a challenge (Bolker and Grenfell, 1995). This has led modelers to focus on disease patterns at the regional, rather than the community, level. Viewed at this scale, spatial dynamics play a crucial role in the persistence of measles, since transport of infection between cities can reintroduce infection following a local stochastic extinction. The degree to which spatial effects can enhance persistence of infection depends in an important way on the synchrony of epidemics in different cities. If epidemics are synchronized, then extinctions are likely to occur simultaneously across a region, eliminating the possibility of reintroduction.

In this chapter, we review the literature on the transmission dynamics of measles, highlighting how statistical analyses and mathematical models have been used to address the questions posed above. We begin by describing the primary data sets that have been used to study the long-term patterns of measles and discuss the role played by the analysis of those data in the development of mathematical models for the dynamics. We then return to our earlier questions to assess the role of mathematical models in developing an understanding of the spatial dynamics of measles within and among communities.

13.2 Data sources

Detailed records of the incidence of measles (i.e., the number of new cases per unit time) are available for many countries, providing high quality temporal and, in many cases, spatiotemporal data (Cliff *et al.*, 1993). Since measles has long been a notifiable disease in many of these locations, incidence records often span a considerable time period. Most studies of the spatiotemporal spread of measles have considered one of three primary data sources: a) weekly reports of measles cases from cities and villages in England and Wales, b) monthly or weekly reports of measles cases in the United States, and c) monthly reports of measles cases from the island of Iceland.

The majority of studies aimed at exploring the long-term persistence of measles within communities and the role of synchrony in regional patterns of incidence have drawn on the England and Wales data. These data have been recorded in reports of the Registrar General, including the Registrar General's Weekly Return. Measles has been a notifiable disease in England and Wales since 1889; data exist for each local authority area from this time and also for selected urban areas from a few decades earlier. These data record measles notifications during disease outbreaks at a fine scale in both time (weekly reports) and space (from large cities, down through small towns, villages and rural areas). Bartlett (1957) used these records to derive his initial estimates of the critical community size needed to maintain measles in an endemic state within a community. The collated national data were also used by Fine and Clarkson (1982) to assess the importance of schools and other seasonal factors on determining the periodicity observed in the incidence data. To facilitate examination of the spatial distribution of measles epidemics in England and Wales, Haggett, Cliff and co-workers collated and analyzed data from southwest England (Haggett, 1972; Cliff et al., 1975, 1993; Cliff and Haggett, 1988), using a data set covering 222 weeks (just over 4 years) and sampled at 179 spatial locations, although some of their analyses focused on subsets of this data (such as 27 locations from the county of Cornwall or 72 locations from the counties of Cornwall and Devon). Later, Grenfell and his group extended this data set across all of England and Wales, first to the largest cities and then to the entire country, eventually obtaining weekly data for the 50 year period between 1944 and 1994 at as many as 1400 spatial locations (Bolker and Grenfell, 1996; Rohani et al., 1999; Grenfell et al., 2001, 2002; Bjørnstad et al., 2002). This extended data set has provided the basis for much of the recent modeling work on the spatiotemporal spread of measles.

Long-term time series of measles incidence data are also available from the United States and have been used occasionally in studies of persistence and synchrony of measles outbreaks. Weekly reporting of measles incidence data began in the United States in 1893, but these reports were patchily distributed until 1925, when all states were required by the US Public Health Service to report their measles cases (Cliff et al., 1992a). Since 1951, weekly reports of measles cases by state and territory have been published in The Centers for Disease Control (CDC)'s *Morbidity and Mortality Weekly Report* (MMWR). These data are collated at the state level and so, while the data set covers a larger geographic extent and population base (several US states have population sizes that are sizeable fractions of the entire population of England and Wales), its spatial resolution is much more coarse than the UK data set. Aspects of this data set, particularly with regard to synchrony, are described by Cliff, Haggett and co-workers (Cliff et al., 1992a,b). Some city-level data is available for the US, and was used by London and Yorke (London and Yorke, 1973; Yorke and London, 1973) to examine seasonal patterns in transmission.

The Iceland measles incidence data are notable in that they provide a highly detailed record of epidemics as they ran their course throughout the island. These data are recorded in *Heilbrigðisskýrslur* (*Public Health in Iceland*), and extend back to 1896. This source, which has been extensively described and studied by Cliff, Haggett and co-workers (Cliff et al., 1981, 1993; Cliff and Haggett, 1988), not only provides numerical data on incidence but also includes written descriptions of the individual epidemics that swept across the country. The level of detail in these accounts is impressive, documenting many specific events that played a role in an outbreak. Cliff and colleagues offer a particularly illustrative example. In 1907 a young girl from Reykjavik visited a village that was experiencing cases of measles. The village was isolated and, in order to limit transmission of the virus, visitors to the village (including the young girl) were confined for two weeks to homes without other young children. After her stay in such a house, the young girl returned home with isolation certificate in hand and mingled with a crowd of people during the Danish King's visit to Reykjavik. It turned out that the girl had acquired a mild case of measles in the village, which, predictably, triggered an epidemic in Reykjavik that then spread rapidly within the city and, from there, to the rest of Iceland.

Because of the small size of its population, measles could not persist in Iceland; instead, a succession of distinct epidemic waves was seen. The majority of measles epidemics on the island began in Reykjavik, as it was the main point of contact to other countries. The clear separation between the epidemic waves, together with the level of detail provided by the Icelandic records, is sufficient to allow the construction of maps that document the routes by which infection spread from community to community, at least during the period up to the Second World War. Using the maps that they constructed, Cliff, Haggett and co-workers (Cliff et al., 1981, 1993) showed that measles exhibited two different patterns of spread on the island. From the capital, the infection first spread hierarchically to the other main urban population centers, after which the disease spread outwards from these locations by diffusion into their surrounding areas. Since measles was not an endemic infection there, we shall not focus our attention on Iceland, although, as we shall see, the repeated invasion waves and extinctions have parallels with the behavior seen in some regions of larger countries such as the UK.

One particularly interesting, and, from the viewpoint of our story, fortunate feature of both the England and Wales and US data sets is that they span the introduction of mass vaccination against measles. This 'natural experiment' (Grenfell and Harwood, 1997) represented a large perturbation to the epidemiological system: the transient behavior that resulted is extremely informative dynamically and the accurate depiction of pre- and post-vaccination dynamics using a single model presents a major challenge. The data sets provide detailed information on the dynamics of the infection in both the pre- and post-vaccine eras.

Finally, very recent work has examined data from a quite different setting: the sub-Saharan African country of Niger (Ferrari et al., 2008). The demographics of this country differ considerably from those of the more commonly studied Western countries. In particular, the birth rate is much higher in Niger, and since, as we shall see below, births are a major driver of measles outbreaks, this has significant implications for the dynamics of the infection.

13.3 Local dynamics: Periodicity and endemic fadeout

Many of the important dynamical characteristics of measles are clearly visible in incidence records (Figure 13.1). The two most striking features are the repeated occurrence of large outbreaks and the importance of population size on the dynamics.

Periodicities in the incidence of many infectious diseases have long been described (Hamer, 1906; Soper, 1929). These early studies, including Hamer's celebrated 1906 paper, identified that the depletion and replenishment of the pool of susceptible individuals is a key mechanism underlying the oscillations. An outbreak can only occur when a relatively large number of susceptibles is present. As the outbreak takes place, susceptibles become infected, so their number falls while the number of infectives rises. Eventually, the number of susceptibles falls to a level where there are insufficiently many remaining to sustain the outbreak and so the number of infectives falls. Over time, the susceptible pool is replenished by births and rises to a level at which another outbreak can occur. The process repeats, leading to recurrent epidemics. In this picture, the time between outbreaks is largely determined by the birth rate of the population. In many cases, periodicities are clear from visual inspection of the data, although more detailed understanding can be gained by the use of time series

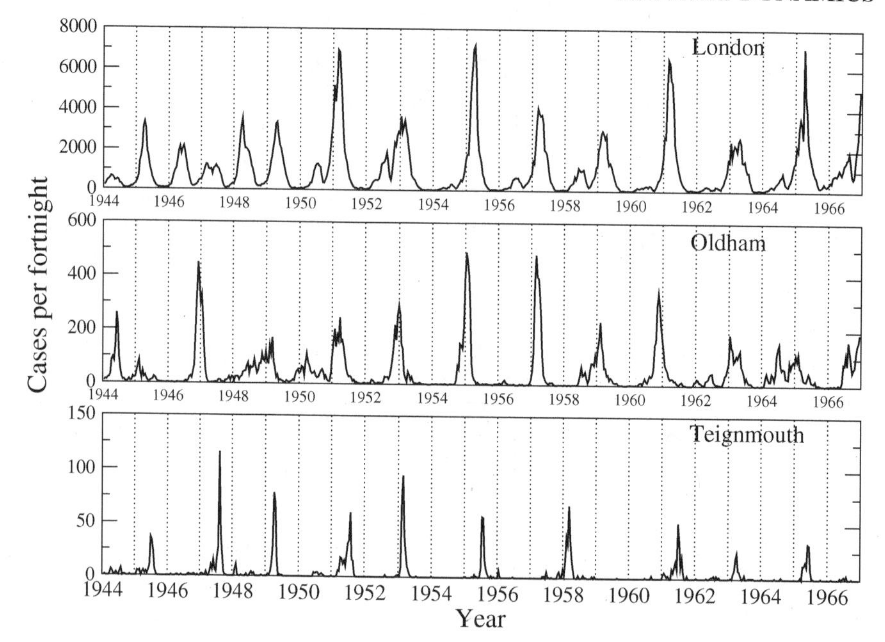

Figure 13.1 Measles incidence (cases per fortnight) for three British towns: Inner London (approximate population size 3.2 million), Oldham (119,000), and Teignmouth (11,000). (The UK measles data used in this figure and in Figure 13.2 were obtained from http://www.zoology.ufl.edu/bolker/measdata.html.)

analyses, such as correlograms or Fourier spectra (Anderson et al., 1984), or, if the time series is nonstationary, wavelet analysis (Grenfell et al., 2001).

Attempts to understand the origins of periodic behavior were one of the motivations behind the development of mathematical models for the transmission dynamics of infectious diseases (Hamer, 1906; Kermack and McKendrick, 1927, 1932, 1933; Soper, 1929). These models include the now familiar deterministic SIR model:

$$\frac{dS}{dt} = \mu N - \mu S - \frac{\beta SI}{N} \tag{13.1}$$

$$\frac{dI}{dt} = \frac{\beta SI}{N} - (\mu + \gamma)I \tag{13.2}$$

$$\frac{dR}{dt} = \gamma I - \mu R. \tag{13.3}$$

Here, each member of the population is assumed to be either susceptible to infection, infectious, or recovered, with the numbers of each being written as S, I, and R. The terms of the model describe the flows between these three compartments as people are born or die, become infected or recover. In the form of the model given here, the infection is assumed to be nonfatal and the population is taken to be in demographic equilibrium, so that the per-capita birth and death rates are equal, with their common value written as μ. The total population size, N, is therefore constant and since we have $S + I + R = N$, we need only track the numbers in two of the three compartments. The recovery rate is taken to be constant, and is written as γ, corresponding an average duration of infection of $1/\gamma$. The transmission parameter, β, is a compound parameter that describes the contact rate at which individuals meet and the probability that infection would occur if one of them was infectious and the other susceptible.

Although the simplest deterministic models do exhibit oscillatory behavior, they fail to produce sustained oscillations. Provided that the infection is sufficiently transmissible, its prevalence in such models settles into a steady state, approached via damped oscillations (Bartlett, 1956). Interestingly, for parameter values that are appropriate for measles, it is observed that the period of these damped oscillations is on the order of two years, in agreement with the period seen in many incidence records.

The key ingredient required for the maintenance of oscillations in the deterministic model framework is seasonality in transmission. Infections such as measles have a low average age at infection: a large number of cases involve children of school age. As a result, schools are an important transmission venue, with contact rates between children rising and falling between school terms and vacations. The resulting seasonal variations in transmission were documented by London and Yorke (London and Yorke, 1973; Yorke and London, 1973) by fitting simple deterministic SIR models to monthly incidence data obtained for several childhood diseases from several American cities. Later studies, including those of Fine and Clarkson (1982) and Grenfell and co-workers (Finkenstädt and Grenfell, 2000; Bjørnstad et al., 2002), provided a more detailed picture of this seasonal variation, either by examining more frequently sampled data (weekly or fortnightly) or by using more sophisticated models and model-fitting procedures. Perhaps the best characterization of seasonality is provided by Finkenstädt and Grenfell (2000), using the discrete time TSIR (Time-series SIR) model to interpret weekly data from the pre-vaccine era in England and Wales (see also Bjørnstad et al., 2002).

Seasonality is typically included in a model by allowing the contact rate to vary according to some function that has a period of one year. Commonly used seasonal terms vary in complexity from a sinusoidal function (Dietz, 1976) through to a detailed weekly (or fortnightly) function that accurately depicts the opening and closing of schools (Schenzle, 1984; Finkenstädt and Grenfell, 2000). Inclusion of seasonality in SIR-type models leads to rich dynamical behavior. In an early systematic exploration, Dietz (1976) showed that weak seasonality typically leads to annual oscillations in prevalence (and hence incidence) and that amplitude magnification is observed, i.e., a relatively small fluctuation in the contact rate can give rise to much larger oscillations in the prevalence of infection. This magnification results from a resonance effect between the annual forcing and the natural period of the damped oscillation. Moderate levels of seasonality typically lead to biennial oscillations (Dietz, 1976). Beyond annual and biennial oscillations, seasonally forced SIR-type models generate a diverse menagerie of more complex dynamical behaviors, including triennial and longer-period oscillations and deterministic chaos (Olsen and Schaffer, 1990). A correspondingly large literature documents this complexity, although the relevance of some of these behaviors to real-world measles outbreaks has been questioned on the grounds that their generation requires levels of seasonality that are far higher than those seen in reality (Pool, 1989; but see also Ferrari et al., 2008).

The recurrent epidemic pattern is clearly impacted by the size of the population under consideration. In large cities or countries, the repeated outbreaks generally occur at regular intervals. For smaller cities or countries, however, the pattern is less regular and periods are seen when there are no cases. In both settings, the number of infectives falls to low levels between outbreaks, but if the population size is large enough then the infection can persist through these periods. In small populations there is a chance that the chain of infection can be broken, leading to extinction of the infection— endemic fadeout. The endemic fadeout phenomenon was characterized by Bartlett in the late 1950s in terms of the critical community size (CCS) (Bartlett, 1956, 1957, 1960). The CCS is the smallest size of an isolated population that can maintain infection without endemic fadeout. For measles, Bartlett showed that the CCS was on the order of two to three hundred thousand people.

Following fadeout in a given locale, further outbreaks can only occur there when infection is reintroduced from elsewhere. If reimportation of infection is reasonably frequent then outbreaks can be triggered as soon as the susceptible pool has been sufficiently replenished. Less frequent importations mean that the wait for the next outbreak can be longer, and so outbreaks occur irregularly. These two situations are characterized as exhibiting type II and III epidemic waves, respectively, while the fadeout-free recurrent epidemics of the largest population centers are known as type I epidemic waves. In type I settings, the time between outbreaks is governed by the rate at which births refill the susceptible pool, while in type II and III settings the inter-outbreak time also depends on the rate at which infection is reintroduced by contact with other populations.

Deterministic models are unable to reproduce either the population size-dependent behavior or the fadeout effects seen in the incidence records. The latter deficiency is a consequence of such models treating the numbers of infectives (and susceptibles) as continuously varying quantities, allowing infection to persist even when the number of infectives falls to a fraction (often a very small fraction) of a single individual (Bolker and Grenfell, 1995). Bartlett addressed this deficiency by incorporating demographic stochasticity within the nonseasonal SIR framework, taking the numbers of susceptibles, infectives, and recovereds to be integers and modeling the discrete transitions that occur as individuals move between classes as probabilistic processes (Bartlett, 1956). Such models can reproduce the fadeout effect and, for biologically realistic parameter values, give reasonable estimates of the observed CCS for measles.

Typically, stochastic models include a term depicting immigration of infective individuals in order to reseed infection following a fadeout. Without this term, which can be thought of as mimicking certain aspects of spatial structure, fadeout would lead to permanent extinction of infection. Bartlett argued that the rate of immigration was likely to scale linearly with population size and showed that this leads to the period of epidemics in populations exhibiting type II waves having the form $a + b/\sqrt{N}$, where a and b are constants and N is the size of the population (Bartlett, 1956, 1957).

Bartlett's nonseasonal stochastic model cannot, however, reproduce the dynamical complexities seen in incidence time series: as discussed above, this requires the inclusion of seasonality. Seasonally forced stochastic models, though, predict critical community sizes that are much greater than those seen in reality unless an infective immigration term is included (Bolker and Grenfell, 1995). This result hints at the importance of spatial structure for persistence of the infection.

13.4 Regional persistence and spatial synchrony

Persistence of measles at the regional level can be seen as a classic metapopulation problem (Levins, 1969), the outcome of which depends on the balance between the frequency of local extinctions due to fadeout, and the rate at which infection is reintroduced from elsewhere to locales that have undergone fadeout. The potential for reintroduction to counteract local fadeout depends on the degree to which the incidence of infection is synchronized between different locales. If outbreaks—and hence fadeouts—are highly synchronized then the infection will tend to undergo simultaneous (or nearly simultaneous) extinction in many locales. A high level of synchrony between outbreaks, therefore, reduces the probability that reintroduction will occur, reducing the persistence of the infection (Bolker and Grenfell, 1995; Ferguson et al., 1997; Grenfell and Harwood, 1997; Keeling, 2000).

The importance of synchrony in understanding persistence of infection has led to numerous studies examining the spatial synchrony of epidemics. These have provided both a description of the patterns seen in spatially resolved incidence records and a fair understanding of the mechanisms that give rise to the observed synchrony (Bartlett, 1956; Murray and Cliff, 1977; Cliff et al., 1992a,b; Bolker and Grenfell, 1996; Lloyd and May, 1996; Rohani et al., 1999; Keeling, 2000; Grenfell et al., 2001; Keeling and Rohani, 2002).

Two distinct patterns of synchrony are seen, depending on the sizes of communities being observed. For a collection of cities (or other locales) that are each above the critical community size, a high degree of synchrony is often seen between their recurrent epidemics. On the other hand, for a collection of smaller-sized towns or locales that surround a large population center, wave-like behavior is often seen as infection spreads outwards from the large city into the surrounding region. This second pattern exhibits lagged synchrony, with phase differences between outbreaks in different locales. We shall discuss these two patterns separately.

13.5 Spatial synchrony among large population centers

Measles outbreaks in the United States, before mass vaccination largely curtailed indigenous transmission, exhibited moderate to high levels of synchrony between states. Cliff, Haggett and co-workers (Cliff et al., 1992a,b) documented this coherence by calculating the correlation between the incidence time series of different states. The monthly incidence data from the period 1962 to 1988 was divided into a number of overlapping portions using a sliding window that spanned 24 months. For each time window, correlation coefficients were calculated between pairs of time series aggregated at three different spatial scales (states, nine standard regions, each made up of a number of states, and three divisions, obtained by amalgamating regions). Two measures of coherence were employed. External coherence was used to quantify the correlation *between* incidences at the *same* spatial scale (e.g., between regions or between states). Internal coherence, on the other hand, was used to measure the extent of correlation within a given region (or division) by calculating the average of the correlation coefficients between its constituent states (or regions). For both measures, high levels of coherence were seen at all spatial scales for the early part of the time series, corresponding to the period before the introduction of mass vaccination. For a majority of states, the highest synchrony seen with another state occurred with one of the immediate neighbors. Interestingly, coherence was seen to fall as measles incidence declined in the wake of immunization (Cliff et al., 1992b).

Using weekly data from the seven largest cities in England and Wales (corresponding fortnightly data are shown in Figure 13.2), Bolker and Grenfell (1996) examined the degree of synchrony between outbreaks in these urban populations over the period 1948-1988. Using correlation coefficients calculated over 4, 10, or 20 year blocks, a high degree of synchrony was noted between outbreaks in the pre-vaccine era, with, again, the onset of mass vaccination leading to a decline in synchrony. The bottom panel of Figure 13.2 shows the results of a simple version of Bolker and Grenfell's analysis: using overlapping six-year windows, the correlation between incidence in each of the pairs of cities was calculated and, at each time point, the average was taken over the 21 resulting correlation coefficients. Biennial dynamics are seen to dominate between about 1953 and the mid sixties, with most of the cities having large outbreaks in odd-numbered years. Correlation was high during this period, but declined noticeably as mass vaccination was introduced. (The slightly lower synchrony in the first few years of the period shown in the figure reflects the fact that some cities were undergoing annual outbreaks, while others exhibited biennial patterns.)

Data were also available at the sub-city (borough) level for London, with borough-level data exhibiting high levels of synchrony. Curiously, a preliminary investigation at this spatial scale by Bolker and Grenfell found that vaccination did not lead to a fall in synchrony (Bolker and Grenfell, 1996). Another analysis, however, suggested that such a fall can be observed if the boroughs' boundary changes are accounted for (Ferguson et al., 1997).

A more detailed analysis at the city level was provided by Rohani et al. (1999), using a data set covering 60 British cities. Even though many of these cities fall below the CCS, the picture of declining synchrony with the onset of vaccination was confirmed. (A particularly interesting observation of this study, although not of direct relevance to us here, was that vaccination need not lead to a decrease in synchrony for all childhood diseases: the dynamics of whooping cough were seen to become more synchronized over time.)

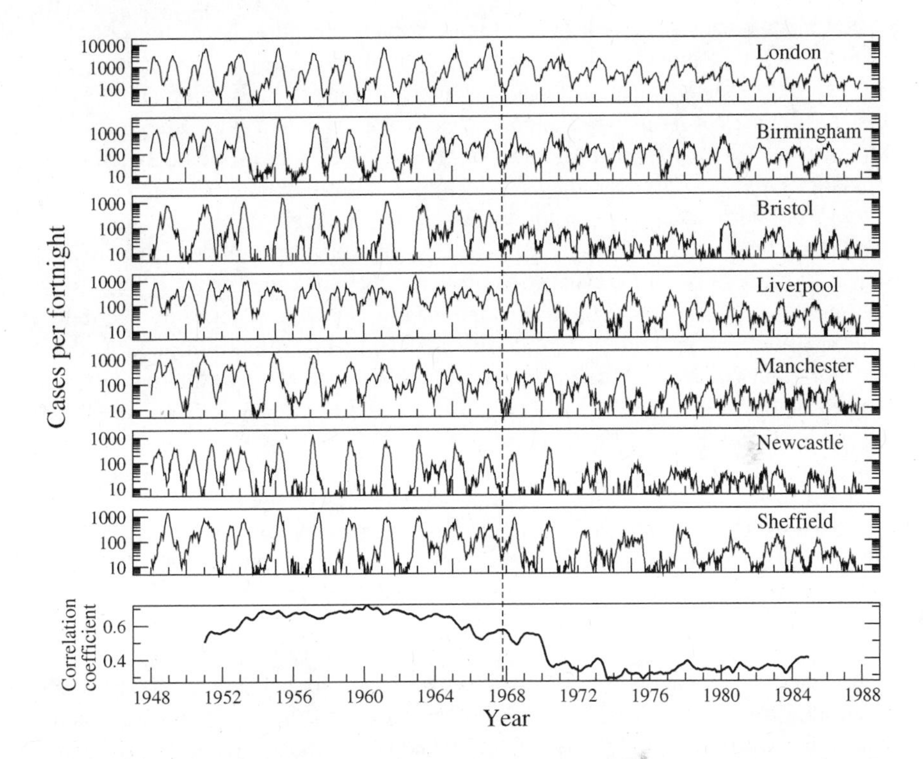

Figure 13.2 Synchrony between measles outbreaks in seven large cities of England and Wales. Upper seven panels show the fortnightly numbers of cases seen in each city over the period 1948-1988, plotted on logarithmic scales. (For London, the scale runs from 20 to 20,000 cases per fortnight, for Birmingham from 5 to 5,000, and for the remaining five graphs the scale runs from 5 to 2,000 cases per fortnight.) The lower panel depicts the mean cross correlation between incidence in the seven cities, calculated using a six-year sliding window and with all incidences first transformed as $\log_{10}(\text{cases} + 1)$. The vertical dashed line denotes the approximate date at which mass vaccination was started in the United Kingdom. Notice that the log scale de-emphasizes the decline in incidence seen in the wake of mass immunization.

Much of the work on the synchrony of outbreaks between large population centers has echoes of classical ecological studies of the spatiotemporal dynamics of oscillatory populations. Perhaps the best known example concerns the abundance of the Canadian lynx (*Lynx canadensis*), described by the records of the fur catches of the Hudson's Bay Company. Elton and Nicholson (1942) described a ten year oscillation in the abundance of the lynx and the surprising degree of synchrony seen in this oscillation across a wide area. To quote Elton and Nicholson: "The most extraordinary feature of this cycle is that it operates sufficiently in line over several million square miles of country not to get seriously out of phase in any part of it." They further note "the remarkable degree of coherence in the cycle in regions thousands of miles apart" (Elton and Nicholson, 1942, pg. 239).

Several mechanisms for synchronizing population cycles have been suggested, but the two that are of most relevance to the epidemiological setting are the Moran effect (Moran, 1953) and dispersal of individuals. Moran (1953), in an attempt to understand the synchronous population cycles of the Canadian lynx, suggested that extrinsic factors, such as meterological or other climatic factors, could synchronize populations in two locations if they shared common local dynamics and the extrinsic factors were correlated between the locations. Specifically, Moran considered two populations whose dynamics were governed by the same linear model and were both subject to random perturbations. The correlation between the resulting population time series was shown to equal the correlation between the time series of the random perturbations. As admitted by Moran, the analysis is much more difficult to carry out in nonlinear settings, but synchronization due to the Moran effect is still seen to occur both in theoretical (Ranta et al., 1995, 1997; Haydon and Steen, 1997) and real-world (Grenfell et al., 1998) settings.

Dispersal of individuals between locations can also lead to synchrony between the dynamics in the different locations (Levin, 1974; Hastings, 1993; Ranta et al., 1995, 1997; Haydon and Steen, 1997). Short range dispersal leads to more closely located populations being more highly synchronized than distantly located populations: synchrony tends to decrease with increasing distance (Ranta et al., 1995).

In the epidemiological setting, seasonality provides a global synchronizing mechanism that is somewhat analogous to the Moran effect: the simultaneous opening of schools in different regions causes outbreaks to start at similar times (provided that infection is present in each region in the first place). The effect of seasonality, however, is not quite so straightforward: because this forcing often leads to cyclic behavior with period two (or more) years, two regions can undergo oscillations in which their large outbreaks occur in different years. Seasonality can then hinder synchronization as the outbreaks in the different regions are forced to occur at particular times of the year, making it difficult for the phase difference to be reduced. In this way, seasonal forcing can lead to stable out of phase behavior. Examples of out-of-phase oscillations being maintained in nearby locations have been documented, such as the British cities of Cambridge and Norwich, which, as discussed by Grenfell et al. (2001), remained out of phase for a 16 year period of their data set.

Dispersal of individuals underlies epidemiological coupling between regions, although, for people at least, this movement more often takes the form of short-term visits than permanent migration. Attempts to understand the impact of epidemiological coupling on synchrony have often employed simple multi-patch (e.g., multi-city) extensions of the basic SIR model, such as

$$\dot{S}_i = \mu N_i - \mu S_i - S_i \sum_{j=1}^{n} \frac{\beta_{ij} I_j}{N_i} \qquad (13.4)$$

$$\dot{I}_i = S_i \sum_{j=1}^{n} \frac{\beta_{ij} I_j}{N_i} - (\mu + \gamma) I_i \,. \tag{13.5}$$

Here, S_i denotes the number of susceptible individuals in patch i, I_j the number of infectious individuals in patch j, and N_i the total size of the population in patch i. The β_{ij} are the transmission parameters for within and between patch transmission.

Movement between patches is not explicitly modeled by these equations: individuals remain in their home patch. Instead, movement is modeled implicitly, with the force of infection experienced by susceptibles in a given patch being depicted as a weighted sum of the levels of infection in each patch. An alternative to this cross-coupled model formulation does model migration explicitly, using equations that are of the form

$$\dot{S}_i = \mu N_i - \mu S_i - \frac{\beta_i S_i I_i}{N_i} + \sum_{j=1}^{n} \Omega_{ij}^S S_j \tag{13.6}$$

$$\dot{I}_i = \frac{\beta_i S_i I_i}{N_i} S_i - (\mu + \gamma) I_i + \sum_{j=1}^{n} \Omega_{ij}^I I_j \,. \tag{13.7}$$

Here, transmission only occurs within a patch, described by the transmission parameters β_i, and the per-capita rates of movement of susceptibles and infectives from patch j to patch i are given by Ω_{ij}^S and Ω_{ij}^I. An extension of this migration formulation separately tracks permanent and temporary residents of a patch (Sattenspiel, 1988; Sattenspiel and Dietz, 1995; Keeling and Rohani, 2002). It turns out that, at least in terms of some qualitative properties, these different formulations exhibit fairly similar behaviors (Ball, 1991; Keeling and Rohani, 2002; Lloyd and Jansen, 2004).

The description of contact or movement between different locations potentially leads to an explosion in the number of model parameters because of the n^2 entries that make up either β_{ij} or Ω_{ij}. Clearly, it would be a difficult task to estimate all of these from epidemiological data alone, particularly since within-patch contacts are likely much more frequent than between-patch contacts. One technique that has been employed is to estimate these quantities from transportation data, such as train or plane passenger movement data (Baroyan and Rvachev, 1967; Rvachev and Longini, 1985). An alternative approach curbs the number of parameters by employing a function that relates the contact between inhabitants of pairs of locations to the population sizes of the locations and the distance between them. These so-called gravity models originated in the geography and sociology literatures (see, for example, Stewart, 1948), and were first employed in an epidemiological setting by Murray and Cliff (1977) in a multi-patch model for the spread of measles in the British city of Bristol. Xia et al. (2004) employed a generalized gravity model that assumes susceptibles in location k experience an additional force of infection due to infectives in location j equal to $\theta N_k^{\tau_1} I_j^{\tau_2} / d_{jk}^{\rho}$. Here d_{jk} denotes the distance between the two locations, N_k is the size of the population at location k, and I_j is the number of infectives at location j. Their gravity model adds just four parameters, θ, τ_1, τ_2, and ρ, the last of which measures the rapidity with which contact between locations declines with increasing separation. Even so, estimation of all four of these parameters may not be straightforward (Xia et al., 2004), although estimates were obtained using the pre-vaccine era England and Wales data set.

Theoretical studies often employ a highly simplified description of coupling, assuming, for instance, that movement only occurs between neighboring patches or that all between-patch transmission parameters take the same value, usually written as some multiple of the within-patch transmission parameter (Bolker and Grenfell, 1995; Lloyd and May, 1996).

Deterministic spatial models typically exhibit the threshold behavior that is familiar from the simplest SIR models: when the basic reproductive number is greater than one, the numbers of infectives in each patch approach an endemic equilibrium (see, for example, Arino and van den Driessche, 2003), typically via damped oscillations. While the spatial model is sufficiently complicated that a general analysis of the dynamics of this behavior is impossible, progress can be made in a number of special cases, such as when there is symmetric coupling between equally-sized identical patches (Bartlett, 1956; Lloyd and May, 1996; Lloyd and Jansen, 2004). The system, either under cross-coupling or migration, can be linearized about its endemic equilibrium and the approach towards this equilibrium can be decomposed into an in-phase mode and a number of out-of-phase modes. The analysis provides the decay rates of each of these modes and it is found that the typical behavior is for the out-of-phase modes to decay much more rapidly than the in-phase mode. Consequently, following a short transient, the system approaches its endemic equilibrium via synchronized damped oscillations.

This analysis of the synchronizing effect of epidemiological coupling is based on a noise-free deterministic system. Stochasticity, if uncorrelated between patches, as demographic stochasticity would be, will tend to desynchronize patches. The synchronizing effect of epidemiological coupling, however, can be strong enough to overcome the impact of noise: synchrony can still be observed in stochastic versions of the spatial model (Lloyd and May, 1996). Synchrony between two patches can be assessed by examining the correlation between the numbers of infectives seen in the two patches at different times. In the case of a two-patch stochastic model, Keeling and Rohani (2002) used moment equations to calculate the correlation coefficient, ρ, between the numbers of infectives in the two patches in terms of the strength of coupling between patches, denoted by ϵ. Their relationship takes the form $\rho = \epsilon/(\xi + \epsilon)$, where the quantity ξ depends on various epidemiological parameters. The important point to take from this relationship is the fact that synchrony (correlation) and coupling strength are positively related.

These model-based analyses of synchrony, however, have limited applicability. For instance, both the deterministic analysis and the moment equation-based analysis predict that synchrony should increase with the onset of mass-vaccination (Lloyd and May, 1996; Root and Lloyd, 2008), making exactly the opposite prediction to what was seen in reality for measles. This failing results from these analyses being based on behavior near the endemic equilibrium of nonseasonal models. Analytic approaches have yet to be extended to seasonal models, but several numerical studies

(Lloyd, 1996; Rohani et al., 1999) have documented a complex interplay between the intrinsic damped oscillatory predator-prey dynamics of measles and the demographic stochasticity and seasonality that act upon the system. Seasonally forced stochastic models can reproduce the decline in synchrony seen with vaccination (Bolker and Grenfell, 1996; Lloyd, 1996; Lloyd and May, 1996; Rohani et al., 1999). Another weakness of the above analyses is that they do not take account of the fadeout effect: their applicability is limited to population centers that are sufficiently large to maintain infection without fadeout.

Despite their weaknesses, the above analyses demonstrate the strong synchronizing effect of spatial coupling. This effect, however, leads to spatial coupling having two conflicting effects on persistence in a collection of populations that are close to the CCS. If spatial coupling is low, it will be unlikely for infection to be reintroduced into a locale in which a fadeout has previously occurred. If it had no other effect on the system, the probability of reintroduction would increase with increasing spatial coupling. But stronger spatial coupling also increases the likelihood that outbreaks are synchronized in different regions and hence increases the chance of simultaneous fadeout, in which case reintroduction cannot occur (Bolker and Grenfell, 1995; Ferguson et al., 1997). Consequently, persistence is greatest at intermediate levels of spatial coupling (Keeling, 2000).

13.6 Moving beyond simple synchrony: Reinvasion waves and phase relationships

For towns that are below the critical community size, dynamics will be dominated by extinction and reinvasion of infection. Consequently, in a region consisting of such populations, the dynamics will share some similarities with invasions in more general ecological contexts (Skellam, 1951; Elton, 1958; Shigesada and Kawasaki, 1997). Ecological invasion theory predicts the occurrence of wave-like behavior, with infection spreading outwards from the location of reintroduction. In the epidemiological context, this spread occurs across a heterogeneous enviroment made up of rural locales and towns of different sizes, and for which “epidemiological distance” need not correspond to geographical distance, given that movement of people between locations depends not just on physical separation but also on social structures (see, for example, section 13.3 of Cliff et al., 1993). Together with the randomness that arises from the small numbers of infectives during the invasion process, heterogeneity will lead to a spatial pattern that is more complex than a simple wave, although signatures of wave-like behavior will still be present.

Using a variety of approaches, Cliff, Haggett and co-workers explored the dynamics of measles outbreaks in the South-West of England, primarily in the counties of Cornwall and Devon (Haggett, 1972; Cliff et al., 1975, 1993; Cliff and Haggett, 1988). Calculation of cross-correlation functions between pairs of time series from the 72 locales and between the time series of locales and the incidence aggregated over the entire region showed lagged relationships: rather than being in perfect synchrony, some towns' incidences peaked before the peak in average incidence, while some towns lagged behind. Maps depicting phase relationships (lead, in-phase, lag, or uncertain) were suggestive of spatial clustering and interpreted as being indicative of spread outwards from several sources (Cliff et al., 1975).

A closer investigation of this behavior represented the region by a graph of nodes and edges, in which nodes depicted locales and nodes were joined by an edge if the corresponding locales were physically adjacent to each other. The spatial lag between two locales, defined as the minimum number of edges that have to be traversed in order to travel between the corresponding nodes on the graph (a quantity known in graph theory as the distance between the nodes), roughly corresponds to geographical distance. The correlation between time series at different locales was found to first decline with spatial lag but then increase for larger lags (Cliff et al., 1975). The initial decrease is as expected, indicating lower contact between more separated regions. The later increase coincided with the spatial lag approaching the average separation between locales and the two major regional cities of Bristol (population 500,000) and Plymouth (population 250,000), in which measles remained persistent over the entire four year observation period.

A more qualitative analysis mapped out locales in which there was either a new outbreak, an ongoing outbreak, a recent fadeout, or no cases. The majority of new outbreaks were found to occur in areas that were adjacent to existing outbreaks, with new outbreaks being less frequently found at increasing spatial lags (less than 5% were found at 3 steps away, and none were found at 4 or more steps away from an existing outbreak). Cliff et al. (1975) concluded that the observed patterns were consistent with a hierarchical spread, with initial outbreaks occurring in the larger urban centers, followed by outward spread from these into the surrounding towns, villages, and rural areas.

The statistical power of Cliff and Haggett's studies was limited by the relatively short length of their time series: their four-year observation window covered just two major measles outbreaks. Grenfell and co-workers' longer time series (Grenfell et al., 2001) gave them more statistical power, but required an analysis that can cope with temporal changes in the dynamics of incidence, i.e., nonstationarity of the time series. Traditional linear techniques, such as autocorrelation or Fourier spectra, assume stationarity of time series. Measles dynamics, however, exhibit significant dynamical changes over time, such as changes in the oscillation period in response to variations in birth rate or level of vaccination (Earn et al., 2000). In such situations, wavelet analysis (Torrence and Compo, 1998; Grenfell et al., 2001) provides a useful tool. In common with Fourier analysis, wavelet analysis is a decomposition approach, but, instead of representing the data in terms of a collection of sine and cosine functions, it uses basis functions that are localized in time (and frequency). Using this approach on the 1944-94 measles incidence series for London, Grenfell et al. (2001) clearly demonstrated the mix of annual and biennial behavior seen in the immediate aftermath of World War II, which gave way to a dominant biennial behavior, and then, in the wake of mass vaccination, a lengthening of the oscillation period and a reduction of its amplitude.

Application of the wavelet approach allows the time series at each spatial location to be decomposed into a number of components (such as a two-yearly component that would correspond to the typical dominant biennial pattern) and provides the phases (phase angles) of these various components at each time point (Grenfell et al., 2001). Comparison of the phases at different locations gives insight into the way in which outbreaks spread across a region. Outbreaks in small towns and rural areas are seen to lag behind those in large cities, and phase differences are seen to generally increase with increasing distance from the city (Grenfell et al., 2001). For example, Grenfell et al. describe a well-defined wave that spreads outwards from London at a speed of around 5 km per week, with similar (also slightly less well defined) waves seen around the other major cities. Calculation of the correlation between phase angles at various locales in terms of their separation gives insight into the spatial extent of the waves spreading outwards from major population centers. The overall spatial coherence was seen to decrease with the onset of mass vaccination, in agreement with the earlier analyses of spatial synchrony, but the wavelet analysis gives the additional insight that the extent of the spatial waves also decreased with this change.

The wavelet analysis of the more complete data set provides a similar, but more definitive, description of epidemic spread to that obtained from the smaller set of data from the South-West of England: a hierarchical pattern of spread, with transmission between the large cities and then outwards into the small towns and rural areas that surround these cities (Grenfell et al., 2001). This pattern is very similar to the one described for the repeated invasion waves seen in Iceland (Cliff et al., 1981). Mechanistic models can reproduce this pattern of hierarchical spread (Grenfell et al., 2001; Xia et al., 2004), and emphasize the ingredients necessary for its generation, in particular its dependence on heterogeneity in the sizes of population centers and the occurrence of fadeouts in smaller-sized communities. Outbreaks in large cities, i.e., those above the CCS, will tend to have a high degree of synchrony with each other, while wave-like behavior will be seen in surrounding regions as a result of repeated cycles of introduction, epidemic, and fadeout.

13.7 Discussion

One of the major challenges for the development of spatial models is providing an appropriate description of movement or contact. As discussed above, this raises issues of parameterization that have been addressed in a number of ways, including the use of transportation data and gravity models. Several new approaches offer additional insight into human movement, including the analysis of data on commuting patterns (Riley and Ferguson, 2006; Viboud et al., 2006) and tracking individuals using cellphones, GPS devices, or even banknotes (Brockmann et al., 2006; González et al., 2008). An important issue that has to be addressed before such data can inform epidemic models is that one must not only know how many people are traveling and the locations between which they are traveling, but also who those people are. In the case of measles, most transmission occurs amongst children so it is more important to know about their movement patterns than those of adults. But for other nonchildhood diseases, spatial coupling may be more strongly determined by the movement of adults. As a result, the movement pattern that influences transmission, i.e., the one that should be incorporated into a model, could be quite different from one disease to another (Viboud et al., 2006).

The development of mathematical models for the transmission dynamics of measles, and the understanding and insights that these have provided into the epidemiological processes at work, has been driven by the availability of remarkably detailed data sets, both in terms of their spatial and temporal resolution and because of the long time period (i.e., the number of epidemic cycles) that they document. Confrontation of their models with the data has forced modelers to progress from simple deterministic models through to complex spatially-structured, seasonally forced stochastic models, with the incorporation of additional complexities whenever the data have highlighted the inadequacies of simpler descriptions. Clearly, this would not have been possible without data collected at appropriate temporal and spatial scales: much information, for instance, is hidden by spatial aggregation, with dynamical complexities visible at small (e.g., city-level) scales being masked by averaging effects when examined at larger (e.g., country-level) scales (Sugihara et al., 1990; Ferrari et al., 2008). As a result, most work has focused on Iceland and England and Wales, providing a good understanding of the dynamics in these two locations. In contrast, the United States has received relatively little attention, presumably in part due to the most readily-available data being spatially-coarse state-level data.

Our discussion here has focused entirely on the dynamics seen in a few Western countries, and since we have seen that behavior depends in an important way on both the demographic and social structure of a population, the dynamics and patterns of spatial spread might be quite different elsewhere. Recent work (Ferrari et al., 2008) has examined the spatial spread of measles in Niger, a sub-Saharan African country that has a much higher birth rate than Western countries, with particular attention on its capital, Niamey. Apart from the higher birth rate, which, by itself would tend to promote an annual outbreak pattern (McLean and Anderson, 1988), the epidemiological system is subject to a much higher level of seasonality than is seen in England and Wales. This seasonality, which is thought to arise from migration from rural to urban areas with the start of the dry season, is strong enough to drive chaotic dynamics in deterministic epidemic models (Olsen and Schaffer, 1990). The strongly forced dynamics lead to a much higher critical community size: model simulation suggests fadeouts would be common even for a population size of five million—much larger than Niamey's 750,000 inhabitants (Ferrari et al., 2008)—and indeed, fadeouts are frequently seen, typically as the rainy season starts. As a result, the incidence records show an erratic outbreak pattern, with large outbreaks, inevitably followed by stochastic fadeout, with a gap of several years before the next outbreak: even with the very high birth rate, the large outbreaks mean that replenishment of the susceptible pool takes a few years. At the national level, measles persists as a result of weak coupling between different regions (both within and outside the country) that exhibit asynchronous outbreaks. Aggregated at this national scale, measles incidence exhibits annual behavior, but this is somewhat misleading in terms of the underlying transmission dynamics because this pattern results from averaging the (asynchronous) outbreaks seen in different locales (Ferrari et al., 2008).

As several authors have eloquently written (e.g., Grenfell and Harwood, 1997), the main themes that emerge from the exploration of the transmission dynamics of measles are familiar ecological ones. Attempts to gain an ever more detailed understanding of this epidemiological system has led its modelers to address the mechanisms that give rise to population cycles, persistence of a population in the face of demographic stochasticity, the complex dynamics that result from seasonal forcing of populations, the dynamical intricacies that result from the interplay between stochasticity and highly nonlinear systems, metapopulation dynamics and its impact on persistence, spatial synchrony of cycling populations, and invasion (and reinvasion) dynamics. As a result of the ecological importance attached to these issues, and particularly in light of the existence of spatiotemporal data at a level of detail that few ecological data sets can rival, it is hardly surprising that ecologists have shown considerable interest in this system. Major contributions to our understanding have been made by geographers, again drawn to the system by the availability of detailed spatial information. Beyond this, the rich dynamics seen in measles outbreaks have attracted much attention from statisticians, mathematicians, and physicists from a nonlinear dynamics viewpoint. Although much of the recent work has been undertaken outside the field, perhaps partly because the development and widespread deployment of measles vaccines have reduced the public health significance of the disease in developed countries (although measles is still a highly signficant disease in many other parts of the world), the importance of the early contributions made by pioneers in the field of epidemiology should not be underestimated or forgotten.

The natural history of measles infection is simple at the individual level, having none of the complexities that arise with the multiple strain structure of infections such as influenza or malaria, and having the simple picture of permanent immunity upon infection. At first sight, at least, measles should be one of the most straightforward infections to describe from a modeler's viewpoint. As we have seen, a number of factors conspire to make the modeler's task anything but straightforward. While the simplicity of the natural history of measles has facilitated the elucidation of its epidemiology (Cliff et al., 1981), the richness and complexity of this epidemiological system remains truly surprising.

13.8 References

R. M. Anderson, B. T. Grenfell, and R. M. May (1984), Oscillatory fluctuations in the incidence of infectious disease and the impact of vaccination: Time series analysis, *J. Hygiene* **93**: 587–608.

J. Arino and P. van den Driessche (2003), The basic reproduction number in a multi-city compartmental model, in *Positive Systems*, ed. by L. Benvenuti, A. De Santis, and L. Farina, Lecture Notes in Computer Science, Vol. 294, Springer-Verlag, Berlin, pp. 135–142.

F. G. Ball (1991), Dynamic population epidemic models, *Math. Biosci.* **107**: 299–324.

O. V. Baroyan and L. A. Rvachev (1967), Deterministic models of epidemics for a territory with a transport network, *Kibernetika (Cybernetics)* **3**: 67–74.

M. S. Bartlett (1956), Deterministic and stochastic models for recurrent epidemics. in *Proceedings of the Third Berkeley Symposium on Mathematical Statistics and Probability*, Vol. 4, ed. by J. Neyman, University of California Press, Berkeley, pp. 81–109.

M. S. Bartlett (1957), Measles periodicity and community size, *J. R. Stat. Soc. A* **120**: 48–70.

M. S. Bartlett (1960), The critical community size for measles in the United States, *J. R. Stat. Soc. A* **123**: 37–44.

O. N. Bjørnstad, B. F. Finkenstädt, and B. T. Grenfell (2002), Dynamics of measles epidemics: Estimating scaling of transmission rates using a time series SIR model, *Ecol. Monogr.* **72**: 169–184.

B. Bolker and B. Grenfell (1995), Space, persistence and dynamics of measles epidemics, *Phil. Trans. R. Soc. Lond. B* **348**: 309–320.

B. M. Bolker and B. T. Grenfell (1996), Impact of vaccination on the spatial correlation and persistence of measles dynamics, *Proc. Natl. Acad. Sci. USA* **93**: 12648–12653.

D. Brockmann, L. Hufnagel, and T. Geisel (2006), The scaling laws of human travel, *Nature* **439**: 462–465.

A. D. Cliff and P. Haggett (1988), *Atlas of Disease Distributions: Analytic Approaches to Epidemiological Data*, Blackwell, Oxford.

A. D. Cliff, P. Haggett, J. K. Ord, K. Bassett, and R. Davies (1975), *Elements of Spatial Structure*, Cambridge University Press, Cambridge.

A. D. Cliff, P. Haggett, J. K. Ord, and J. R. Versey (1981), *Spatial Diffusion: An Historical Geography of Epidemics in an Island Community*, Cambridge University Press, Cambridge.

A. D. Cliff, P. Haggett, and D. F. Stroup (1992a), The geographic structure of measles epidemics in the northeastern United States, *Am. J. Epidemiol.* **136**: 592–602.

A. D. Cliff, P. Haggett, D. F. Stroup, and E. Cheney (1992b), The changing geographical coherence of measles morbidity in the United States, 1962-88, *Stat. Med.* **11**: 1409–1424.

A. Cliff, P. Haggett, and M. Smallman-Raynor (1993), *Measles: An Historical Geography of a Major Human Viral Disease*, Blackwell, Oxford.

K. Dietz (1976), The incidence of infectious diseases under the influence of seasonal fluctuations, *Lecture Notes in Biomathematics*, **11**: 1–15.

D. J. D. Earn, P. Rohani, B. M. Bolker, and B. T. Grenfell (2000), A simple model for complex dynamical transitions in epidemics, *Science* **287**: 667–670.

C. Elton and M. Nicholson (1942), The ten-year cycle in numbers of the lynx in Canada, *J. Anim. Ecol.* **11**: 215–244.

C. S. Elton (1958), *The Ecology of Invasion by Animals and Plants*, Methuen, London.

N. M. Ferguson, R. M. May, and R. M. Anderson (1997), Measles: Persistence and synchronicity in disease dynamics, in *Spatial Ecology: The Role of Space in Population Dynamics and Interspecific Interactions*, ed. by D. Tilman and P. Kareiva, Princeton University Press, Princeton, pp. 137–157.

M. J. Ferrari, R. F. Grais, N. Bharti, A. J. K. Conlan, O. N. Bjørnstad, L. J. Wolfson, P. J. Guerin, A. Djibo, and B. T. Grenfell (2008), The dynamics of measles in sub-Saharan Africa, *Nature* **451**: 679–684.

P. E. M. Fine and J. A. Clarkson (1982), Measles in England and Wales - I: An analysis of factors underlying seasonal patterns, *Int. J. Epidemiol.* **11**: 5–14.

B. F. Finkenstädt and B. T. Grenfell (2000), Time series modelling of childhood diseases: A dynamical systems approach, *Appl. Statist.* **49**: 187–205.

M. C. González, C. A. Hidalgo, and A.-L. Barabási (2008), Understanding individual human mobility patterns, *Nature* **453**: 779–782.

B. Grenfell and J. Harwood (1997), (Meta)population dynamics of infectious diseases, *Trends Ecol. Evol.* **12**: 395–399.

B. T. Grenfell, K. Wilson, B. F. Finkenstädt, T. N. Coulson, S. Murray, S. D. Albon, J. M. Pemberton, T. H. Clutton-Brock, and M. J. Crawley (1998), Noise and determinism in synchronized sheep dynamics, *Nature* **394**: 674–677.

B. T. Grenfell, O. N. Bjørnstad, and J. Kappey (2001), Travelling waves and spatial hierarchies in measles epidemics. *Nature* **414**: 716–723.

B. T. Grenfell, O. N. Bjørnstad, and B. F. Finkenstädt (2002), Dynamics of measles epidemics: Scaling noise, determinism, and predictability with the TSIR model, *Ecol. Monogr.* **72**: 185–202.

P. Haggett (1972), Contagious processes in a planar graph; an epidemiological application, in *Medical Geography; Techniques and Field Studies*, ed. by N. D. McGlashan, Methuen, London, pp. 307–324.

W. H. Hamer (1906), Epidemic disease in England- the evidence of variability and of persistency of type, *Lancet* **i** (Vol. **167**, no. 4307): 733–739.

A. Hastings (1993), Complex interactions between dispersal and dynamics: Lessons from coupled logistic equations, *Ecology* **74**: 1362–1372.

D. T. Haydon and H. Steen (1997), The effects of large- and small-scale random events on the synchrony of metapopulation dynamics: A theoretical analysis, *Proc. R. Soc. Lond. B* **264**: 1375–1381.

M. J. Keeling (2000), Metapopulation moments: Coupling, stochasticity and persistence. *J. Anim. Ecol* **69**: 725–736.

M. J. Keeling and P. Rohani (2002), Estimating spatial coupling in epidemiological systems: A mechanistic approach, *Ecol. Lett.* **5**: 20–29.

W. O. Kermack and A. G. McKendrick (1927), A contribution to the mathematical theory of epidemics, *Proc. R. Soc. Lond. A* **115**: 700–721.

W. O. Kermack and A. G. McKendrick (1932), Contributions to the mathematical theory of epidemics. II- The problem of endemicity, *Proc. R. Soc. Lond. A* **138**: 55–83.

W. O. Kermack and A. G. McKendrick (1933), Contributions to the mathematical theory of epidemics. III- Further studies of the problem of endemicity, *Proc. R. Soc. Lond. A* **141**: 94–122.

S. A. Levin (1974), Dispersion and population interactions, *Am. Nat.* **108**: 207–228.

R. Levins (1969), Some demographic and genetic consequences of environmental heterogeneity for biological control, *Bull. Ent. Soc. Am.* **15**: 237–240.

A. L. Lloyd and V. A. A. Jansen (2004), Spatiotemporal dynamics of epidemics: Synchrony in metapopulation models, *Math. Biosci.* **188**: 1–16.

A. L. Lloyd (1996), *Mathematical Models for Spatial Heterogeneity in Population Dynamics and Epidemiology*, PhD thesis, University of Oxford.

A. L. Lloyd and R. M. May (1996), Spatial heterogeneity in epidemic models, *J. Theor. Biol.* **179**: 1–11.

W. P. London and J. A. Yorke (1973), Recurrent outbreaks of measles, chickenpox and mumps. I. Seasonal variation in contact rates, *Am. J. Epidemiol.* **98**: 453–468.

A. R. McLean and R. M. Anderson (1988), Measles in developing countries. Part I. Epidemiological parameters and patterns, *Epidemiol. Infect.* **100**: 111–133.

P. A. P. Moran (1953), The statistical analysis of the Canadian lynx cycle. II. Synchronization and meteorology, *Aust. J. Zool.* **1**: 291–298.

G. D. Murray and A. D. Cliff (1977), A stochastic model for measles epidemics in a multi-region setting. *Trans. Inst. Brit. Geog.* **2**: 158–174.

L. F. Olsen and W. M. Schaffer (1990), Chaos versus noisy periodicity: Alternative hypotheses for childhood epidemics, *Science* **249**: 499–504.

R. Pool (1989), Is it chaos, or is it just noise? *Science* **243**: 25–28.

E. Ranta, V. Kaitala, J. Lindstrom, and H. Linden (1995), Synchrony in population dynamics, *Proc. R. Soc. Lond. B* **262**: 113–118.

E. Ranta, V. Kaitala, J. Lindstrom, and E. Helle (1997), The Moran effect and synchrony in population dynamics, *Oikos* **78**: 136–142.

S. Riley and N. M. Ferguson (2006), Smallpox transmission and control: Spatial dynamics in Great Britain, *Proc. Natl. Acad. Sci. USA* **103**: 12637–12642.

P. Rohani, D. J. D. Earn, and B. T. Grenfell (1999), Opposite patterns of synchrony in sympatric disease metapopulations, *Science* **286**: 968–971.

A. M. Root and A. L. Lloyd (2008), Spatial synchrony in epidemic models (Manuscript in preparation).

L. A. Rvachev and I. M. Longini Jr. (1985), A mathematical model for the global spread of influenza, *Math. Biosci.* **75**: 1–22.

L. Sattenspiel (1988), Spread and maintenance of a disease in a structured population, *Am. J. Phys. Anthropol.* **77**: 497–504.

L. Sattenspiel and K. Dietz (1995), A structured epidemic model incorporating geographic mobility among regions, *Math. Biosci.* **128**: 71–91.

D. Schenzle (1984), An age-structured model of pre- and post-vaccination measles transmission, *IMA J. Math. Appl. Med. Biol.* **1**: 169–191.

N. Shigesada and K. Kawasaki (1997), *Biological Invasions: Theory and Practice*, Oxford University Press, Oxford.

J. G. Skellam (1951), Random dispersal in theoretical populations, *Biometrika* **38**: 196–218.

H. E. Soper (1929), The interpretation of periodicity in disease prevalence, *J. R. Stat. Soc. A* **92**: 34–61.

J. Q. Stewart (1948), Demographic gravitation: Evidence and applications, *Sociometry* **11**: 31–58.

G. Sugihara, B. Grenfell, and R. M. May (1990), Distinguishing error from chaos in ecological time series, *Phil. Trans. R. Soc. Lond. B* **330**: 235–251.

C. Torrence and G. P. Compo (1998), A practical guide to wavelet analysis, *Bull. Am. Met. Soc.* **79**: 61–78.

C. Viboud, O. N. Bjørnstad, D. L. Smith, L. Simonsen, M. A. Miller, and B. T. Grenfell (2006), Synchrony, waves, and spatial hierarchies in the spread of influenza, *Science* **312**: 447–451.

Y. Xia, O. N. Bjørnstad, and B. T. Grenfell (2004), Measles metapopulation dynamics: A gravity model for epidemiological coupling and dynamics, *Am. Nat.* **164**: 267–281.

J. A. Yorke and W. P. London (1973), Recurrent outbreaks of measles, chickenpox and mumps. II. Systematic differences in contact rates and stochastic effects, *Am. J. Epidemiol.* **98**: 469–482.

CHAPTER 14

Rules of thumb for the control of vector-borne diseases in a spatial environment

Matthew D. Potts
University of California at Berkeley

Tristan Kimbrell
Temple University

Abstract. While recent infectious disease modeling efforts have started to explore the impact of spatiotemporal heterogeneity on disease dynamics, few models have been developed to investigate the implications of spatial heterogeneity on the optimality of different disease control strategies. In this chapter, we take a first step towards exploring under what situations spatially targeted vector control strategies are superior to aspatial strategies. Inspired by the dengue fever disease system, we develop a patch model constructed with a series of coupled ordinary differential equations to study the spatiotemporal time course of a dengue fever epidemic and the impact of different control strategies. We focus on two different movement patterns of hosts and vectors. In what we term the unlimited movement case, we assume that the vast majority of hosts and vectors are not constrained by distance and randomly move to a different patch during each time step. In the second case, which we term the limited movement case, we assume that only a very small percentage of hosts and vectors move to a neighboring patch during each time step. We compare three different rule of thumb control strategies: constant, spot, and ring control and investigate the impact of a time lag between the onset of an epidemic and the implementation of control efforts. We find that for the unlimited movement case, constant control is always optimal in terms of preventing the greatest number of hosts from becoming infected, while for the limited movement case we find that the time when control efforts begin relative to the onset of the epidemic is the key determinant of the optimality of different control strategies. For the limited movement case, when control efforts start immediately constant control is best, but when there is a time lag the spatially targeted spot or ring control is better. Taken as a whole our results suggest that the degree of spatial heterogeneity among infected hosts at the time control is implemented is a key factor determining which control strategy is optimal. When the distribution of infected hosts is spatially homogenous constant control is best. When the distribution of infected hosts is spatially heterogeneous a spatially targeted control strategy is best. We discuss the policy implications of our results for the implementation of control efforts in a real world setting and the greater need for more spatially explicit data on vector borne disease incidence and dynamics.

14.1 Introduction

While epidemiology has a long history of using mathematical models to study infectious diseases (Ross 1911, MacDonald 1957) and mathematical models have been instrumental in identifying the concept of epidemic thresholds, understanding the population dynamics of vector and host species, and designing control strategies, it is only relatively recently that mathematical models of infectious diseases have incorporated aspects of spatial heterogeneity to explore the role of space in disease dynamics and control.

Empirical studies of various disease systems have consistently demonstrated that real world populations are not spatially homogenous (Lajmanovich and York 1976, Dietz 1988, Sattenspiel and Dietz 1995, Galvani and May 2005). Heterogeneity may exist in the spatial distribution of vectors (Getis et al. 2003) and their dispersal patterns (Russell et al. 2005), in the probability of being bitten by an infected vector (Kelly and Thompson 2000), or in the infectiousness of hosts due to genetic or behavioral factors (Woolhouse et al. 1997, Lloyd-Smith et al. 2005).

Recent mathematical models of infectious disease have incorporated spatial heterogeneity using a number of different approaches including metapopulation models (e.g., Rodríguez and Torres-Sorando 2001, Dobson 2003, Luz et al. 2003, Lloyd and Jansen 2004, Favier 2005), network models (e.g., Newman 2003, Verdasca et al. 2005), and diffusion models (e.g., Raffy and Tran 2005, Tran and Raffy 2006).

These new spatial infectious disease models illustrate the multitude of ways different aspects of spatial heterogeneity affect disease dynamics. A few examples include: Gudelj and White's (2004) work on behavioral effects which shows that if a disease causes infected individuals to behave differently in how they move through space than noninfected individuals, then the behaviors may have large impacts on the ability of a disease to spread, and on the resulting spatial distribution of the population; the Favier et al. (2005) study on the effect that host and vector patch structure have on dengue disease dynamics, which demonstrates that by including heterogeneous patch structure the model more closely approximates real dengue epidemics in Easter Island, Belém, and Brasília; and Bjornstad and Grenfell's (2008) study on the spatiotemporal time course of measles epidemics in the United Kingdom which illustrates that the waiting time between epidemics is strongly determined by regional prevalence, spatial coupling, and the density of local susceptibles.

However, a gap in this growing modeling literature is the impact of spatial heterogeneity on the design of control efforts. While a few papers exist (Fulford et al. 2002, Gaff and Gross 2007, Asano et al. 2008), there has yet to be a systematic exploration of the influence of spatial heterogeneity on the design of efficient and effective control strategies. The need to understand how spatial heterogeneity may affect existing control efforts and inform on the design of new control efforts is especially acute for vector-borne infectious diseases in which both the host and disease vector often move. For many of these diseases no effective vaccine exists. The need for new control efforts is especially great for the dengue fever disease system.

Dengue is a viral disease that causes more illness and death than any other arbovirus and is endemic in more than 100 countries. Worldwide, there are approximately 2.500 billion people at risk of infection, and the World Health Organization (WHO) estimates that there are about 50-100 million cases per year (WHO 2002, PAHO 2002). *Aedes aegypti* is the main vector. It is a mosquito that lives in close association with humans in urban and sub-urban environments, ingesting preferably human blood and breeding in artificial containers (Gubler 1998, Service 1992). Dengue is generally considered a disease of urban areas and its epidemiology is highly related to the biology of the mosquito vector and human behavior, as well as the environment and the virus itself.

The incidence of dengue has increased significantly over the past 25 years (Gubler 2005), and it has been classified as an "emerging or uncontrolled disease" (TDR 2005). In the Americas, strong control campaigns eliminated *Ae. aegypti* from most of Central and South America during the 1950s, but discontinuation of the program led to re-infestation during the 1970s and 1980s and re-emergence of dengue (Gubler 1998). Globalization, population growth, and uncontrolled or unplanned urbanization have all been major factors influencing the current pandemic (Kuno 1995). These demographic and social changes, as well as a lack of effective mosquito control, have facilitated the spread and permanence of *Ae. aegypti* and dengue virus in many areas of the world (Gubler 1998).

Since there is no effective vaccine for dengue, vector control is the main approach for control and prevention. Although insecticide spraying has been used extensively, larval source reduction (eliminating or cleaning water-filled containers that can harbor *Ae. aegypti* larvae) is considered the most effective way of reducing and controlling the mosquito populations (Gubler 1998). These vertical control methods have had poor sustainability, as have community-based approaches with extensive health education and community outreach. Few places have achieved and documented successful source reduction efforts (Focks et al. 2000), and the increasing spread and incidence of dengue suggests that the current measures employed are generally ineffective, inappropriate, or are being applied incorrectly (WHO/TDR 2002, Ooi et al. 2006).

Given the failure of existing control efforts and the realization that spatial heterogeneity significantly influences diseases dynamics, there is a pressing need to develop infectious disease models that explore how different spatial control strategies perform in preventing outbreaks of vector-borne diseases. In this chapter, we take a first step towards understanding how spatial heterogeneity affects the design of disease control strategies for the prevention and mitigation of epidemic outbreaks of vector-borne disease in a spatial context.

Specifically, we numerically explore how the relative rates of movement of hosts and vectors influences the relative effectiveness of different spatially explicit vector-control strategies on a 10×10 grid of patches. The patch network may be imagined to be small villages in a rural setting or city blocks in an urban setting. To highlight the importance of spatial heterogeneity we focus on two extreme cases of host and vector movement: i) hosts and vectors are not movement limited and 75% of them move to another patch in each time step; and ii) host and vectors are highly movement limited with only 1% of hosts and vectors moving to a nearest neighbor patch in each time step. From hereafter i) is referred to as the unlimited movement case and ii) is referred to as the limited movement case.

To be as socially realistic as possible, we focus on rule-of-thumb control strategies because such strategies are most likely to be implemented by agencies tasked with disease control. The three strategies we explore are: a) constant, where the same control effort is applied in all patches, b) spot, where all control efforts are applied in the patch with the most infected hosts; and c) ring, where control is applied in the focal patch with the highest number of infected hosts as well as in the surrounding patches. We assume that the difference in cost in implementing the different control strategies is negligible. In addition, we explore how failing to immediately recognize a disease outbreak or a delay in implementing control strategies affects which rule-of-thumb control strategy is optimal.

For all control types the same total amount of control is used over the course of the whole epidemic. We judge the control method that produces the fewest number of total infected hosts over the course of the whole epidemic to be the best.

We use parameters taken from studies of dengue fever to parameterize our model. Dengue fever has been reported in the medical literature since 1779 (Rigau-Pérez et al. 1998), and many of the most important parameters thought to affect the dynamics of the disease have been empirically determined.

The layout of the rest of the chapter is as follows. In Section 14.2, we describe our coupled ordinary differential equation model in detail and give the model parameters. The effectiveness of different rule of thumb control strategies is presented in Section 14.3 and discussed in Section 14.4. The chapter concludes with Section 14.5 which suggests some areas of further research.

14.2 Model specification

To model the dynamics of the infectious disease we started with the classic SI ordinary differential equation framework for the vectors and SEIR ordinary differential equation framework for the hosts. However, since we are modeling disease dynamics on a network of interconnected patches, we modified the mean-field SEIR model so that both vectors and hosts may move between patches. Dengue fever has four serotypes that can infect humans; in this chapter we assume that only a single serotype is present in the system.

For an $N \times N$ network of patches six coupled ordinary differential equations specify disease dynamics in each patch at time t. For an arbitrary patch i, j the equations are as follows:

$$
\begin{aligned}
\frac{dV_S(i,j)}{dt} &= \mu \sum_{k,l=1}^{N} \mathrm{M}_{V_I} V_I(k,l) - \sum_{k,l=1}^{N} \mathrm{M}_{V_S} V_S(k,l) \Bigg[c(i,j) \\
&+ \frac{qb \sum_{k,l=1}^{N} \mathrm{M}_{H_I} H_I(k,l)}{\sum_{k,l=1}^{N} \mathrm{M}_{H_S} H_S(k,l) + \sum_{k,l=1}^{N} \mathrm{M}_{H_E} H_E(k,l) + \sum_{k,l=1}^{N} \mathrm{M}_{H_I} H_I(k,l) + \sum_{k,l=1}^{N} \mathrm{M}_{H_R} H_R(k,l)} \Bigg] \\
\frac{dV_I(i,j)}{dt} &= -\sum_{k,l=1}^{N} \mathrm{M}_{V_I} V_I(k,l) \big[c(i,j) + \mu \big] \\
&+ \frac{qb \sum_{k,l=1}^{N} \mathrm{M}_{V_S} V_S(k,l) \sum_{k,l=1}^{N} \mathrm{M}_{H_I} H_I(k,l)}{\sum_{k,l=1}^{N} \mathrm{M}_{H_S} H_S(k,l) + \sum_{k,l=1}^{N} \mathrm{M}_{H_E} H_E(k,l) + \sum_{k,l=1}^{N} \mathrm{M}_{H_I} H_I(k,l) + \sum_{k,l=1}^{N} \mathrm{M}_{H_R} H_R(k,l)} \\
\frac{dH_S(i,j)}{dt} &= \\
&- \frac{wb \sum_{k,l=1}^{N} \mathrm{M}_{V_I} V_I(k,l) \sum_{k,l=1}^{N} \mathrm{M}_{H_S} H_S(k,l)}{\sum_{k,l=1}^{N} \mathrm{M}_{H_S} H_S(k,l) + \sum_{k,l=1}^{N} \mathrm{M}_{H_E} H_E(k,l) + \sum_{k,l=1}^{N} \mathrm{M}_{H_I} H_I(k,l) + \sum_{k,l=1}^{N} \mathrm{M}_{H_R} H_R(k,l)} \\
\frac{dH_E(i,j)}{dt} &= -\rho \sum_{k,l=1}^{N} \mathrm{M}_{H_E} H_E(k,l) \\
&+ \frac{wb \sum_{k,l=1}^{N} \mathrm{M}_{V_I} V_I(k,l) \sum_{k,l=1}^{N} \mathrm{M}_{H_S} H_S(k,l)}{\sum_{k,l=1}^{N} \mathrm{M}_{H_S} H_S(k,l) + \sum_{k,l=1}^{N} \mathrm{M}_{H_E} H_E(k,l) + \sum_{k,l=1}^{N} \mathrm{M}_{H_I} H_I(k,l) + \sum_{k,l=1}^{N} \mathrm{M}_{H_R} H_R(k,l)} \\
\frac{dH_I(i,j)}{dt} &= \rho \sum_{k,l=1}^{N} \mathrm{M}_{H_E} H_E(k,l) - \theta \sum_{k,l=1}^{N} \mathrm{M}_{H_I} H_I(k,l) \\
\frac{dH_R(i,j)}{dt} &= \theta \sum_{k,l=1}^{N} \mathrm{M}_{H_I} H_I(k,l)
\end{aligned}
\tag{14.1}
$$

In the equations above, $V_S(i,j)$ is the number of susceptible vectors in patch i, j, $V_I(i,j)$ is the number of infected vectors, $H_S(i,j)$ is the number of susceptible hosts, $H_E(i,j)$ is the number of exposed hosts who are not yet infectious, $H_I(i,j)$ is the number of infectious hosts, and $H_R(i,j)$ is the number of recovered hosts. The parameter μ is the natural daily mortality rate of the vector, b is the bite rate of the vector per day, q is the probability of a vector becoming infected after feeding on an infected host, w is the probability of a susceptible host becoming infected after being bitten by an infected vector, $1/\rho$ is the number of days the virus is latent in the exposed host, and $1/\theta$ is the number of days the host is infectious.

Each state variable has its own associated movement matrix. For example, the matrix $\mathbf{M}_{V_S}$ is the movement matrix for susceptible vectors. The rate of vectors and hosts moving from a focal patch to other patches in the system was held constant during a model run.

We explored two extreme cases of host and vector movement: i) hosts and vectors are not movement limited and 75% of them move to another patch in each time step (unlimited movement); and ii) host and vectors are highly movement limited with only 1% of hosts and vectors moving to a nearest neighbor patch in each time step (limited movement). For all state variables for arbitrary patches k, l for i) $M_{kl} = 0.25$ for $k = l$ and $M_{kl} = .75/(N^2 - 1)$ for $k \neq l$ and for ii) $M_{kl} = 0.99$ for $k = l$; $M_{kl} = .0025$ for $k \neq l$ and $\|k - l\| = 1$, $M_{kl} = 0$ otherwise. In addition for case ii) we assumed that the boundaries were reflecting.

The variable $c(i, j)$ is the amount of control applied in patch i, j. When $c(i, j) = 0$ in all patches, we assume that the total vector population is constant, that vectors die at the rate of natural mortality, and that dead vectors regardless of disease status are instantly replaced by new adult susceptible vectors. However, when $c(i, j) \neq 0$, we assume that control measures kill both susceptible and infected adult vectors above the rate of natural mortality in patch i, j thereby reducing the total number vectors in the system.

We explored three different rule of thumb control strategies. a) Constant control in all of the patches. Under this control strategy c equals a constant in all patches of the model. b) Spot control, applied in only the patch with the highest number of infected hosts. As the patch containing the most number of infected hosts changes, the patch in which spot control is applied changes accordingly. c) Ring control, applied in the focal patch with the highest number of infected hosts, as well as in the surrounding patches. In the ring control strategy, the focal patch and the twelve patches two grid steps away were treated (unless the focal patch was near an edge, in which case correspondingly fewer patches were treated). In all of the control strategies examined, the same amount of total control was always applied to the system. Thus, when spot control was applied in one patch, c was N^2 times larger in that patch than the amount of control applied in one patch when the constant control strategy was examined.

In practice, control strategies may not begin at exactly the same time as infected hosts enter a system. There may be a significant lag between the emergence of infected hosts and the start of control. As a consequence, we examined how lags in the start of the three control strategies influenced the total number of infected hosts in the system. If a lag occurred, the total amount of control used in the system was still the same as when there was no lag. Thus, if a lag occurred, the rate of control used in a patch in the remaining time of the model run was greater than if there was no lag.

Table 14.1 shows the values of the parameters used in the various model runs. The model was initialized with 1000 susceptible vectors and hosts in each patch, and no infected vectors or hosts. In the case of limited vector and host movement the model was first run for 11,000 time steps to let the distribution of hosts and vectors come to a steady state due to the reflecting boundary condition. This was not necessary for the cases where vector and hosts moved freely. Ten infected hosts were then introduced into a corner patch. The model was then run for 2000 time-steps, which was sufficient time for the disease epidemic to travel through the population. The amount of control used was varied from none to the 0.03% increase in the natural vector mortality rate across all patches for all time steps for the constant control case and the equivalent total effect under the two other control strategies. Vector control was started at 0 and 600 time steps after the emergence of the first infected hosts to explore the effect of time lags in initiating control on the optimality of different control strategies.

Table 14.1. List of parameters used in most runs of the model.

Parameter	**Description**	**Value**
N^2	Number of patches	100
μ	Natural daily mortality rate of vector	0.11
b	Bite rate of vector per day	0.5
q	Probability of a vector feeding on an infected host becoming infected	0.38
w	Probability of a susceptible host bitten by an infected vector becoming infected	0.38
$1/\rho$	Days virus is latent in host	5
$1/\theta$	Days host is infectious	6

The equations were programmed in C++ and simulated using a Runge-Kutta-Fehlberg 4, 5 method. A typical model run on a modern desktop computer took no more than a few minutes.

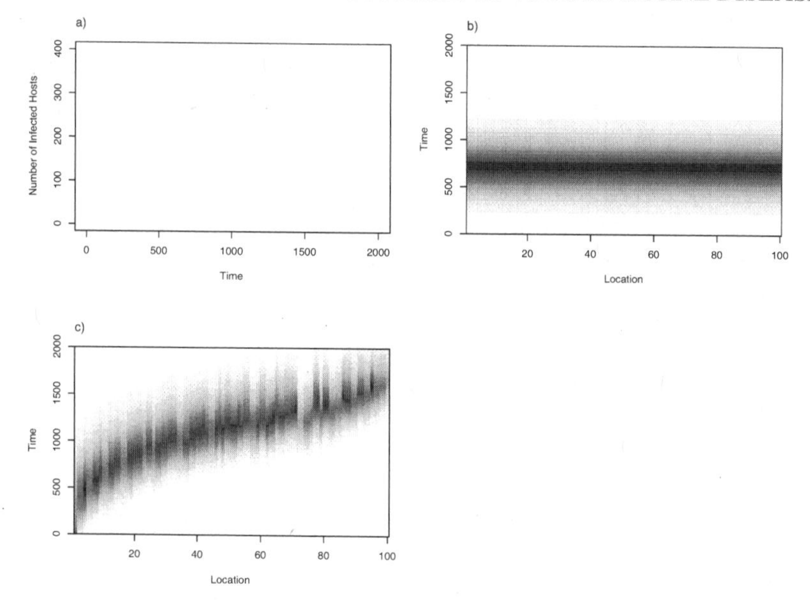

Figure 14.1 Spatiotemporal dynamics of an epidemic in the absence of control. Panel a) is a time series of the number of infected hosts over the course of a single epidemic. The solid line corresponds to the unlimited movement case and the dashed line corresponds to limited movement case. Panels b) and c) illustrate the spatiotemporal time course of the nonspatial and spatial cases respectively. The 10×10 patch network at each time slice has been transformed into a vector with location 0 corresponding to patch $(0, 0)$ and location 100 corresponding to patch $(10, 10)$. The darker the pixel the greater the number of infected hosts.

14.3 Results

No control

In the absence of any control efforts the two different movement rules for hosts and vectors led to a similar total number of infected hosts during the course of an epidemic. Out of the population of 100,000 total hosts 17,120 became infected for the unlimited movement case and 16,286 hosts became infected for the limited movement case. Thus, the unlimited movement case led to approximately 5% more infections than the limited movement case.

However, both the temporal and spatiotemporal aspects of the epidemic differed greatly between the two movement cases. Unlimited movement led to a much more rapidly developing epidemic that peaked earlier with a higher number of infected hosts at the peak (Figure 14.1a). In the limited movement case, the epidemic developed much more slowly, lasted longer, and had fewer infected hosts at the peak. Plotting the entire spatiotemporal course of the epidemic makes these temporal differences even more pronounced (Figures 1b, 1c). The unlimited movement case led to an epidemic that was homogenous in space. Infected hosts slowly built up through the whole population, with the peak number of infected hosts occurring at the same time in all the patches and then dying down at the same rate across all the patches. In contrast, the limited movement case led to an epidemic that was very heterogeneous in space. Infected hosts initially built up near where the first infected individuals were introduced and then progressed as a wave across the patch network.

Table 14.2. Total number of infected individuals for different dispersal kernels and control strategies. Best control strategy is indicated in *italics*. Panels a) & b) give the results for the unlimited movement case while panels c) & d) give the results for limited movement case. For panels a) and c) there is no time lag before starting control efforts while for panels b) and d) there is a 600-time step lag after the first infected hosts before starting control efforts. In the absence of control, scenarios a) and b) result in a total of 17,120 infected hosts while c) and d) results in 16,286 infected hosts.

a)

		Control Type		
		Constant	**Spot**	**Ring**
Control Amount	**0.0001**	*9955*	9978	9959
	0.0002	*3951*	4000	3958
	0.0003	*1613*	1648	1618

b)

		Control Type		
		Constant	**Spot**	**Ring**
Control Amount	**0.0001**	*15560*	15565	15561
	0.0002	*14355*	14371	14358
	0.0003	*13404*	14432	13408

c)

		Control Type		
		Constant	**Spot**	**Ring**
Control Amount	**0.0001**	*4379*	7424	7460
	0.0002	*1240*	5939	4787
	0.0003	*640*	5459	3769

d)

		Control Type		
		Constant	**Spot**	**Ring**
Control Amount	**0.0001**	10543	*10354*	10478
	0.0002	6796	7367	*6404*
	0.0003	5164	6172	*5154*

Impact of control

To highlight the impact of control we first restrict our discussion to the constant control strategy with no time lag from the emergence of the first infected hosts and the initiation of control efforts. For both movement cases even small amounts of control had a significant impact on the total number of infected hosts during the course of an

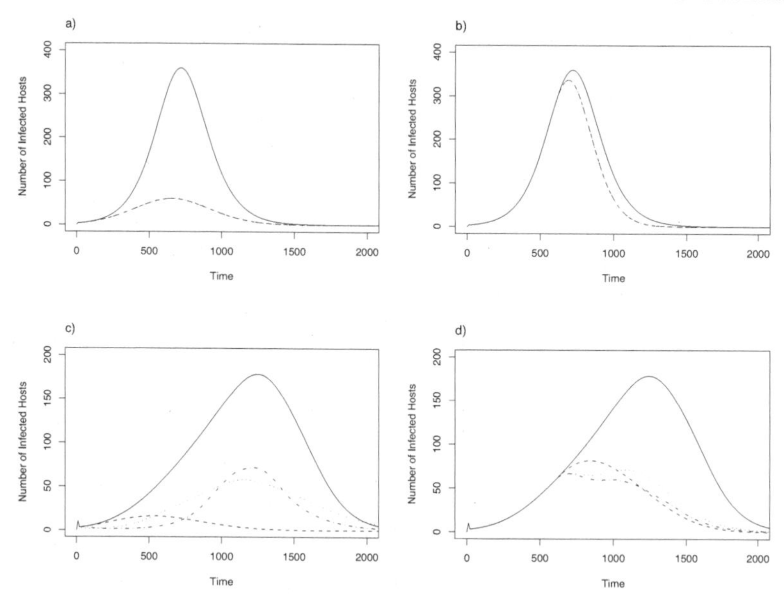

Figure 14.2 Time series of number of infected hosts for different hosts and vector movement rates, vector control strategies and time lag before initiating vector control. The control amount for all cases is 0.0002. Panels a) & b) illustrate the unlimited movement case while panels c) & d) illustrate the limited movement case. For panels a) and c) there is no time lag before starting control efforts while for panels b) and d) there is a 600-time step lag after the first infected hosts before starting control efforts. —: no control; - - - -: constant control; . . . : spot control; · — · — ·—: ring control.

epidemic (Table 14.2). A 0.01% increase in the intrinsic vector mortality rate for the unlimited movement case cut the number of infected hosts nearly in half to 9,955, while an increase of vector morality rate by 0.03% led to the number of infected individuals dropping by nearly 90%. The results for the limited movement case were even more dramatic. A 0.01% increase in the intrinsic vector morality rate reduced the number of cases by 75% while a 0.03% increase cut the number of cases by more than 95%.

The effects of control on the temporal and spatiotemporal aspects of the epidemic for the two movement cases were equally pronounced. For both cases with a control amount of 0.0002, Figures 14.2a and 14.2c clearly illustrate a dramatic drop in the number of infected individuals at each time step over the course of the epidemic. However, the length of the epidemic with control differed for the two movement cases. With unlimited movement the length of the epidemic was more or less the same as without control (Figure 14.2a), and although the spatial intensity was greatly decreased, all patches still had infected individuals (Figure 14.4a). For the limited movement case, however, application of control dramatically decreased the length of the epidemic (Figure 14.2c) and decreased its spatial intensity (Figure 14.6a). In fact,

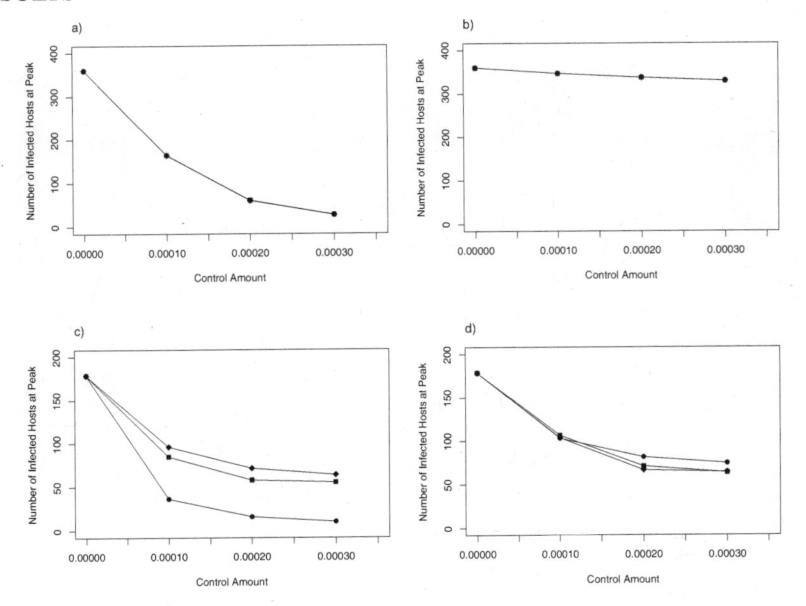

Figure 14.3 Total number of infected hosts at epidemic peak for different host and vector movement rates, vector control strategies, vector control amounts, and time lag before initiating vector control. Panels a) & b) illustrate the unlimited movement case while panels c) & d) illustrate the limited movement case. For panels a) and c) there is no time lag before starting control efforts while for panels b) and d) there is a 600-time step lag after the first infected hosts before starting control efforts. • - constant control; □ - spot control; and ◇ - ring control.

with control the majority of patches did not have a single infected host.

Optimality of different spatial control strategies

When treatment is initiated immediately after the first host becomes infected, regardless of the type of host and vector movement and amount of control used, the constant control strategy is superior. For the unlimited movement case the superiority of constant control over the other control strategies was very small regardless of the amount of control used. The difference in the total number of infected hosts between the different control amounts and control strategies was less than 3% (Table 14.2a). In addition, there was no difference between the control strategies in terms of the total number of hosts infected at the peak or the number of days for which there were more than 10 infected hosts, a measure of the total length of the epidemic (Figure 14.3a, 14.4a). Finally, the temporal (Figure 14.2a) and spatiotemporal course of the epidemic for the different control strategies were very similar (Figure 14.5a, b, c) though the spot control (Figure 14.5b) leads to a slightly greater number of infected hosts at the peak.

The superiority of constant control over spot and ring was much more pronounced

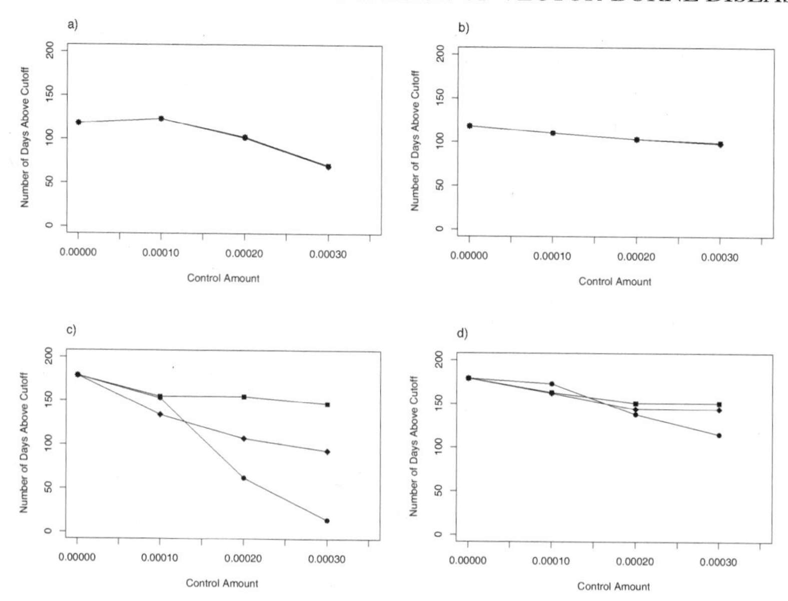

Figure 14.4 Number of days the total number of infected hosts is above 10 individuals for different host and vector movement rates, vector control strategies, vector control amounts, and time lag before initiating vector control. Panels a) & b) illustrate the unlimited movement case while panels c) & d) illustrate the limited movement case. For panels a) and c) there is no time lag before starting control efforts while for panels b) and d) there is 600-time step lag after the first infected hosts before starting control efforts. • : constant control; □ :spot control; and ◇:ring control.

for the limited movement case. With a control amount of 0.001, constant control produced an average of 58% as many infected hosts as compared to the ring or spot control strategies. For a higher control amount (0.0003) the superiority of constant control was even greater leading to only an average of 14% of the number of infected hosts as compared to the other two control strategies (Table 14.2c). In addition, constant control greatly reduces the number of hosts infected at the peak (Figure 14.2c & 14.3c) as well as the duration of the epidemic (Figure 14.2c & 14.4c). Finally, looking at the spatiotemporal course of the infection, it is clear that the spot and ring control strategies are far worse than constant control in terms of the length and spatial extent of the epidemic (Figures 14.6a, b & c).

Impact of time lags on optimality of different spatial control strategies

Delaying the start of control efforts greatly reduced the efficacy of control in all situations. The delay greatly increased the total number of hosts infected over the course of the epidemic, the length of the epidemic, and spatiotemporal intensity of the epidemic. In the unlimited movement cases where the epidemic developed and

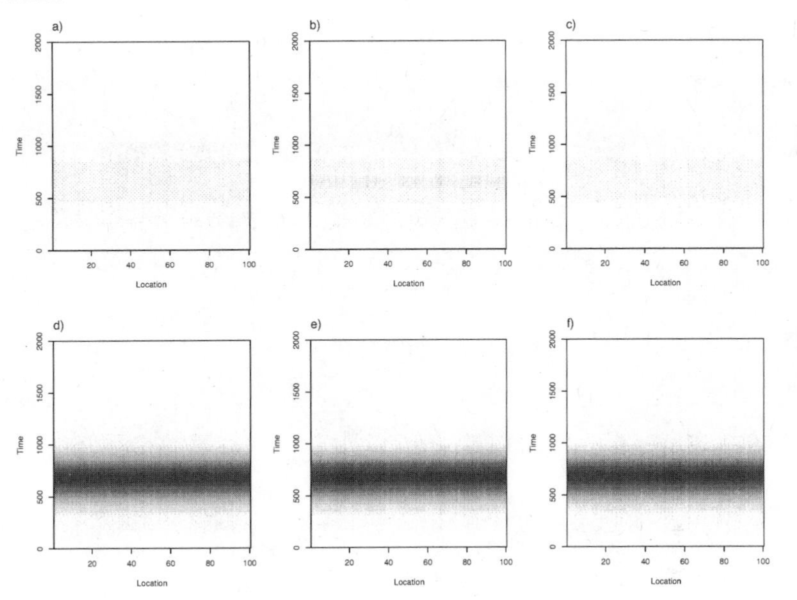

Figure 14.5 Spatiotemporal time series of an epidemic for different vector control strategies, and time lag before imitating control efforts for the unlimited movement case and 0.0002 total amount of control. Panel a), b), & c) corresponding to no time lag while d), e), & f) correspond to a 600-time step lag before starting control efforts. Control strategies employed are as follows: a),d): constant; b),e): spot; c),f) ring.

intensified very quickly, regardless of the amount of control or control strategy used, the total number of infected hosts was only reduced by 20% compared to no control efforts (Table 14.2b). Likewise, the temporal and spatiotemporal time course of the epidemic was also very similar to the no control case (Figure 14.2b & Figures 14.5b, c, d). However, similar to the situation where control was initiated immediately, constant control was superior but only slightly so. The superiority was so slight that the number of hosts infected at the peak (Figure 14.3b) and total length of the epidemic (Figure 14.4b) were indistinguishable for the different control strategies.

The impact of a delay in initiating control efforts for the limited movement case had a much more pronounced and diverse impact on the efficacy of control efforts as compared to the unlimited movement case. While as with the unlimited movement case the overall number of infected hosts increased with a time delay in initiating control the impact was less and much more dependent on the amount of control used. For a small amount of control (0.0001) the total number of infected hosts was reduced by 36% as compared to no control cases and for a large amount of control (0.0003) the total number of infected hosts was reduced by 70% as compared to the no control case (Table 14.2d). In addition, there was an interaction between the amount of control used and the control strategy employed and the temporal and spatiotemporal time course of the epidemic. For example, for a low amount of control (0.0001) spot

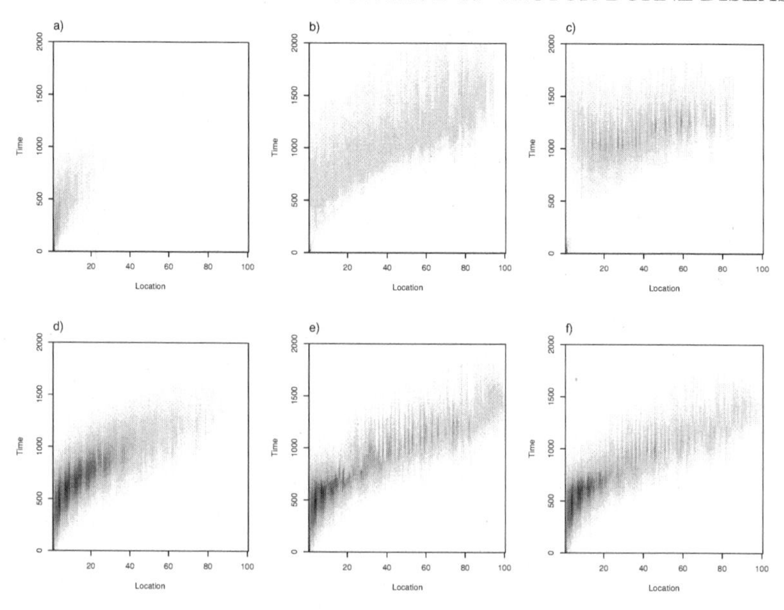

Figure 14.6 Spatiotemporal time series of an epidemic for different vector control strategies and time lag before initiating control efforts for the limited movement case and 0.0002 total amount of control. Panel a), b), & c) corresponding to no time lag while d), e), & f) correspond to a 600-time step lag before starting control efforts. Control strategies employed are as follows: a),d): constant; b),e): spot; c),f): ring.

and ring strategies led to a lower number of infected hosts at the peak (Figure 14.3d) and shorter overall epidemic (Figure 14.4d).

Most importantly, though, with a time lag in starting control for the limited movement case, constant control was no longer superior in terms of the total number of infected hosts over the course of the epidemic no matter the amount of vector control used (Table 14.2d). For low amounts of control a spot strategy was slightly superior and for greater amounts of control the ring strategy was slightly superior. In addition, the temporal and spatiotemporal course of the epidemic differed across strategies. For a control amount of 0.0002, constant control led to a shorter, less spatially extensive, initially more intense epidemic (Figure 14.2d & Figure 14.6c) as compared to ring or spot control strategies (Figure 14.2d and Figure 14.6c & 14.6d).

14.4 Discussion

The object of our modeling effort was to explore the impact of spatiotemporal heterogeneity on the relative superiority of different spatial control strategies. The unlimited movement case approximates a mean-field model that is spatiotemporal homogenous. For this case, vectors and hosts are more or less randomly mixed in every time step meaning that infected hosts and vectors quickly spread throughout the whole population and throughout the epidemic the number of infected hosts is more or less spatially homogenous. Thus, it is not surprising that an aspatial control strategy is optimal since it is an effective way to quickly reduce vector populations across the whole community simultaneously. However, it is a bit surprising that the spatially targeted spot and ring control strategies had a similar impact in terms of reducing the totally number of infected hosts. The rapid random mixing of hosts and vectors means that the effect of any control effort rapidly propagates through the entire population.

The limited movement case tends towards the opposite end of the spectrum with extreme spatial heterogeneity in the system. In this case we might have expected that spatially targeted control strategies would perform much better than an aspatial control strategy. We did not find this. When control efforts began shortly after disease emergence a constant control strategy was much better than any spatially targeted control strategy. The reason for this is that constant control kills vectors throughout the system, whereas targeted control only kills vectors in a limited area. Thus, as targeted control is continually applied in a limited area, there is eventually a diminishing return, and fewer total vectors are killed with target control than with constant control. This would seem to indicate that quickly reducing vector populations across the whole community is the fastest way to end an epidemic when the host and vectors are limited in movement. However, when there is a delay in implementing control measures in the limited movement case, spatially targeted control strategies are slightly better than the aspatial approach. The time lag case creates extreme heterogeneity in the spatial distribution of infected hosts before control efforts are implemented and therefore targeted control strategies make more sense.

Putting this all together, the results of the modeling effort indicate that what is most important in choosing a control effort is to understand the spatial distribution of infected hosts and vectors in the system when the time control efforts are initiated. When the spatial distribution of infected hosts is spatially homogenous then an aspatial control strategy is optimal. However when the spatial distributions of infected hosts are highly spatially heterogeneous then spatially targeted control strategies are optimal. It is important to note that this finding appears to hold regardless of the number of infected hosts as the time control is initiated and the movement rates of hosts and vectors.

Our findings have interesting implications for the implementation of vector-control strategies in a real world setting. First and foremost, from the discussion above it is clear that what is most important is to understand the spatial distribution of infected hosts at the time that a decision is made concerning the implementation of a particular vector-control strategy. This calls for better and more detailed disease surveillance efforts that provide real-world information on the time and location of infected hosts. In addition, these enhanced surveillance efforts might also lead to the ability to reduce control efforts when the risk of outbreak is low and thereby concentrate control efforts when the risk of outbreak is greatest. This has two distinct advantages. Many epidemic disease control efforts, especially for dengue, rely on individual citizens reducing the number of breeding sites in their homes and yards. Continual control efforts of this sort have low sustainability leading to reduced efficacy over long periods of time. Warning citizens to act when the risk of outbreak is the greatest is likely to lead to much higher compliance. Secondly, resources for vector control are often limited. Therefore using them when their temporal or spatial effectiveness is greatest is most efficient.

The second policy implication is how the choice of an objective function influences the optimality of different control strategies. In our model, we judged the optimality of a control effort from the standpoint of which strategy led to the fewest number of cases. For diseases with high morbidity rates this may make the most sense but for diseases that do not have high morbidity rates but require substantial medical care it may be socially optimal to focus on other aspects of the spatiotemporal time course of the epidemic. For these types of diseases the key objective may be to reduce the total number of cases at any one time rather than the total number of cases over the whole course of the epidemic. In our model, we saw that in the limited movement case spot and ring produced a comparable number of cases (Figure 14.2c) but spot control led to a longer but less intense epidemic as compared to ring control. Thus, it may be important to include other benefits of control other than the absolute reduction in the number of infected hosts over the course of the whole epidemic as objectives of control.

The other aspect of the impacts of control that should be taken into account when formulating a proper objective function is the cost of control measures. In our model, we assumed that constant, spot, and ring control measures had the same cost to implement. This is probably not true in a real world setting. It is likely that constant control could end up costing much more to implement than ring or spot control. Constant control requires repeated visits to all patches while ring and spot control are much more spatiotemporally targeted. The interesting part of our results is that when there is a significant time lag in initiating treatment then there was very little difference between the three control strategies in terms of the absolute number of total infected hosts, meaning that costs and feasibility of implementation may be appropriate guiding factors in terms of the choice of control efforts.

14.5 Conclusion

Our findings point to two key areas of future research. First, there is a need to obtain better parameters for our model. While we gathered a number of parameters from the literature the parameters most likely influencing model results (host movement, vector movement, and relative patch size) are not well known. While there has been considerable work on human movement through space (e.g., Hufnagel et al. 2004; Crépey and Barthélemy 2007) there has been very little research on how vectors move through space. To develop optimal control strategies, it will be necessary to gain a better understanding of the range of these parameters.

There are two ways in which this might be accomplished. The first is more direct observational studies of host and vector movement. The second and perhaps more logistically feasible is to employ a pattern-oriented modeling (POM) approach (Grimm *et al.* 2005). POM provides a strategy to deal with two problems concerning the modeling of large multifaceted systems: complexity and uncertainty. POM leverages the fact that patterns contain "coded" information concerning underlying model processes and structure. Even our relatively simple models indicate that different movement rates lead to readily observed temporal and spatiotemporal differences in the dynamics of epidemics. Thus, the greatest need in terms of empirical data collection may lie in developing detailed spatially explicit databases of disease incidence.

The second key area of future research is to place our optimization in a proper optimal control framework. The theory and methods to optimize systems of coupled ordinary differential equation are well developed (Lenhart and Workman 2007). Using a proper optimal control framework would readily allow us to incorporate a range of costs and benefits of different control strategies and determine how far away our rule of thumb control strategies are from the optimal ones for a given disease system.

14.6 Acknowledgments

This research was made possible by funding from NSF Division of Biological Infrastructure Program (DBI-0628908) and the University of Miami, FL. We thank Chris Cosner for helpful discussions during the development of the modeling approach.

14.7 References

E.L. Asano, J. Gross, S. Lenhart, and L.A. Real (2008), Optimal control of vaccine distribution in a rabies metapopulation model, *Mathematical Biosciences and Engineering* **5**:219-238.

O.N. Bjornstad and B.T. Grenfell (2008), Hazards, spatial transmission and timing of outbreaks in epidemic metapopualtions, *Environ. Ecol. Stat.* **15**:265-277.

P. Crépey, and M. Barthélemy (2007), Detecting robust patterns in the spread of epidemics: A case study of influenza in the United States and France, *American Journal of Epidemiology* **166**:1244-1251.

K. Dietz (1988), Mathematical models for transmisson and control of malaria, in *"Principles and Practice of Malariology,"* ed. by W. Wernsdofer and Y. McGregor, Churchill Livingstone, Edinburgh, pp. 1091-1133.

A. Dobson (2003), Metalife!, *Science* **301**:1488-1490.

C. Favier, D. Schmit, C.D.M.M.L. Graf, B. Cazelles, N. Degallier, B. Mondet, and M.A. Dubois (2005), Influence of spatial heterogeneity on an emerging infectious disease: The case of dengue epidemics, *Proc. Biol. Sci.* **272**:1171-1177.

D.A. Focks, R.J. Brenner, J. Hayes, and E. Daniels (2000), Transmission thresholds for dengue in terms of Aedes aegypti pupae per person with discussion of their utility in source reduction efforts, *American Journal of Tropical Medicine & Hygiene* **62**:11-18.

G.R. Fulford, M.G. Roberts, and J.A.P. Heesterbeek (2002), The metapopulation dynamics of an infectious disease: Tuberculosis in possums, *Theoretical Population Biology* **61**:15-29.

H.D. Gaff and L.J. Gross (2007), Modeling tick-borne disease: A metapopulation model, *Bulletin of Mathematical Biology* **69**:265-288.

A.P. Galvani and R.M. May (2005), Epidemiology: Dimensions of superspreading, *Nature* **438**:293-295.

A. Getis, A.C. Morrison, K. Gray, and T.W. Scott (2003), Characteristics of the spatial pattern of the dengue vector, *Aedes aegypti*, in Iquitos, Peru, *American Journal of Tropical Medicine & Hygiene* **69**:494-505.

V. Grimm, E. Reilla, T. Berger, F. Jeltsch, W.M. Mooij, S.F. Railsback, H. Thulke, J. Weiner, T. Wiegand, and D. DeAngelis (2005), Pattern-oriented modeling of agent-based complex systems: Lessons from ecology, *Science* **310**:987-91.

D.J. Gubler (1998), Dengue and dengue hemorrhagic fever, *Clinical Microbiology Reviews* **11**:480-496.

D.J. Gubler (2005), The emergence of epidemic dengue fever and dengue hemorrhagic fever in the Americas: A case of failed public health policy, *Panamerican Journal of Public Health* **17**:221-224.

I. Gudelj and K.A.J. White (2004), Spatial heterogeneity, social structure and disease dynamics of animal populations, *Theoretical Population Biology* **66**:139-149.

L. Hufnagel, D. Brockmann, T. Geisel, and R. May (2004), Forecast and control of epidemics in a globalized world, *Proceedings of the National Academy of Sciences of the United States of America* **101**: 15124-15129.

D.W. Kelly and C.E. Thompson (2000), Epidemiology and optimal foraging: Modeling the ideal free distribution of insect vectors, *Parasitology* **120**:319-327.

G. Kuno (1995), Review of the factors modulating dengue transmission, *Epidemiol Rev.* **17**:321-335.

A. Lajmanovich and J.A. Yorke (1976), A deterministic model of gonorrhea in a nonhomgeous population, *Math Biosciences* **28**:221-236.

S. Lenhart and J.T. Workman (2007), *Optimal Control Applied to Biological Models,* Chapman & Hall/CRC, New York.

A.L. Lloyd and V.A.A. Jansen (2004), Spatiotemporal dynamics of epidemics: Synchrony in metapopulation models, *Mathematical Biosciences* **188**:1-16.

J.O. Lloyd-Smith, S.J. Schreiber, P.E. Kopp, and W.M. Getz (2005), Superspreading and the effect of individual variation on disease emergence, *Nature* **438**:355-359.

P.M. Luz, C.U.T. Codeao, E. Massad, and C.J. Struchiner (2003), Uncertainties regarding dengue modeling in Rio de Janeiro, Brazil, *Memorias do Instituto Oswaldo Cruz* **98**:871-878.

G. Macdonald (1957), *The Epidemiology and Control of Malaria,* Oxford University Press, London.

M.E.J. Newman (2003), The structure and function of complex networks, *SIAM Review* **45**:167-256.

E.E. Ooi, K.T. Goh and D.J. Gublert (2006), Dengue prevention and 35 years of vector control in Singapore, *Emerging Infection Diseases* **12**: 887-893.

(PAHO), Pan American Health Organization, Rockefeller Foundation, Pediatric Dengue Vaccine Initiative (2002), Report: Workshop on Dengue Burden Studies. Washington DC, USA. Accessed January 2006. Available from: http://www.paho.org/English/AD/DPC/CD/burden-dengue-11-2002.htm.

M. Raffy and A. Tran (2005), On the dynamics of flying insects populations controlled by large scale information, *Theor. Popul. Biol.* **68**:91-104.

J.G. Rigau-Pérez, G.G. Clark, D.J. Gubler, P. Reiter, E.J. Sanders, and A.V. Vorndam (1998), Dengue and dengue haemorrhagic fever, *The Lancet* **352**:971-977.

D.J. Rodriguez and L. Tores-Sorando (2001), Models of infectious diseases in spatially heterogeneous environments, *Bull. Math. Biol.* **63**: 547-571.

R. Ross (1911), *The Prevention of Malaria,* Murray, London.

R.C. Russell, C.R. Williams, R.W. Sutherst, and S.A. Ritchie (2005), *Aedes (Stegomyia) albopictus* – a dengue threat for southern Australia? *Communicable Diseases Intelligence* **29**:296-298.

L. Sattenspiel and K. Dietz (1995), A structured epidemic model incorporating geographic-mobility among regions, *Mathematical Biosciences* **128**:71-91.

M.W. Service (1992), Importance of ecology in *Aedes aegypti* control, *Southeast Asian Journal of Tropical Medicine & Public Health* **23**:681-690.

(TDR), Special Programme for Research and Training in Tropical Diseases (2005), TDR Diseases. Accessed January 2006. Available from: http://www.who.int/tdr/diseases/default.htm

A. Tran and M. Raffy (2006), On the dynamics of dengue epidemics from large-scale information, *Theoretical Population Biology* **69**:3-12.

J. Verdasca, M.M.T. da Gama, A. Nunes, N.R. Bernardino, J.M. Pacheco, and M. C. Gomes (2005), Recurrent epidemics in small world networks, *J. Theor. Biol.* **233**:553–561.

(WHO), World Health Organization (2002), Dengue and dengue hemorrhagic fever. Fact sheet 117. Accessed January 2006. Available from: http://www.who.int/mediacentre/factsheets/fs117/en/

M.E. Woolhouse, C. Dye, J.F. Etard, T. Smith, J.D. Charlwood, G.P. Garnett, P. Hagan, J.L. Hii, P.D. Ndhlovu, R.J. Quinnell, C.H. Watts, S.K. Chandiwana, and R.M. Anderson. (1997), Heterogeneities in the transmission of infectious agents: Implications for the design of control programs, *Proceedings of the National Academy of Science USA* **94**:338-342.

CHAPTER 15

Modeling spatial spread of communicable diseases involving animal hosts

Shigui Ruan
University of Miami

Jianhong Wu
York University

Abstract. In this chapter, we review some previous studies on modeling spatial spread of specific communicable diseases involving animal hosts. Reaction-diffusion equations are used to model these diseases due to movement of animal hosts. Selected topics include the transmission of rabies in fox populations (Källen et al., 1984; Källen et al., 1985; Murray et al., 1986), dengue (Takahashi et al., 2005), West Nile virus (Lewis et al., 2006; Ou & Wu, 2006), hantavirus spread in mouse populations (Abramson and Kenkre, 2002), Lyme disease (Caraco et al., 2002), and feline immunodeficiency virus (FIV) (Fitzgibbon et al., 1995; Hilker et al., 2007).

15.1 Introduction

Spatial spread of communicable diseases is closely related to the spatial heterogeneity of the environment and the spatial-temporal movement of the hosts. Mathematical modeling of disease spread normally starts with the consideration of the transmission dynamics within a population which is homogeneous in terms of host structures and environmental variation, and then follows by the examination of the impact on the transmission dynamics of the refined and detailed biological/epidemiological structures and patterns of spatial dispersal/diffusion of the hosts.

Epidemic theory for homogeneous populations has shown that the *basic reproductive number*, which may be considered as the fitness of a pathogen in a given population, must be greater than unity for the pathogen to invade a susceptible population (Anderson and May, 1991; Brauer and Castillo-Chavez, 2000; Diekmann and Heesterbeek, 2000; Edelstein-Keshet, 1988; Jones and Sleeman, 2003; Murray, 2003;

Thieme, 2003). It is natural to ask how spatial movement of the hosts affects the spatial-temporal spread pattern of the disease if the basic reproduction number for an otherwise homogeneous population exceeds unity.

Answers to the above question obviously depend on the manner in which hosts move into, out of, and within the considered geographical region. For example, adding an immigration term so that infective individuals enter the system at a constant rate clearly allows the persistence of the disease, because if it dies out in one region then the arrival of an infective from elsewhere can trigger another epidemic. Indeed, a constant immigration term has a mildly stabilizing effect on the dynamics and tends to increase the minimum number of infective individuals observed in the models (Bolker and Grenfell, 1995). Spread of diseases in a heterogeneous population has also been intensively studied using patchy or metapopulation models. These models are formulated under the assumption that the host population under consideration can be divided into multipatches so that the host population within a patch is considered as homogeneous, and the heterogeneity is associated with the rates with which individuals move from one patch to another (Arino and van den Dreissche, 2006).

Another popular way to incorporate the spatial movement of hosts into epidemic models is to assume some types of host random movements, leading to reaction-diffusion equations. See, for example, Busenberg and Travis (1983), Capasso (1978), Capasso and Wilson (1997), De Mottoni et al. (1979), Gudelj et al. (2004), Fitzgibbon et al. (2007), Webb (1981). This strand of theoretical developments built on the pioneering work of Fisher (1937), who used a logistic-based reaction-diffusion model to investigate the spread of an advantageous gene in a spatially extended population. With initial conditions corresponding to a spatially localized introduction, such models predict the eventual establishment of a well-defined invasion front which divides the invaded and uninvaded regions and moves into the uninvaded region with a constant velocity. The velocity at which an infection wave moves is set by the rate of divergence from the (unstable) disease-free state and can be determined by linear methods (Murray, 2003).

Most reaction-diffusion (or reduced/related space-dependent integral) epidemic models are space-dependent extensions of the classical Kermack-McKendrik (Kermack and McKendrik 1927) deterministic compartmental model for a directly transmitted viral or bacterial agent in a closed population consisting of susceptibles, infectives, and recovereds. Their model leads to a nonlinear integral equation which has been studied extensively. The deterministic model of Bartlett (1956) predicts a wave of infection moving out from the initial source of infection. Kendall (1957) generalized the Kermack-McKendrik model to a space-dependent integro-differential equation. Aronson (1977) argued that the three-component Kendall model can be reduced to a scalar one and extended the concept of asymptotic speed of propagation developed in Aronson and Weinberger (1975) to the scalar epidemic model. The Kendall model assumes that the infected individuals become immediately infectious and does not take into account the fact that most infectious diseases have an incubation period. This incubation period was considered by Diekmann (1978, 1979) and Thieme (1977a, 1977b, 1979), using a nonlinear (double) integral equation model. For further study on velocity of spatial spread, we refer to Mollison (1991), van den Bosch et al. (1990), the monograph of Rass and Radicliffe (2003), and references cited therein. Most of these studies concern the existence of traveling waves, and their relation to the disease propagation/spread rate. For additional studies, see Ai and Huang (2005), Cruickshank et al. (1999), Hosono and Ilyas (1995), Kuperman and Wio (1999), Zhao and Wang (2004), etc.

Despite these studies on reaction-diffusion epidemic models, however, there are very few studies on modeling spatial spread of specific diseases using partial differential equation models. In this chapter, we review some previous studies on modeling spatial spread of specific communicable diseases using reaction-diffusion equations. Selected topics include the transmission of rabies in fox population (Källen et al., 1984; Källen et al., 1985; Murray et al., 1986), dengue (Takahashi et al., 2005), West Nile virus (Lewis et al., 2006; Ou and Wu, 2006), hantavirus spread in mouse populations (Abramson and Kenkre, 2002), Lyme disease (Caraco et al., 2002), and feline immunodeficiency virus (FIV) (Fitzgibbon et al., 1995; Hilker et al., 2007).

15.2 Rabies

The celebrated studies by Källen (1984), Källen et al. (1985), and Murray et al. (1986) about the spatial spread of rabies among foxes show the feasibility and usefulness of utilizing a simple reaction-diffusion model for the description of transmission dynamics and spread patterns of specific diseases and for the qualitative evaluation of various space-relevant control strategies. These studies give a fine example of how to build a reaction-diffusion model based on the known ecology of the host behavior and the detailed epidemiology of the disease progression, how to use known data and facts to determine model parameter values, how to calculate the speed of propagation of the epizootic front and the threshold for the existence of an epidemic, and how to use models to quantify and evaluate space-relevant control strategies. They also demonstrate the trade-off between simplicity and the number of parameters that have to be estimated from field studies. It is therefore natural that we start with a brief introduction of these studies to illustrate some of the basic ideas and techniques involved in reaction-diffusion models for disease spread.

Rabies, a viral infection of the central nervous system, is transmitted by direct contact. The dog is the principal transmitter of the disease to man, and it is a particularly horrifying disease for which there is no known case of a recovery once the disease has reached the clinical stage. The aforementioned studies examined the rabies epidemic, which started in 1939 in Poland and moved steadily westward at a rate of 30-60 km per year. The red fox was the main carrier, and victim, of the rabies epidemic under consideration, although most mammals are thought to be susceptible to the disease and although an epidemic, which was mainly propagated by racoons, was also moving rapidly up the east coast of America during that period and subsequently.

The basic model of Källen et al. (1985) is built on the assumptions that foxes are the main carriers of rabies in the rabies epizootic considered, the rabies virus is normally transmitted by bite, and rabies is fatal in foxes. It also assumes that susceptible foxes are territorial, but once the virus enters the central nervous system it induces behavioral changes in its host and, in particular, if it enters the limbic system the foxes become aggressive, lose their sense of direction and territorial behavior, and wander about in a more or less random way.

Let $S(x,t)$ and $I(x,t)$ be the total number of susceptible foxes and the total number of infective foxes, respectively, in the space-time coordinate (x,t) and ignore the incubation period at the moment. Then the model formulated in a one-dimensional unbounded domain takes the form (Källen et al., 1985)

$$\begin{aligned}\frac{\partial S}{\partial t} &= -\beta S(x,t)I(x,t),\\ \frac{\partial I}{\partial t} &= D\frac{\partial^2 I}{\partial x^2} + \beta S(x,t)I(x,t) - \mu I(x,t),\end{aligned} \tag{15.1}$$

where β is the transmission coefficient, μ^{-1} is the life expectancy of an infective fox, and D is the diffusion coefficient.

The basic reproduction number of the corresponding ODE model is $R_0 = \beta S_0/\mu$, with S_0 being the initial susceptible population (with homogeneous environment). If $R_0 < 1$ then the mortality rate is greater than the rate of recruitment of new infectives, and hence the infection is expected to die out quickly. We thus obtain the minimum fox density $S_c := \mu/\beta$ below which rabies cannot persist. It was indeed proven (Källen, 1984) that if $R_0 < 1$, $I(\cdot,0) \geq 0$ has bounded support, and if $S(x,0) = S_0$ for $x \in R$, then $I(x,t) \to 0$ as $t \to \infty$ uniformly on R.

The case where $R_0 > 1$ indicates the persistence of the disease in a spatially homogeneous setting. The spatial diffusion then propagates the disease so that a small localized introduction of rabies evolves into a traveling wave with a certain wave speed, that is, a solution with $I(x,t) = f(z)$, $S(x,t) = g(z)$ with the wave variable $z = x - ct$ so that the wave forms (profiles) f and g are determined by the asymptotic boundary value problem

$$\begin{aligned}&Df'' + cf' + \beta fg - \mu f = 0, cg' - \beta fg = 0;\\ &f(\pm\infty) = 0,\ g(+\infty) = S_0, g(-\infty) = S_\infty,\end{aligned}$$

where primes denote differentiation with respect to z, S_∞ gives the number of susceptible foxes that remain after the infective wave has passed, and this number is found by solving the final size equation

$$S_\infty/S_0 - R_0^{-1}\ln(S_\infty/S_0) = 1.$$

The existence of traveling waves with speeds larger than $c_0 = 2\sqrt{1 - R_0^{-1}}$ is established by Källen (1984) and Källen et al. (1985), and the importance of the traveling wave with the minimal wave speed c_0 is shown by Källen (1984). Namely, it was shown that if $I(\cdot,0)$ has compact support, then for every $\delta > 0$ there exists N so that $I(x + c_0 t - \ln t/c_0, t) \leq \delta$ for every $t > 0$ and for all $x > N$. Therefore, if a fox travels with speed $c(t) = c_0 - (c_0 t)^{-1}\ln t$ towards $+\infty$ (in space) to the right of the support of $I(\cdot,0)$, the infection will never overtake the fox. In other words, the asymptotic speed of the infection must be less than $c(t)$. As a consequence, if $I(x,t)$ takes the form of a traveling wave for large t, it must do so for the one with the minimal speed c_0.

Estimating such a propagation speed is feasible once we know the relevant parameter values. In (Murray et al., 1986), R_0 was set to 2 according to the observed mortality rate $65 - 80\%$ during the height of the epizootic. The diffusion coefficient D is estimated to be 60 km^{2}yr^{-1}, using the average territory of a fox and the mean time a fox stays in its territory. This yields the minimal wave speed near 50 km per year, in good agreement with the empirical data from Europe.

The diffusion model provides a useful framework to evaluate some spatially related control measures such as the possibility of stopping the spread of the disease by creating a rabies 'break' ahead of the front through vaccination to reduce the susceptible population to a level below the threshold for an epidemic to occur. Based on parameter values relevant to England, the model suggests that vaccination has considerable advantages over severe culling. Using a classical logistic model for the growth of susceptible foxes, one can explain the tail part of the wave, and in particular, the oscillatory behavior. Indeed, Anderson et al. (1981) speculated that the periodic outbreak is primarily an effect of the incubation period, and Dunbar (1983) and Murray et al. (1986) obtained some qualitative results that show sustained oscillations if the classical logistic model is used and the carrying capacity of the environment is sufficiently large.

It was noted that juvenile foxes leave their home territory in the fall, traveling distances that typically may be 10 times a territory size in search of a new territory. If a fox happens to have contracted rabies around the time of such long-distance movement, it could certainly increase the rate of spread of the disease into uninfected areas (see Murray et al. (1986)). To address this impact of the age-dependent diffusion of susceptible foxes, Ou and Wu (2006) started with a general model framework in population biology and spatial ecology wherein the individual's spatial movement behaviors depend on its maturation status, and they illustrated how delayed reaction-diffusion equations with nonlocal interactions arise naturally. For the above mentioned spatial spread of rabies by foxes, they showed how the distinction of territorial patterns between juvenile and adult foxes yields a class of partial differential equations involving delayed and non-local terms that are implicitly defined by a hyperbolic-parabolic equation. They then demonstrated how incorporating this distinction into the model leads to a formula describing the relation of the minimal wave speed and the maturation time of foxes. Their work involves $I(t,a,x)$ and $S(t,a,x)$ as the population density at time t, age $a \geq 0$, and spatial location $x \in R$ for the infective and the susceptible foxes, respectively, and τ as the maturation time which is assumed to be a constant. It was shown that the total population of the infective foxes $J(t,x) = \int_0^\infty I(t,a,x)\,da$ and the density of the adult susceptible foxes

$M(t,x) = \int_\tau^\infty S(t,a,x)da$ satisfy

$$\frac{\partial J}{\partial t} = D_I \frac{\partial^2 J}{\partial x^2} + \beta M(t,x)J(t,x) - d_I J(t,x) + \beta J(t,x) \int_0^\tau S(t,a,x)da,$$
$$\frac{\partial M}{\partial t} = -\beta M(t,x)J(t,x) - d_S M(t,x) + S(t,\tau,x),$$

where D_I is the diffusive coefficient, d_I is the death rate for the infective foxes, β is the transmission rate, the constant d_S is the death rate for the susceptible foxes, and $S(t,a,x)$ with $0 \leq a \leq \tau$ can be solved implicitly in terms of (J, M) by considering

$$\begin{cases} \left(\frac{\partial}{\partial t} + \frac{\partial}{\partial a}\right) S(t,a,x) = D_Y \frac{\partial^2}{\partial x^2} S(t,a,x) - \beta S(t,a,x)J(t,x) - d_Y S(t,a,x), \\ S(t,0,x) = b(M(t,x)), \end{cases}$$

where D_Y and d_Y are the diffusive and death coefficients for the immature susceptible foxes and $b(\cdot)$ is the birth function of the susceptible foxes.

It was shown in Ou and Wu (2006) that some of the key issues related to the spatial spread can be addressed, despite the difficulty in obtaining an explicit analytic formula of $S(t,a,x)$ in terms of the historical values of M at all spatial locations. For example, the minimal wave speed can be shown to be a decreasing function of the maturation period. This result coincides in principle with the speculation by Anderson et al. (1981) and Murray et al. (1986), and gives a more precise qualitative description of the influence of maturation time on the propagation of the disease in space.

15.3 Dengue

Dengue fever (DF) and dengue hemorrhagic fever (DHF) are caused by one of four closely related, but antigenically distinct, virus serotypes (DEN-1, DEN-2, DEN-3, and DEN-4) of the genus Flavivirus. Infection by one of these serotypes provides immunity to only that serotype for life, so persons living in a dengue-endemic area can have more than one dengue infection during their lifetime. DF and DHF are primarily diseases of tropical and sub-tropical areas, and the four different dengue serotypes are maintained in a cycle that involves humans and the Aedes mosquito. Here, *Aedes aegypti*, a domestic, day-biting mosquito that prefers to feed on humans, is the most common Aedes species. Infections produce a spectrum of clinical illness ranging from a nonspecific viral syndrome to severe and fatal hemorrhagic disease. Important risk factors for DHF include the strain of the infecting virus, as well as the age, and especially the prior dengue infection history of the patient (CDC, 2007a).

Winged female *Aedes aegypti* in search of human blood or places for oviposition are the main reason for local population dispersal and the slow advance of a mosquito infestation. On the other hand, wind currents may also result in an advection movement of large masses of mosquitoes and consequently cause a quick advance of infestation. The study (Takahashi et al., 2005) we describe here focuses on an urban scale of space, wherein a (local) diffusion process due to autonomous and random search movements of winged *Aedes aegypti* is coupled to constant advection which may be interpreted as the result of wind transportation.

Takahashi et al. (2005) considered only two sub-populations: the winged form (mature female mosquitoes) and an aquatic population (including eggs, larvae and pupae), with mortality rates μ_1 and μ_2. The spatial density of the winged *A. aegypti* and aquatic population at point x and time t are denoted by $M(x,t)$ and $A(x,t)$, respectively. The specific maturation rate of the aquatic form into winged female mosquitoes is γ, which is saturated by a term describing a carrying capacity K_1; that is, $\gamma A(1 - M/K_1)$. Similarly, the rate of oviposition by female mosquitoes, which is the only source of the aquatic form, is proportional to their density but is also regulated by a carrying capacity effect dependent on the occupation of the available breeders; that is, $rM(1 - A/K_2)$. Since the focus is on the *A. aegypti* dispersal as a result of a random (and local) flying movement, macroscopically represented by a diffusion process with coefficient D, coupled to a wind advection caused by a constant velocity flux ν, we obtain naturally the coupled system of reaction-diffusion equations

$$\begin{aligned} \frac{\partial M}{\partial t} &= D\frac{\partial^2 M}{\partial x^2} - \frac{\partial(\nu M)}{\partial x} + \gamma A(1 - \frac{M}{K_1}) - \mu_1 M, \\ \frac{\partial A}{\partial t} &= rM(1 - \frac{A}{K_2}) - (\mu_2 + \gamma)A. \end{aligned} \tag{15.2}$$

Traveling wave solutions representing an invasion process (linking two stationary and spatially homogeneous solutions) were formally investigated under the assumption that the invasion speeds obtained for the two sub-populations are equal. This assumption was justified by the following biological argument: Suppose that there are distinct subpopulations linked with the wave speed for the winged population larger than that for the aquatic population. If we wait long enough there will be some distant interval where the (faster) mosquito population will reach values close to the saturation level with practically no aquatic population for as long as we want. That would contradict the vital dynamics, since in that interval a large population of mosquitoes would lay eggs at an enormous rate because (almost) no saturation effect exists without a sizable aquatic population. A similar argument works if the wave speed for the winged population is smaller than that for the aquatic population. Consequently, from a practical point of view, we should only expect a time delay between the wavefronts and a constant spatial gap, not an expanding one.

Existence and uniqueness of a positive spatially homogeneous equilibrium is guaranteed if the mortality rate μ_1 is less than the oviposition rate r and if the basic reproduction number $R_0 = \frac{r\gamma}{(\gamma+\mu_2)\mu_1}$ is larger than 1. The traveling wave with the minimal wave speed was shown numerically to have the strong stability and attractivity property, and hence an effective strategy for controlling the *A. aegypti* dispersal based on the above model is to ensure the minimal wave speed is as small as possible. In relation to this containment strategy, a numerical examination of dependence of the wave speed on a few vital model parameters was carried and it was shown that an application of insecticide against the winged (mosquito) phase is much more effective as an infestation containment strategy than insecticide application against its aquatic phase. This should not be surprising, since the winged form is the one responsible for the *A. aegypti* movement. However, it was also shown that a saturation effect is very apparent and massive insecticide application to increase the mosquito mortality rate beyond a certain value will show very little improvement in wave speed reduction. In addition, it was shown that insecticide application against the aquatic form is not very effective for wave control, but if a chemical attack against the winged form is coupled with the elimination of infested water-holding containers, the results are surprisingly effective.

The study of the wavefront speed dependence on advection, i.e., wind transportation, is interesting from a prediction point of view, and numerical analysis shows that the wavefront speed varies linearly with the advection velocity but not in the same way as in the classical Fisher model. Since the advection only carries the winged form, and the mosquitoes need some time to oviposit, the dependence of the wavefront speed in the model on the advection velocity is not as strong as in Fisher's model. Although advection by natural causes cannot be controlled, the above discussion may be useful for the prediction of patterns of *A. aegypti* invasion in urban areas exposed to strong and constant winds.

Notice that model (15.2) only considers mosquito movement. More realistic models need to include both host and vector populations. Some related models can be found in Favier et al. (2005) and Tran and Raffy (2006).

15.4 West Nile virus

West Nile virus (WNV) was first isolated from a febrile adult woman in the West Nile District of Uganda in 1937. The ecology was characterized in Egypt in the 1950s. The virus became recognized as a cause of severe human meningitis or encephalitis (inflammation of the spinal cord and brain) in elderly patients during an outbreak in Israel in 1957. Equine disease was first noted in Egypt and France in the early 1960s. WNV first appeared in North America in 1999, with encephalitis reported in humans and horses. The subsequent spread in North America is an important milestone in the evolving history of this virus (CDC, 2007b).

West Nile virus belongs to a family of viruses called Flaviviridae. It is spread by mosquitoes that have fed on the blood of infected birds. West Nile virus is closely related to the viruses that cause Dengue fever, Yellow fever, and St. Louis encephalitis. People, horses, and most other mammals are not known to develop infectious-level viremias very often, and thus are probably "dead-end" or incidental-hosts (CDC, 2007b; PHAC, 2007).

Lewis et al. (2006) investigated the spread of WNV by spatially extending the non-spatial dynamical model of Wonham et al. (2004) to include diffusive movements of birds and mosquitoes, resulting in a system of 7 reaction-diffusion equations. A reduced 2-equation model takes the form

$$\begin{aligned}\frac{\partial I_V}{\partial t} &= \epsilon\frac{\partial^2 I_V}{\partial x^2} + \alpha_V\beta_R\frac{I_R}{N_R}(A_V - I_V) - d_V I_V,\\ \frac{\partial I_R}{\partial t} &= D\frac{\partial^2 I_R}{\partial x^2} + \alpha_R\beta_R\frac{N_R - I_R}{N_R}I_V - \gamma_R I_R,\end{aligned} \tag{15.3}$$

where d_V is the adult female mosquito death rate, γ_R is the bird recovery rate from WNV, β_R is the biting rate of mosquitoes on birds, α_V and α_R are the WNV transmission probability per bite to mosquitoes and birds, respectively, ϵ and D are the diffusion coefficients for mosquitoes and birds, respectively, $I_V(x,t)$ and $I_R(x,t)$ are the numbers of infectious (infective) female mosquitos and birds at time t and spatial location $x \in R$, N_R is the number of live birds, and A_V is the number of adult mosquitoes.

Phase-plane analysis of the spatially homogeneous system shows that a positive (endemic) equilibrium (I_V^*, I_R^*) exists if and only if the basic reproduction number R_0 is larger than 1, where

$$R_0 = \sqrt{\frac{\alpha_V\alpha_R\beta_R^2 A_V}{d_V\gamma_R N_R}}.$$

Moreover, this endemic equilibrium, if it exists, is globally asymptotically stable in the positive quadrant.

For the spatially varying model, the vector field is cooperative, therefore an application of the general result in (Li et al., 2005) ensures that there exists a minimal speed of traveling fronts c_0 such that for every $c \geq c_0$, the system has a non-increasing traveling wave solution $(I_V(x-ct), I_R(x-ct))$ with speed c, linking (I_V^*, I_R^*) to $(0,0)$. The cooperative nature of the vector field ensures that the minimal wave speed c_0 coincides with the spread rate in the sense that if the initial values of $(I_V(\cdot,0), I_R(\cdot,0)), I_V(\cdot,0) + I_R(\cdot,0) > 0$, have compact support and are not identical to either equilibrium, then for small $\epsilon > 0$,

$$\begin{aligned}&\lim_{t\to\infty}\left\{\sup_{|x|\geq(c_0+\epsilon)t}||(I_V(x,t), I_R(x,t))||\right\} = 0,\\ &\lim_{t\to\infty}\left\{\sup_{|x|\leq(c_0-\epsilon)t}||(I_V(x,t), I_R(x,t)) - (I_V^*, I_R^*)||\right\} = 0.\end{aligned}$$

In addition, this c_0 is linearly determined and thus could be explicitly calculated from model parameters. In particular, using real data estimated from Wonham et al. (2004) on the original 7-dimensional system, it was shown that a diffusion coefficient of about 5.94 is needed in the model to achieve the observed spread rate of about $1000 km/year$ in North America.

The work in Liu et al. (2006), using a patchy model based on the framework of Bowman et al. (2004), seems to indicate the spread speed may be different if the movement of birds has a preference direction.

One important biological aspect of the hosts in many epidemiological models, namely the stage structure, seems to have received little attention, although structured population models have been intensively studied in the context of population dynamics and spatial ecology, and the interaction of stage-structure with spatial dispersal has drawn considerable attention in association with the theoretical development of the so-called reaction-diffusion equations with nonlocal delayed feedback (see the survey of Gourley and Wu (2006) and the references therein). The developmental stages of hosts have a profound impact on the transmission dynamics of vector borne diseases. In the case of West Nile virus the transmission cycle involves both mosquitoes and birds, the crow species being particularly important. Nestling crows are crows that have hatched but are helpless and stay in the nest, receiving more-or-less continuous care from the mother for up to two weeks and less continuous care thereafter. Fledgling crows are old enough to have left the nest (they leave it after about five weeks), but they still cannot fly very well. After three or four months these fledglings will be old enough to obtain all of their food by themselves. Consequently, adult birds, fledglings, and nestlings are all very different from a biological and an epidemiological perspective, and a realistic model needs to take these different stages into account. For example, in comparison with grown birds, the nestlings and fledglings have much higher disease induced death rate, much poorer ability to avoid being bitten by mosquitoes, and much less spatial mobility.

Gourley et al. (2007) derived a structured population model in terms of a system of delay differential equations describing the interaction of five subpopulations, namely susceptible and infected adult and juvenile reservoirs and infected adult vectors, for a vector borne disease with particular reference to West Nile virus. Spatial movement was then incorporated into this model to yield an analogue reaction-diffusion system with nonlocal delayed terms. This permits a consideration of some specific conditions for the disease eradication and sharp conditions for the local stability of the disease-free equilibrium, as well as a formal calculation of the minimal wave speed for the traveling waves and subsequent comparison with field observation data.

15.5 Hantavirus

Hantaviruses are rodent-borne zoonotic agents that result in hemorrhagic fever with renal syndrome or hantavirus pulmonary syndrome. Hemorrhagic fever with renal syndrome was first reported in 1951 when an outbreak occurred among military personnel involved in the Korean War (Lee and van der Groen, 1989) and now has been identified in Asia and Europe (Shi, 2007). In 1993, hantavirus pulmonary syndrome was identified from an outbreak in New Mexico, USA (Schmaljohn and Hjelle, 1997). Since then, it has been discovered in various regions of southwestern US and in other countries in the Americas. Each hantavirus is generally associated with a primary rodent host. Human infection occurs primarily through the inhalation of aerosolized saliva and excreta of infected rodents. The case fatality rate for hantavirus pulmonary syndrome in the United States is 37% (CDC, 2002a). Hantaviruses pathogenic to humans in the United States include Sin Nombre virus hosted by the deer mouse (*Peromyscus maniculatus*) (Mills et al., 1999), New York virus hosted by the white-footed mouse (*Peromyscus leucopus*) (Song et al., 1994), Black Creek Canal virus hosted by the cotton rat (*Sigmodon hispidus*) (Glass et al., 1998), and Bayou virus hosted by the rice rat (*Oryzomys palustris*) (McIntyre et al., 2005).

In the last few years, several mathematical models have been used to investigate the temporal and spatial dynamics of various hantavirus reservoir species and their relation to the human population. Allen et al. (2003) proposed ordinary differential equation models to study hantavirus infection (Black Creek Canal virus) and arenavirus infection (Tamiami virus) in cotton rats. The two viruses differ in their modes of infection; the first virus is horizontally transmitted, whereas the second is primarily vertically transmitted. Sauvage et al. (2003) considered Puumala virus infection in bank voles (*Clethrionomys glareolus*). Their model is a system of ordinary differential equations for rodents infected with hantavirus in two different habitats: optimal and suboptimal. The population is subdivided into susceptible and infected juveniles and adults. Allen et al. (2006) developed two new mathematical models for hantavirus infection in male and female rodents. The first model is a system of ordinary differential equations while the second model is a system of stochastic differential equations.

Taking the random movement of the rodent population into account, Abramson and Kenkre (2002) and Abramson et al. (2003) used partial differential equation models to study Sin Nombre virus in deer mice. Suppose that the whole mice population is composed of two classes, susceptible and infected, represented by M_S and M_I, respectively, with $M_S + M_I = M$. Since the virus does not affect properties such as the mortality of the mice, the death rate is assumed to be the same for both susceptible and infected mice. It is also not transmitted to newborns, so that no mice are born in the infected state. The infection is transmitted from mouse to mouse through individual contacts, such as fights. The dispersal of mice is modeled as a diffusion process. Finally, intra-species competition for resources indicates a logistic population growth. The model is described by the following equations:

$$\begin{aligned} \frac{\partial M_S}{\partial t} &= D\frac{\partial^2 M_S}{\partial x^2} + bM - cM_S - \frac{M_S M}{K} - aM_S M_I, \\ \frac{\partial M_I}{\partial t} &= D\frac{\partial^2 M_I}{\partial x^2} - cM_I - \frac{M_I M}{K} + aM_S M_I. \end{aligned} \tag{15.4}$$

All parameters characterizing the different processes affecting the mice are supposed constant, except the carrying capacity K of the mouse population, which we will sometimes write $K = K(x,t)$ explicitly to indicate the dependence on the location and time which allows for diversity in habitats and temporal phenomena. The birth rate b characterizes a source of susceptible mice only. The death rate, common to both subpopulations, is c. The contagion rate is the parameter a. Finally, a diffusion coefficient D characterizes a diffusive transport mechanism for the mice.

The sum of the two equations in (15.4) reduces to a Fisher type equation for the whole population

$$\frac{\partial M}{\partial t} = D\frac{\partial^2 M}{\partial x^2} + (b-c)M\left[1 - \frac{M}{(b-c)K}\right]. \tag{15.5}$$

Abramson and Kenkre (2002) showed that, as a function of K, the system undergoes a bifurcation between a stable state with only susceptible mice (and $M_I = 0$) to a stable state with both subpopulations positive. The value of the critical carrying capacity is a function of other parameters and is given by $K_c = \frac{b}{a(b-c)}$. This critical value does not depend on D, and the same bifurcation is observed either in a spatially independent system ($D = 0$) or in a homogeneous extended one in the presence of diffusion. In a nonhomogeneous situation, for moderate values of the diffusion coefficient, the infected subpopulation remains restricted to those places where $K(x,t) > K_c$, becoming extinct in the rest.

Yates et al. (2002) found that the outbreaks of hantavirus pulmonary syndrome in southwest US in 1993 and again in 1998-2000 were associated with the El Niño-southern oscillation phenomenon, which produced increased amounts of fall-spring precipitation in the arid and semi-arid regions of New Mexico and Arizona and in turn initiated greater production of rodent food resources. Consequently, rodent population increased dramatically, and at high densities, rodents began dispersing across the landscape and coming into contact with humans in homes and businesses. This suggests that a 'wave' of virus infection was following the 'wave' of rodent dispersal.

Let $z_1 = x - v_S t$ and $z_2 = x - v_I t$, where v_S and v_I are the speeds of the susceptible and infected waves, respectively. The wave form equations are

$$\begin{aligned} D\frac{d^2 M_S}{dz_1^2} + v_S\frac{dM_S}{dz_1} + bM - cM_S - \frac{M_S M}{K} - aM_S M_I = 0, \\ D\frac{d^2 M_I}{dz_2^2} + v_I\frac{dM_I}{dz_2} - cM_I - \frac{M_I M}{K} + aM_S M_I = 0. \end{aligned} \tag{15.6}$$

There are two interesting scenarios.

(i) Initially the system is at a state of low carrying capacity (below K_c) and the population consists of uninfected mice only at the stable equilibrium. When the environment changes so that $K > K_c$, the population will be out of equilibrium: the susceptible mice population will evolve towards a new equilibrium and a wave of infected mice will invade the susceptible population. Analysis at the unstable equilibrium $(K(b-c), 0)$ implies that traveling wave speed satisfies $v \geq 2\sqrt{D[-b + aK(b-c)]}$.

(ii) Initially the system is empty of mice. Consider a system with $K > K_c$ and with $M_S = M_I = 0$ in almost all of its range, but with a small region where $M_S > 0$ and $M_I > 0$. A wave of both mouse populations will develop and invade the empty region. The wave speed of the susceptible is $v_S \geq 2\sqrt{D(b-c)}$ and the wave speed of the infected $v_I \geq 2\sqrt{D[-b + aK(b-c)]}$.

The density of susceptible mice rises from zero and lingers near the positive unstable equilibrium before tending to the stable one.

Barbera et al. (2008) generalized the Abramson-Kenkre reaction-diffusion model to a hyperbolic reaction-diffusion model for the hantavirus infection in mouse populations and investigated traveling wave solutions related to the spread of the infection in the landscape. For further studies on modeling spatial spread of hantavirus, we refer to Giuhhioli et al. (2005), Kenkre et al. (2007), and the references cited therein.

15.6 Lyme disease

In 1975, a group of children in the Lyme, Connecticut, area were originally diagnosed as having juvenile rheumatoid arthritis (Steere et al., 1977). Subsequently it became apparent that this occurrence was actually a delayed manifestation of a tick-transmitted multisystem disease for which some manifestations had been reported previously in Europe (Steere et al., 2004). In 1976, the disease was recognized as a seperate entity and named Lyme disease (Steere, 1989). In 1981 the spirochetal bacterium *Borrelia burgdorferi* from the deer tick *Ixodes scapularis* was identified (Burgdorfer et al., 1982) and cultured from patients with early Lyme disease (Steere et al., 1983). Lyme disease is now the most commonly reported tick-borne illness in the US, Europe, and Asia (Dannis et al., 2002; CDC, 2002b; Zhang et al., 1998).

New cases of Lyme disease appear at unabated rates in endemic regions, the geographic distribution of the incidence of Lyme disease has expanded rapidly, and the spread of the disease involves direct interactions among no fewer than four species (Ostfeld et al., 1995). The hematophagous vector is the deer tick *Ixodes scapularis*. Larval and nymphal ticks feed primarily on the white-footed mouse *Peromyscus leucopus* but will attack a variety of hosts; inadvertent nymph bites can infect humans with the spirochete. Adult ticks feed preferentially on white-tailed deer *Odocoileus virginianus*.

Caraco et al. (1998) proposed an ODE model focusing on these four species and let infection in humans follow as a consequence of the community's population dynamics. According to Caraco et al. (1998), *Ixodes scapularis* exhibits a two-year life cycle. 89% of newly hatched larvae attack white-footed mice. Larvae that obtain a blood meal drop off their host and then overwinter as nymphs. At the beginning of the second year, nymphs quest for a blood meal (the second of the life cycle). If they succeed, the nymphs mature to the adult stage. Adult females feed almost exclusively on white-tailed deer and mate there. Females eventually drop off the deer they have parasitized, lay about 2000 fertile eggs nearby, and die. It is estimated that 20 – 33% of nymphs in infected areas (ticks that have previously taken a single blood meal) are infected, and that 50% of questing adults (those that have already taken two blood meals) are infected with the spirochete. Interestingly, it is the tick to mouse to tick enzootic cycle of infection that maintains the spirochete. Seasonality helps drive the cycle. Nymphs infected last year appear first as warmer weather begins; these ticks pass the spirochete to susceptible mice. After summer has arrived larvae hatch, quest for a blood meal, and acquire the spirochete when they attack an infected mouse. These individuals then become quiescent as infected nymphs, completing the cycle of infection.

Since deer move fecund adult ticks, their dispersal influences the spatial pattern of tick larvae. But deer cannot be infected and do not disperse the pathogen. Furthermore, *Borrelia* cannot survive outside of its hosts. Mice usually disperse juvenile ticks, and dispersal of infectious mice can introduce the spirochete into tick populations. So the spatial advance of infection must be driven by dispersal of mice and other hosts to juvenile ticks (Van Buskirk and Ostfeld, 1998). Caraco et al. (2002)

modeled the advance of the natural infection cycle as a reaction-diffusion process. The model may help identify factors influencing the rate at which the disease spreads and predict the velocity at which spirochete infection advances spatially.

The model treats population densities at locations (x, y) in a two-dimensional domain Ω. Parameters for birth, death, infection, and developmental advance do not depend on spatial location. Diffusion approximates dispersal via random motion. It is assumed that the dynamics and dispersal of mice are independent of infestation/infection status. To limit the number of variables, dispersal of nymphs is ignored. At equilibrium population densities, nymph dispersal does not affect the spread of Lyme disease. Dispersal of larvae is important; spatial dispersion of replete larvae governs the pattern in the risk of Lyme disease when these animals quest as nymphs. Therefore, dispersal of larvae while they feed on mice is considered. Adult ticks reproduce and disperse diffusively; dispersal of adults mimics movements of deer while ticks mate (deer are not modeled explicitly). Natality and mortality among black-legged ticks are apparently independent of *Borrelia* infection (Van Buskirk and Ostfeld, 1998). The model requires six state variables for the reaction-diffusion dynamics; among them, three subsidiary variables are required to model the tick's population structure.

Mice reproduce in a density-dependent manner and incur density-independent mortality. Since mice are born uninfected, the equation for susceptible-mouse density $M(x, y, t)$ includes birth, death, acquisition of the spirochete from infectious-nymph bites, and dispersal:

$$\frac{\partial M}{\partial t} = D_M\left(\frac{\partial^2 M}{\partial x^2} + \frac{\partial^2 M}{\partial y^2}\right) + r_M(M+m)\left(1 - \frac{M+m}{K_M}\right) - \mu_M M - \alpha\beta M n, \tag{15.7}$$

where D_M is the diffusion coefficient for mice with unit (distance)^2/time; r_M is the intrinsic birthrate; K_M is the spatially homogeneous carrying capacity; μ_M is the individual mortality rate among mice; α is the attack rate of juvenile ticks questing for mice; $\beta(0 < \beta < 1)$ is a mouse's susceptibility to pathogen infection when bitten by an infectious nymph.

The density of pathogen-infected mice $m(x, y, t)$ increases as susceptible mice are bitten by infectious nymphs and decreases through mortality. The equation for infected mice includes infection, death, and dispersal:

$$\frac{\partial m}{\partial t} = D_M\left(\frac{\partial^2 m}{\partial x^2} + \frac{\partial^2 m}{\partial y^2}\right) - \mu_M m + \alpha\beta M n. \tag{15.8}$$

The subsidiary variable $L(A, a)$ is the density of questing larvae which declines through mortality and attacks on mice, where $A(x, y, t)$ and $a(x, y, t)$ are density of uninfected adult ticks and pathogen-infected adult ticks, respectively. It is assumed that larval hatching rate depends nonlinearly on adult tick density. Then, at each point (x, y) :

$$\frac{dL}{dt} = r(A+a)[1 - c(A+a)] - \mu_L L - \alpha L(M+m), \tag{15.9}$$

where r is the tick's per capita reproduction at low density; μ_L is the mortality rate among questing larvae; and c represents crowding among reproducing ticks. Larvae must hatch at a positive rate when $(A+a) > 0$, so c is small. Essentially, c is inversely proportional to deer density, which is assumed a constant and treated implicitly.

The density of larvae infesting susceptible mice $V(x, y, t)$ varies in successful attack, completion of the first blood meal, death, and dispersal while they infest mice:

$$\frac{\partial V}{\partial t} = D_M\left(\frac{\partial^2 V}{\partial x^2} + \frac{\partial^2 V}{\partial y^2}\right) - (\sigma + \mu_V)V + \alpha ML, \tag{15.10}$$

where σ is the rate at which larvae infesting mice complete their meal, and μ_V is the mortality rate among larvae infesting mice. Since the duration of a larval meal seldom exceeds a few days, $\sigma > \mu_V$. The assumptions concerning the density of larvae infesting pathogen-infected mice, $v(x, y, t)$, are similar. We substitute the density of infectious mice (m) for susceptible-mouse density (M) and obtain $\partial v/\partial t$.

The subsidiary variable $N(V, v)$ is the density of susceptible questing nymphs at (x, y, t), which increases as larvae complete their first meal without acquiring the spirochete. The larvae may have infested a susceptible mouse or attacked an infectious mouse and avoided infection. As they die, bite humans, and attack mice, $N(V, v)$ decreases. Combining processes yields

$$\frac{dN}{dt} = \sigma[V + (1 - \beta_T)v] - N[\gamma + \alpha(M + m) + \mu_N], \tag{15.11}$$

where β_T $(0 < \beta_T < 1)$ is a tick's susceptibility to infection when feeding on an infected mouse. The mortality rate among questing nymphs is μ_N, and γ is the rate at which nymphs bite humans.

The subsidiary variable $n(v)$ is the density of questing infectious nymphs at (x, y, t). Infectious nymphs must have attacked a mouse infected with *Borrelia* as larvae and then acquired the pathogen. Their density at any location (x, y) varies as

$$\frac{dn}{dt} = \beta_T \sigma v - n[\gamma + \alpha(M + m) + \mu_N], \tag{15.12}$$

where the term γn represents the local risk of Lyme disease to humans.

The density of uninfected adult ticks $A(x, y, t)$ changes through attacks of those nymphs on mice, death of adults, and dispersal:

$$\frac{\partial A}{\partial t} = D_H\left(\frac{\partial^2 A}{\partial x^2} + \frac{\partial^2 A}{\partial y^2}\right) - \mu_A A + \alpha N[M + (1 - \beta_T)m], \tag{15.13}$$

where μ_A is the density-independent mortality rate among adult ticks. The diffusion coefficient D_H models dispersal of adult ticks while they infest deer.

The density of pathogen-infected adult ticks $a(x, y, t)$ increases as infected nymphs attack any mouse and as susceptible nymphs attack infected mice and acquire *Borrelia* during their second blood meal. Adding death and dispersal yields

$$\frac{\partial a}{\partial t} = D_H\left(\frac{\partial^2 a}{\partial x^2} + \frac{\partial^2 a}{\partial y^2}\right) - \mu_A a + \alpha[(M + m)n + \beta_T mN]. \tag{15.14}$$

To analyze this model, Caraco et al. (2002) first identified three aspatial equilibria: extinction of the system, positive abundance of ticks and mice in the absence of spirochete, and proportional infection of both mice and ticks. Then they studied how adult tick mortality and juvenile attack rate influence the velocity at which infection spreads in the diffusion model. Their results indicate that as vector mortality rates vary, the disease spread velocity is roughly proportional to the density of infectious vectors, and thus proportional to the local risk of zoonotic infection. However, as the rate at which juvenile ticks attack hosts varies, the spread velocity of infection may increase or decrease. In both cases, the disease spread velocity remains proportional to the frequency of infection among hosts.

15.7 Feline immunodeficiency virus (FIV)

In 1987, the isolation of a T-lymphotropic virus possessing the characteristics of a lentivirus from pet cats in Davis, California was reported (Pedersen et al., 1987). The virus is a member of the family of retroviruses and causes an acquired immunodeficiency syndrome in cats. It shares many physical and biochemical properties with human immunodeficiency virus (HIV) and was therefore named feline immunodeficiency virus (FIV). Today FIV has been detected worldwide. The prevalences vary, ranging from 2% in Germany and 16% in the United States to 33% in the United Kingdom and 44% in Japan (Hartmann, 1998).

FIV can be isolated from blood, serum, plasma, cerebrospinal fluid, and saliva of infected cats. The infection is much more common in males than females since the transmission mode is through bites inflicted during fights and biting is more apt to occur between male cats (Yamamoto et al., 1989). Veneral transmission from infected males to females is possible. In experimental studies, infection has been shown to occur not only via a vaginal route, but also via rectal mucous membrane (Moench et al., 1993).

Though there is no evidence that FIV can spread to humans, it is important to study its epidemiology for a variety of reasons. Its spread mimics the spread of HIV within the human population and it is possible that subsequent mutations of FIV could produce a virus capable of infecting humans. Courchamp et al. (1995) constructed a deterministic model to study the circulation of FIV within populations of domestic cats. Since all sexually transmitted diseases can be transferred from males to males, from females to females, and from males to females and vice versa, Fitzgibbon et al. (1995) proposed a criss-cross infection model to describe the spread of FIV. Their model uses Fickian diffusion to account for the geographic spread of the disease and introduces age of the disease within an individual as a structural variable.

Divide the feline population sexually into male and female classes. Each of these classes is in turn subdivided into susceptible and infective subclasses. Consider four state variables u, w, v, z representing population densities of susceptible males, infective males, susceptible females, and infective females, respectively. Assume that the infection spreads from infective males to susceptible males and females and from infective females to susceptible males and females, with different infection rates. Let $\Omega \subset R^n (1 \le n \le 3)$ be a bounded region which lies locally on one side of its boundary $\partial\Omega$, which is sufficiently smooth. The criss-cross epidemic model without age structure is, for $x \in \Omega, t > 0$:

$$\begin{aligned} \frac{\partial u}{\partial t} &= d_1 \Delta u - k_1 uw - k_2 uz, \\ \frac{\partial w}{\partial t} &= d_2 \Delta w + k_1 uw + k_2 uz - \lambda_1 w, \\ \frac{\partial v}{\partial t} &= d_3 \Delta v - k_3 vw - k_4 vz, \\ \frac{\partial z}{\partial t} &= d_4 \Delta z + k_3 vw + k_4 vz - \lambda_2 z \end{aligned} \tag{15.15}$$

with Neumann boundary conditions

$$\frac{\partial u}{\partial n} = \frac{\partial w}{\partial n} = \frac{\partial v}{\partial n} = \frac{\partial z}{\partial n} = 0, \quad x \in \partial\Omega, \quad t > 0 \tag{15.16}$$

and initial conditions

$$\begin{aligned} u(x,0) = u_0(x) \ge 0, \quad & v(x,0) = v_0(x) \ge 0, \\ w(x,0) = w_0(x) \ge 0, \quad & z(x,0) = z_0(x) \ge 0, \quad x \in \Omega, \end{aligned} \tag{15.17}$$

where $k_i (i = 1, ..., 4)$ are the infection rates of the four subclasses; λ_1 and λ_2 are the removal rate of the infective males and females, respectively; $d_i (i = 1, ..., 4)$ are the diffusion rates of the four subclasses. All parameters are positive constants. The Neumann boundary conditions imply that all populations remain confined to the region Ω for all time.

The analysis of Fitzgibbon et al. (1995) indicates that the infective population is always ultimately extinguished. Thus, the model applies to a short term development of FIV, which extinguishes because of a lack of new susceptibles. From their ODE model, Courchamp et al. (1995) claim that FIV is endemic in domestic feline populations. The reason is that the model of Courchamp et al. incorporates a logistic growth nonlinearity for the total population, whilst the the model of Fitzgibbon et al. does not include demographic population dynamics of the feline population.

Hilker et al. (2007) extended the model of Courchamp et al. (1995) to the reaction-diffusion system version. Let $S(x,t)$ and $I(x,t)$ denote the densities of susceptible and infectious cats in the location $x \in \Omega$ (in km) and at time $t > 0$ (in years), so that $P(x,t) = S + I$ is the density of the cat population (in number of individuals per km^2). The model takes the form

$$\begin{aligned} \frac{\partial S}{\partial t} &= D_S \Delta S - \sigma \frac{SI}{P} + \beta(P)P - \mu(P)S, \\ \frac{\partial I}{\partial t} &= D_I \Delta I + \sigma \frac{SI}{P} - \mu(P)I - \alpha I, \end{aligned} \tag{15.18}$$

where D_S and D_I (km^2 per year) are the diffusion rates of the susceptibles and infectives, respectively; σ is the transmission coefficient; and α is the disease related death rate. The fertility function $\beta(P) \ge 0$ and the mortality function $\mu(P) \ge 0$ are assumed to be density-dependent; the intrinsic per-capita growth rate is $g(P) =$

$\beta(P)-\mu(P)$. If $\beta(P)=b>0, \mu(P)=m+rP/K, r=b-m, m>0$, one obtains the well-known logistic per-capita growth rate $g(P)=r(1-P/K)$. If

$$\beta(P)=\begin{cases} a[-P^2+(K_++K_-+e)P+c], & 0\leq P\leq K_++K_- \\ \text{nonnegative and nonincreasing}, & \text{otherwise} \end{cases}$$

and $\mu(P)=a(eP+K_+K_-+c)$, then the per-capita growth rate

$$g(P)=a(K_+-P)(P-K_-)$$

describes the strong Allee effect in the vital dynamics. This type of function can be used to model the fact that cat is a very opportunistic predator and is one of the worst invasive species threatening many indigenous species.

For the model with logistic growth, numerical simulation indicates that a traveling infection wave emerges and advances with a constant speed. In its wake, the population settles down to the endemic state. In the model with Allee effect, the emergence and propagation of a traveling wave can be observed as well. However, if the transmission coefficient is further increased so that the nontrivial state disappears, two different scenarios are possible: (a) front reversal with eventual host extinction (see Fig. 2c, Hilker et al. 2007) and (b) a transient (and spatially restricted) epidemic before disease-induced extinction (see Fig. 10, Hilker et al. 2007). In both cases, the propagation of traveling pulse-like epidemics will wipe out the host population.

Recently, the spatial spread of some infectious diseases, including FIV and FeLV (Feline Leukemia Virus), among animal populations distributed on heterogeneous habitats has been extensively studied. We refer to Fitzgibbon et al. (2001), Fitzgibbon and Langlais (2008), Malchow et al. (2008), and the references cited therein.

15.8 Summary

We have summarized a few models developed for specific diseases which involve animal hosts and have significant implication to human health: rabies, dengue, West Nile virus, hantavirus, Lyme disease, and feline immunodeficiency virus. A common feature of these diseases is the involvement of a certain animal carrier and at least a subgroup of individuals in the animal population that may move more or less randomly in space. This feature leads naturally to the addition of diffusion and perhaps advection terms to classical compartmental models.

Most studies introduced here started with the assumption that the disease is capable of invading the susceptible population in a spatially homogeneous environment, and these studies then considered the issue of spatial spread patterns and disease propagation speeds under various conditions of spatial movement of the host population. A particular object is the existence of traveling wave fronts and the minimal wave speed of such fronts that is believed to coincide with the disease spread speed.

Spatial diffusion may interact with structural heterogeneity, for example, maturation status of the host population. How this interaction leads to particular spatio-temporal patterns of disease spread and the implication for the design of containment strategies was a key issue of some of the studies discussed here. Further work in this area is discussed in two other recent review articles (Ruan, 2007; Gourley et al., 2008).

Diseases involving multiple species or higher dimensional space may also permit different propagation speeds for different species and dimensions, and pose great mathematical challenges for analysis. Furthermore, parameterizing spatial models from epidemiological or biological data (Noble, 1974; Murray et al., 1986) is difficult but crucial in studying the spatial spread of diseases.

So, despite the substantial recent progress in the study of spatial spread of diseases using reaction-diffusions equations, the implications of spatial structure in epidemiological models are still far from clear, and the statement in Murray (2003) remains: "the geographic spread of epidemics is less well understood and much less well studied than the temporal development and control of diseases and epidemics."

15.9 Acknowledgments

Research of S. Ruan was partially supported by NSF grants DMS-0412047, DMS-0715772, and NIH grants P20-RR020770 and R01-GM083607. J. Wu would like to acknowledge support from Natural Sciences and Engineering Research Council of Canada, Mathematics for Information Technology and Complex Systems, the Canada Research Chairs Program, Ontario Ministry of Health and Long-term Care, and Public Health Agency of Canada.

We are very grateful to Thomas Caraco, William Fitzgibbon, Frank Hilker, V. M. Kenkre, James D. Murray, Sergei Petrovskii, Pauline van den Driessche, and Glenn Webb not only for permitting us to adapt their results but also for their careful reading, valuable comments, and helpful suggestions on an earlier version of this chapter. Interestingly, Professor Murray told us that a close school friend of his wife was one of those children who contacted Lyme disease in Old Lyme long before it was recognized as a tick-borne disease and now has terrible physical problems since the disease was not treated at the time and cannot be treated retroactively.

15.10 References

G. Abramson and V. M. Kenkre (2002), Spatiotemporal patterns in hantavirus infection, *Phys. Rev. E* **66**: 011912-1Ű5.

G. Abramson, V. M. Kenkre, T. L. Yates, and R. R. Parmenter (2003), Traveling waves of infection in the hantavirus epidemics, *Bull. Math. Biol.* **65**: 519-534.

S. Ai and W. Huang (2005), Travelling waves for a reaction-diffusion system in population dynamics and epidemiology, *Proc. Roy. Soc. Edinburgh* **A135**: 663-675.

L. J. S. Allen, M. Langlais, and C. Phillips (2003), The dynamics of two viral infections in a single host population with applications to hantavirus, *Math. Biosci.* **186**: 191-217.

L. J. S. Allen, R. K. McCormack, and C. B. Jonsson (2006), Mathematical models for hantavirus infection in rodents, *Bull. Math. Biol.* **68**: 511-524.

R. M. Anderson, H. C. Jackson, R. M. May, and A. M. Smith (1981), Population dynamics of fox rabies in Europe, *Nature* **289**: 765-771.

R. M. Anderson, and R. M. May (1991), *Infectious Diseases of Humans: Dynamics and Control,* Oxford University Press, Oxford.

J. Arino and P. van den Driessche (2006), Metapopulation epidemic models: a survey, in *Nonlinear Dynamics and Evolution Equations*, ed. by H. Brunner, X.-Q. Zhao and X. Zou, Fields Institute Communications **48**, Amer. Math. Soc., Providence, RI, pp. 1-12.

D. G. Aronson (1977), The asymptotic speed of propagation of a simple epidemic, in *Nonlinear Diffusion,* ed. by W. E. Fitzgibbon and H. F. Walker, Research Notes in Math. **14**, Pitman, London, pp. 1-23.

D. G. Aronson and H. F. Weinberger (1975), Nonlinear diffusion in population genetics, combustion, and nerve pulse propagation, in *Partial Differential Equations and Related Topics,* ed. by J. A. Goldstein, Lecture Notes in Math. **446**, Springer, Berlin, pp. 5-49.

E. Barbera, C. Currò, and G. Valenti (2008), A hyperbolic reaction-diffusion model for the hantavirus infection, *Math. Meth. Appl. Sci.* **31**: 481-499.

M. S. Bartlett (1956), Deterministic and stochastic models for recurrent epidemics, *Proc. 3rd Berkeley Symp. Math. Stat. Prob.* **4**: 81-109.

B. M. Bolker and B. T. Grenfell (1995), Space, persistence, and dynamics of measles epidemics, *Phil. Trans. R. Soc. Lond.* **B237**: 298-219.

C. Bowman, A. Gumel, P. van den Driessche, J. Wu, and H. Zhu (2005), A mathematical model for assessing control strategies against West Nile virus, *Bull. Math. Biol.* **67**: 1107-1133.

F. Brauer and C. Castillo-Chavez (2000), *Mathematical Models in Population Biology and Epidemiology,* Springer-Verlag, New York.

W. Burgdorfer, A. G. Barbour, S. F. Hayes, J. L. Benach, E. Grunwaldt, and J. P. Davis (1982), Lyme disease – a tick-borne spirochetosis? *Science* **216**: 1317-1319.

S. N. Busenberg and C. C. Travis (1983), Epidemic models with spatial spread due to population migration, *J. Math. Biol.* **16**: 181-198.

T. Caraco, G. Gardner, W. Maniatty, E. Deelman, and B. K. Szymanski (1998), Lyme disease: Self-regulation and pathogen invasion, *J. Theoret. Biol.* **193**: 561-575.

T. Caraco, S. Glavanakov, G. Chen, J. E. Flaherty, T. K. Ohsumi, and B. K. Szymanski (2002), Stage-structured infection transmission and a spatial epidemic: A model for Lyme disease, *Am. Nat.* **160**: 348-359.

V. Capasso (1978), Global solution for a diffusive nonlinear deterministic epidemic model, *SIAM J. Appl. Math.* **35**: 274-284.

V. Capasso and R. E. Wilson, Analysis of reaction-diffusion system modeling man-environment-man epidemics, *SIAM. J. Appl. Math.* **57**: 327-346.

CDC (2002a), Hantavirus pulmonary syndrome – United States: Updated recommendations for risk reduction, *Morb. Mortal. Wkly. Rep.* **51**: 1Ű12.

CDC (2002b), Lyme disease: United States, *Morb. Mortal. Wkly. Rep.*, **51**: 29-31.

CDC (2007a), Dengue Fever, http://www.cdc.gov/NCIDOD/DVBID/DENGUE/.

CDC (2007b), West Nile virus, http://www.cdc.gov/ncidod/dvbid/westnile/background.htm.

F. Courchamp, D. Pontier, M. Langlais, and M. Artois (1995), Population dynamics of feline immunodeficiency virus within populations of cats, *J. Theoret. Biol.* **175**: 553-560.

I. Cruickshank, W. S. Gurney, and A. R. Veitĉh (1999), The characteristics of epidemics and invasions with thresholds, *Theor. Pop. Biol.* **56**: 279-92.

P. de Monttoni, E. Orlandi, and A. Tesei (1979), Asymptotic behavior for a system describing epidemics with migration and spatial spread of infection, *Nonlinear Anal.* **3**: 663-675.

D. T. Dennis and E. B. Hayes (2002), Epidemiology of Lyme Borreliosis, in *Lyme Borreliosis: Biology, Epidemiology and Control,* ed. by O. Kahl, J.S. Gray, R.S. Lane, and G. Stanek,

CABI Publishing, Oxford, United Kingdom, pp. 251-280.

O. Diekmann (1978), Thresholds and travelling waves for the geographical spread of infection, *J. Math. Biol.* **6**: 109-130.

O. Diekmann (1979), Run for life: A note on the asymptotic spread of propagation of an epidemic, *J. Differential Equations* **33**: 58-73.

O. Diekmann and J. A. P. Heesterbeek (2000), *Mathematical Epidemiology of Infective Diseases: Model Building, Analysis and Interpretation*, Wiley, New York.

S. R. Dunbar (1983), Travelling wave solutions of diffusive Lotka-Volterra equations, *J. Math. Biol.* **17**:11-32.

L. Edelstein-Keshet (1988), *Mathematical Models in Biology*, Birkhäuser Mathematics Series, McGraw-Hill Inc., Toronto.

C. Favier, D. Schmit, C. D. M. Müller-Graf, B. Cazelles, N. Degallier, B. Mondet, and M. A. Dubois (2005), Influence of spatial heterogeneity on an emerging infectious disease: the case of dengue epidemics, *Proc. R. Soc. Lond.* **B272**: 1171-1177.

R. A. Fisher (1937), The wave of advance of advantageous genes, *Ann. Eugenics* **7**: 353-369.

W. E. Fitzgibbon and M. Langlais (2008), Simple models for the transmission of microparasites between host populations living on non coincident spatial domain, in *Structured Population Models in Biology and Epidemiology*, ed. by P. Magal and S. Ruan, Lecture Notes in Math. **1936**, Springer-Verlag, Berlin, pp. 115-164.

W. E. Fitzgibbon, M. Langlais, and J.J. Morgan (2001), A mathematical model of the spread of feline leukemia virus (FeLV) through a highly heterogeneous spatial domain, *SIAM J. Math. Anal.* **33**: 570-588.

W. E. Fitzgibbon, M. Langlais, and J.J. Morgan (2007), A mathematical model for indirectly transmitted diseases, *Math. Biosci.* **206**: 233-248.

W. E. Fitzgibbon, M. Langlais, M. E. Parrott, and G. F. Webb (1995), A diffusive system with age dependency modeling FIV, *Nonlin. Anal.* **25**: 975-989.

L. Giuggioli, G. Abramson, V. M. Kenkre, G. Suzán, E. Marcé, and T. L. Yates (2005), Diffusiono and home range parameters from rodent population measurements in Panama, *Bull. Math. Biol.* **67**: 1135-1149.

G. E. Glass, W. Livingstone, J. N. Mills, W. G. Hlady, J. B. Fine, W. Biggler, T. Coke, D. Frazier, S. Atherley, P. E. Rollin, T. G. Ksiazek, C. J. Peters, and J. E. Childs (1998), Black creek canal virus infection in Sigmodon hispidus in southern Florida, *Am. J. Trop. Med. Hyg.* **59**: 699-703.

S. A. Gourley, R. Liu, and J. Wu (2007), Some vector borne diseases with structured host populations: extinction and spatial spread, *SIAM J. Appl. Math.* **67**:408-432.

S. A. Gourley, R. Liu, and J. Wu (2008), Spatiotemporal patterns of disease spread: Interaction of physiological structure, spatial movements, disease progression and human intervention, in *Structured Population Models in Biology and Epidemiology*, ed. by P. Magal and S. Ruan, Lecture Notes in Math. **1936**, Springer-Verlag, Berlin, pp. 165-208.

S. A. Gourley and J. Wu (2006), Delayed non-local diffusive systems in biological invasion and disease spread, in *Nonlinear Dynamics and Evolution Equations*, ed. by H. Brunner, X.-Q. Zhao and X. Zou, Fields Institute Communications **48**, Amer. Math. Soc., Providence, RI, pp.137-200.

I. Gudelj, K. A. J. White, and N. F. Britton (2004), The effects of spatial movement and group interactions on disease dynamics of social animals, *Bull. Math. Biol.* **66**: 91-108.

K. Hartmann (1998), Feline immunodeficiency virus infection: An overview, *Veterin. J.* **155**: 123-137.

F. M. Hilker, M. Langlais, S. V. Petrovskii, and H. Malchow (2007), A diffusive SI model with Allee effect and apllication to FIV, *Math. Biosci.* **206**: 61-80.

Y. Hosono and B. Ilyas (1995), Traveling waves for a simple diffusive epidemic model, *Math. Models Methods Appl. Sci.* **5**: 935-966.

D. S. Jones and B. D. Sleeman (2003), *Differential Equations and Mathematical Biology*, Chapman & Hall/CRC, Boca Raton, FL.

A. Källen (1984), Thresholds and travelling waves in an epidemic model for rabies, *Nonlinear Anal.* **8**: 651-856.

A. Källen, P. Arcuri, and J. D. Murray (1985), A simple model for the spatial spread and control of rabies, *J. Theor. Biol.* **116**: 377-393.

D. G. Kendall (1957), Discussion of 'Measles periodicity and community size' by M. S. Bartlett, *J. Roy. Stat. Soc.* **A120**: 64-76.

V. M. Kenkre, L. Giuggioli, G. Abramson and G. Camelo-Neto (2007), Theory of hantavirus infection spread incorporating localized adult and itinerant juvenile mice, *Eur. Phys. J. B* **55**: 461-470.

W. O. Kermack and A. G. McKendrik (1927), A contribution to the mathematical theory of epidemics, *Proc. Roy. Soc.* **A115**: 700-721.

J. Koopman (2004), Modeling infection transmission, *Ann. Rev. Public Health* **25**: 303-326.

M. N. Kuperman and H. S. Wio (1999), Front propagation in epidemiological models with spatial dependence, *Physica A* **272**: 206-222.

H. W. Lee and G. van der Groen (1989), Hemorrhagic fever with renal syndrome, *Prog. Med. Virol.* **36**: 92-102.

M. A. Lewis, J. Renclawowicz, and P. van den Driessche (2006), Traveling waves and spread rates for a West Nile Virus model, *Bull. Math. Biol.* **68**: 3-23.

B. Li, H. Weinberger and M. Lewis (2005), Spreading speed as slowest wave speeds for cooperative systems, *Math. Biosci.* **196**: 82-98

R. Liu, J. Shuai, J. Wu, and X. Zou (2006), Modelling spatial spread of West Nile virus and impact of directional dispersal of birds, *Math. Biosci. Eng.* **3**: 45-160.

H. Malchow, S. V. Petrovskii, and E. Venturino (2008), *Spatiotemporal Patterns in Ecology and Epidemiology: Theory, Models, and Simulation*, Chapman & Hall/CRC, Boca Raton, FL.

N. E. McIntyre, Y. K. Chu, R. D. Owen, A. Abuzeineh, N. De La Sancha, C. W. Dick, T. Holsomback, R. A. Nisbet, and C. Jonsson (2005), A longitudinal study of Bayou virus, hosts, and habitat, *Am. J. Trop. Med. Hyg.* **73**(6): 1043-1049.

J. N. Mills, T. L. Yates, T. G. Ksiazek, C. J. Peters, and J. E. Childs (1999), Long-term studies of hantavirus reservoir populations in the southwestern United States: rationale, potential and methods, *Emerg. Infect. Dis.* **5**(1): 95-101.

T. R. Moench, K. J. Whaley, T. D. Mandrell, B. D. Bishop, and C. J. Witt (1993), The cat/feline immunodeficiency virus model for transmucosal transmission of AIDS: nonoxynol-9 contraceptive jelly blocks transmission by an infected cell inoculum, *AIDS* **7**: 797-8O2.

D. Mollison(1991), Dependence of epidemic and population velocities on basic parameters, *Math. Biosci.* **107**: 255-287.

J. D. Murray (2003), *Mathematical Biology II: Spatial Models and Biomedical Applications*, Springer-Verlag, Berlin.

J. D. Murray and W. L. Seward (1992), On the spatial spread of rabies among foxes with immunity, *J. Theor. Biol.* **156**: 327-348.

J. D. Murray, E. A. Stanley, and D. L. Brown (1986), On the spatial spread of rabies among foxes, *Proc. R. Soc. Lond.* **B229**(1255): 111-150.

J. V. Noble (1974), Geographic and temporal development of plagues, *Nature* **250**: 726-729.

M. O'Callagham and A. G. Murray (2002), A tractable deterministic model with realistic latent periodic for an epidemic in a linear habitat, *J. Math. Biol.* **44**: 227-251.

R. S. Ostfeld, O. M. Cepeda, K. R. Hazler, and M. C. Miller (1995), Ecology of Lyme disease: habitat associations of ticks *Ixodes scapularis* in a rural landscape, *Ecol. Appl.* **4**: 242-250.

C. Ou and J. Wu (2006), Spatial spread of rabies revisited: influence of age-dependent diffusion on nonlinear dynamics, *SIAM J. Appl. Math.* **67**: 138-164.

N. C. Pedersen, E. W. Ho, M. L. Brown, and J. K. Yamamoto (1987), Isolation of a T-lymphotropic virus from domestic cats with an immunodeficiency-like syndrome, *Science* **235**: 790-793.

PHAC(2007), West Nile virus general overview, $http://www.phac-aspc.gc.ca/wn-no/gen\_e.html$

L. Rass and J. Radcliffe (2003), *Spatial Deterministic Epidemics*, Math. Surveys Monogr. **102**, Amer. Math. Soc., Providence, RI.

S. Ruan (2007), Spatial-temporal dynamics in nonlocal epidemiological models, in *Mathematics for Life Science and Medicine*, ed. by Y. Takeuchi, K. Sato and Y. Iwasa, Springer-Verlag, New York, pp. 97-122.

F. Sauvage, M. Langlais, N. G. Yoccoz, and D. Pontier (2003), Modelling hantavirus in fluctuating populations of bank voles: The role of indirect transmission on virus persistence, *J. Anim. Ecol.* **72**: 1-13.

C. Schmaljohn and B. Hjelle (1997), Hantaviruses: A global disease problem, *Emerg. Infect. Dis.* **3**: 95-104.

J. Shi (2007), Studies on epidemics of hemorrhagic fever with renal syndrome in China, *Chin. J. Zoonoses* **23**: 296-298.

J. W. Song, L. J. Baek, D. C. Gajdusek, R. Yanagihara, I. Gavrilovskaya, B. J. Luft, E. R. Mackow, and B. Hjelle (1994), Isolation of pathogenic hantavirus from white-footed mouse (*Peromyscus leucopus*), *Lancet* **344**: 1637.

A. C. Steere (1989), Lyme disease, *N. Engl. J. Med.* **321**:586-596.

A. C. Steere, J. Coburn, and L. Glickstein (2004), The emergence of Lyme disease, *J. Clin. Intestigat.* **113**: 1093-1101.

A. C. Steere, R. L. Grodzicki,A. N. Kornblatt, J. E. Craft, A. G. Barbour, W. Burgdorfer, G. P. Schmid, E. Johnson, and S. E. Malawista (1983), The spirochetal etiology of Lyme disease, *N. Engl. J. Med.* **308**: 733-740.

A. C. Steere, S. E. Malawista, D. R. Snydman, R. E. Shope, W. A. Andiman, M. R. Ross, and F. M. Steele (1977), Lyme arthritis: an epidemic of oligoarticular arthritis in children and adults in three Connecticut communities, *Arthritis Rheum.* **20**: 7-17.

L. T. Takahashi, N. A. Maidana, W. C. Ferreira, Jr., P. Pulino, and H. M. Yang (2005), Mathematical models for the *Aedes aegypti* dispersal dynamics: travelling waves by wing and wind, *Bull. Math. Biol.* **67**: 509-528.

H. R. Thieme (1977a), A model for the spatial spread of an epidemic, *J. Math. Biol.* **4**: 337-351.

H. R. Thieme (1977b), The asymptotic behavior of solutions of nonlinear integral equations, *Math. Z.* **157**: 141-154.

H. R. Thieme (1979), Asymptotic estimates of the solutions of nonlinear integral equation and asymptotic speeds for the spread of populations, *J. Reine Angew. Math.* **306**: 94-121.

H. R. Thieme (2003), *Mathematics in Population Biology*, Princeton University Press, Princeton.

A. Tran and M. Raffy (2006), On the dynamics of dengue epidemics from large-scale information, *Theoret. Pop. Biol.* **69**: 3-12.

J. van Buskirk and R. S. Ostfeld (1995), Controlling Lyme disease by modifying the density and species composition of tick hosts, *Ecol. Appl.* **5**: 1133-1140.

F. van den Bosch, J. A. J. Metz, and O. Diekmann (1990), The velocity of spatial population expansion, *J. Math. Biol.* **28**: 529-565.

G. F. Webb (1981), A reaction-diffusion model for a deterministic diffusive epidemic, *J. Math. Anal. Appl.* **84** (1981), 150-161.

M. J. Wonham, T. de-Camino-Beck, and M. A. Lewis (2004), An epidemiological model for West Nile virus: invasion analysis and control applications, *Proc. R. Soc. Lond.* **B271**: 501-507.

J. K. Yamamoto, H. Hansen, E. W. Ho et al. (1989), Epidemiologic and clinical aspects of feline immunodeficiency virus infection in cats from the continental United States and Canada and possible mode of transmission, *J. Amer. Veterin. Med. Associat.* **194**: 213-220.

T. L. Yates, J. N. Mills, C. A. Parmenter, T. G. Ksiazek, R. P. Parmenter, J. R. Vande Castle, C. H. Calisher, S. T. Nichol, K. D. Abbott, J. C. Young, M. L. Morrison, B. J. Beaty, J. L. Dunnum, R. J. Baker, J. Salazar-Bravo, and C. J. Peters (2002), The ecology and evolutionary history of an emergent disease, Hantavirus Pulmonary Syndrome, *Bioscience* **52**: 989-998.

Z. Zhang, K. Wan and J. Zhang (1998), Studies on epidemiology and etiology of Lyme disease in China, *Chin. J. Epidemiol.* **19**: 263-266.

X.-Q. Zhao and W. Wang (2004), Fisher waves in an epidemic model, *Dis. Contin. Dynam. Systems* **4B**: 1117-1128.

CHAPTER 16

Economically optimal management of a metapopulation

James N. Sanchirico
University of California at Davis

James E. Wilen
University of California at Davis

Abstract. An informational revolution is underway in the natural sciences that is generating a comprehensive picture of meso- and micro-scale phenomena of the biosphere that appear as previously unrecognized spatial patterns in natural and man-influenced landscapes. These patterns, in turn, pose new questions about spatial-dynamic processes at various scales in coupled human and natural systems. Many of these questions are spawning new paradigms that focus explicitly on space and the manner in which the dynamics of patterns are generated. With all of this new information, a natural question to ask is: what are the implications for managing natural resources? And, how does the answer to that question depend on the spatial characteristics of the resource, such as spatial heterogeneity and dispersal processes? To provide qualitative insights into these questions, we develop a stylized bioeconomic model of a metapopulation. We use the model to numerically investigate optimal spatial-dynamic harvesting policies for a metapopulation and the consequences of implementing policies that treat the resource as if it is uniformly distributed over a homogenous environment. The latter is representative of the current framework used to develop a suite of policies for managing marine resources. Our findings highlight the need to better understand the economic-ecological implications of spatial-dynamic processes and to better incorporate these processes into management decisions.

16.1 Marine metapopulations

Over the last 20 years, the assembly and synthesis of an enormous amount of new information on ocean processes has greatly improved our understanding of oceanographic processes that operate on spatial and temporal scales ranging from small to large (Hilborn et al. 2003). We are also learning more about how oceanographic processes along with habitat heterogeneity affect the spatial distribution of marine populations. At one end of the spectrum are processes like the Pacific Decadal Oscillation (PDA) and North Atlantic Oscillation that involve long periods where wind, sea surface, and temperature are favorable to certain fish assemblages, followed by "flips" in conditions that then favor other assemblages (Mantua et al. 1997). At an intermediate scale are processes like El Ninos that operate on an intra-annual basis and that dominate the strength and locations of upwelling conditions (Cury et al. 1995) . At the smaller scale are nearshore and short term coastal events such as the springtime currents and winds that blow and then relax periodically.

These short and long-term external oceanographic and environmental drivers interact with larval production, dispersal, and settlement processes (Caley et al. 1997) in complex ways. For example, some populations of species, such as red sea urchins on the west coast of North America, have their entire annual recruitment determined by wind conditions and oceanographic currents during a window of a few days or week (Smith and Wilen 2003). Larvae are not necessarily passive in this process and the degree to which individuals are essentially retained either by means of active biological mechanisms (like changing buoyancy) and/or external drivers in their natal or local habitat is becoming better understood (e.g., Cowen et al. 2000; Warner and Cowen 2002; Swearer et al. 2002; Thorrold et al., 2002, Cowen et al. 2006). However, understanding the rates and nature of dispersal of different age classes within a species is still a topic about which far too little is known (e.g., Shanks et al. 2003) and movement rates appear to be highly variable among even closely related species. The post-settlement success of the recruits also depends on scale and context-dependent demographic factors and population regulation mechanisms that may operate at different strengths and different points in the life cycle (Steele 1997; Quinn and Deriso 1999).

The upshot of all this is that the new science is revealing more information about the patchy distribution of marine populations and the determinants of that patchy distribution. Because of this information revolution, the conventional aspatial paradigm for describing the nature of marine systems is giving way to a new spatial paradigm. One parsimonious conceptual framework for this new paradigm is the concept of a metapopulation, which is a system consisting of local populations occupying discrete habitat patches with significant demographic connectivity between patches (Kritzer and Sale 2004; Sale et al. 2006). As Joan Roughgarden, a prominent mathematical ecologist stated, the "m[M]etapopulation concept is here to stay in marine ecology. Science demands it, fisheries management needs it, and it is the last hope for marine conservation. It marks the most important milestone of marine ecology in more than 50 years" (page xix Sale et al. 2006).

We adopt the metapopulation framework and develop a bioeconomic model that treats space explicitly in the form of three discrete patches (Levin 1992). Fish populations may disperse between the patches via various mechanisms at a variety of rates. Focusing on a small number of discrete patches may seem restrictive, but, as Hastings and Botsford (2003) argue, such an approach can approximate models with space treated continuously. Our framework is especially suited to investigate how patch or habitat heterogeneity interacts with dispersal and connectivity to affect the potential benefits from applying spatially explicit policies.

We employ the model used in Sanchirico et al. (2008). We summarize its basic properties first. Assume that there are three habitats or patches with population sizes x_i with $i = 1, 2, 3$. The populations are assumed to grow at rates $F_i(x_1, x_2, x_3)$, which depend on the population sizes in the system. In particular, the population dynamics within each patch are:

$$\frac{dx_1}{dt} = F_1(x_1, x_2, x_3) - h_1 = r_1x_1(1 - x_1) + d_{13}x_3 + d_{12}x_2 - d_{11}x_1 - h_1 \quad (16.1)$$

$$\frac{dx_2}{dt} = F_2(x_1, x_2, x_3) - h_2 = r_2x_2(1 - x_2) + d_{21}x_1 + d_{23}x_3 - d_{22}x_2 - h_2 \quad (16.2)$$

$$\frac{dx_3}{dt} = F_3(x_1, x_2, x_3) - h_3 = r_3x_3(1 - x_3) + d_{31}x_1 + d_{32}x_2 - d_{33}x_3 - h_3 \quad (16.3)$$

where h_i is the catch rate in patch i and d_{ij} is the dispersal rate between patch i and patch j. The intrinsic growth rates, r_i, can vary due to differences in habitat quality across the system.

For a single isolated patch (for example, let $x_2 = 0$ and $x_3 = 0$), the population x_1 increases logistically with a growth rate r_1 but loses individuals at a density-independent rate d_{11}. When additional patches are included, we assume that the settlement rate is independent of the density in the arrival patch (denoted by the term $d_{12}x_2$ and $d_{13}x_3$). This type of process can be thought of as adult settlement, where adult survivorship is assumed unaffected by density dependent mechanisms.

While this model is very general, there are some constraints on the dispersal parameters that should be imposed to ensure that what leaves one area is greater than or equal to what arrives in another. (This restriction does not allow for the possibility of biomass entering the system from patches other than the three we consider.) In particular, we assume that there is no mortality and that what leaves patch i for patch j arrives in patch j.

We permit d_{ii} and d_{jj} to vary, which allows us, for example, to investigate different types of connectivity. For example, we benchmark our results by investigating a closed or independent system where all of the $d_{ij} = 0$. We also consider a system where adult biomass can get from every patch to every other patch (denoted fully-integrated) where $d_{ij} > 0$. Finally, we investigate a source-sink system, where patch one is assumed to be the source and the dispersal is unidirectional to the sink patches ($d_{j1} > 0$ with $j = 1, 2, 3$ with the remaining d_{ij} set to zero). We illustrate the three metapopulation structures in Figure 16.1.

The model is capable of depicting the variety of behavioral characteristics of a metapopulation as well as connectivity patterns stemming from typical oceanographic features. A discrete model of this type can also depict a range of productivity assumptions in a system of individual patches. Some patches may have higher biological productivity than others, while others may have no inherent productivity. See Sanchirico et al. (2008) for additional types of connectivity nested within this model.

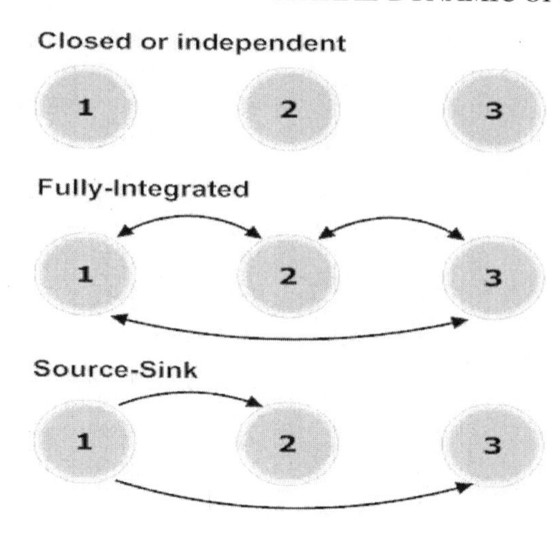

Figure 16.1 Metapopulation structures where the arrows illustrate potential dispersal flows.

16.2 Optimal control model

In this section, we show how one might choose fishing effort rates in each patch and in each period to optimally guide the metapopulation system toward a profit (rent)-maximizing equilibrium. A considerable literature exists (Scott 1954; Crutchfield and Zellner 1962; Brown 1970; Plourde 1970; Burt and Cummings 1970; Smith 1968; Clark 1976; Clark and Munro 1975) that outlines the so-called sole owner optimum, which is the case we are considering, for renewable resource use in an aspatial context. The main conclusions are as follows. First, as long as parameters are constant, there is an optimal long run steady state to which the optimal dynamic trajectories should converge. Second, it is desirable to get to this steady state as quickly as possible. If there are "adjustment costs" the optimal approach will be a more gradual asymptotic approach.

For a review of the literature that investigates the exploitation of a metapopulation from the sole owner perspective, see Sanchirico and Wilen (2007) and the citations therein. We extend these analyses by: (1) using a three patch metapopulation system as opposed to the standard two patch analysis; (2) investigating the role of different forms of connectivity; (3) characterizing the full spatial-dynamic solution rather than just focusing on the optimal long-run steady-sate, and (4) calculating the cost of ignoring the spatial characteristics of the resource in management decisions. A related paper is by Neubert and Herrera (2008) who use a continuous in time and space formulation to look at fishing effort patterns.

We assume, as does most of the previous work, that the regulator is knowledgeable, understanding population dynamics and dispersal mechanics of the biology, with perfect foresight. The objective function of the social-planner is to maximize the present discounted value (which is essentially the sum of current plus all future profits taking into account that the farther into the future profits accrue, the less they are worth) of fishing rents or profits across the patches by choosing the level of fishing effort in each patch in each period $(E_1(t), E_2(t), E_3(t))$. Mathematically, we represent the objective function as:

$$\boldsymbol{J}(E_1(t), E_2(t), E_3(t)) = \int_0^\infty \exp(-\delta t)\{\sum_{i=1}^{3}[p_i h_i(t) - c_i E_i(t)]\}dt \qquad (16.4)$$

where $h_i(t) = q_i E_i(t) x_i(t)$ is the harvest rate in patch i, q_i is the catchability coefficient in patch i, p_i is the price received at the dock from fish in patch i, c_i is a fishing cost parameter in patch i, and δ is the social discount rate (Clark 1990). We allow prices to vary across space due to differences in product quality, but for simplicity assume prices remain constant over time. We also assume that prices for the fish are not responsive to changes in harvest levels, which is an appropriate assumption when the fishery in question makes only a small contribution to a global fish market. Costs and cactchability coefficients can vary due to oceanographic and geographic characteristics of the patches.

The maximization of equation (16.4) is subject to the metapopulation dynamics represented in equations (16.1), (16.2), and (16.3), a set of initial population levels, and additional control constraints. For example, fishing effort in each habitat is constrained between E_i^{min} and E_i^{max}: $E_i^{min} \leq E_i(t) \leq E_i^{max}$.

Sanchirico and Wilen (2005) and Sanchirico et al. (2008) discuss the steps for solving this problem. Essentially, the basic procedure is as follows. First, we note that the optimal fishing effort level is determined by maximizing the Hamiltonian (**H**), which is defined as:

$$\mathbf{H} = \exp(-\delta t)\{\sum_{i=1}^{3}[p_i h_i(t) - c_i E_i(t)]\} + \lambda_1(F_1 - h_1) + \lambda_2(F_2 - h_2) + \lambda_3(F_3 - h_3) \qquad (16.5)$$

where λ_i are the shadow prices or adjoint variables that represent the marginal value of an additional unit of patch biomass on the present discounted fishery profits (Kamien and Schwartz 1991).

Since the problem is linear in the controls, we can first rearrange the Hamiltonian to isolate the control variables. Once this is done, we observe that there are switching functions in the Hamiltonian that are:

$$\sigma_i(t) \equiv \frac{\partial H}{\partial E_i} = exp(-\delta t)(p_i q_i x_i - c_i) - \lambda_i \qquad (16.6)$$

These switching functions are the time-varying coefficients that multiply each of the controls in the rearranged Hamiltonian. By the Pontryagin necessary conditions, each control must be chosen to maximize the Hamiltonian at each instant. Since controls enter the Hamiltonian linearly, the optimal levels of the control instruments in each patch i must satisfy:

$$E_i^* = \begin{cases} E_i^{max} & \text{when} \quad \sigma_i(t) > 0 \\ E_i^s = singular & \text{when} \quad \sigma_i(t) = 0 \\ E_i^{min} & \text{when} \quad \sigma_i(t) < 0 \end{cases} \tag{16.7}$$

When a switching function is positive, the optimal control for that patch is set at its maximum and when the switching function is negative the control must be set at its minimum allowable value. If the switching function is zero, the control must be set at its "singular value."

These conditions have well-known interpretations of tradeoffs between current marginal profits of another unit of harvest in a patch with the shadow value of that same marginal unit; when the difference is positive it pays to harvest and when negative it pays to "invest" the marginal unit in biomass and future profits. In this system, the twist is that a particular patch's shadow value reflects not only the role of a unit of biomass on future own-patch profits, but also the role that the patch plays in generating future harvest profits in other patches due to dispersal and other ecological and oceanographic interconnections.

In addition to the switching functions, the six other necessary conditions include the biomass state equations (16.1), (16.2), and (16.3) along with the adjoint equations (16.8). With respect to the latter, Pontryagin's Principle states that for an optimal solution $(x_1^*, x_2^*, x_3^*, E_1^*, E_2^*, E_3^*)$ there exist adjoint variables λ_1, λ_2, and λ_3 such that

$$\dot{\lambda}_i \equiv \frac{d\lambda_i}{dt} = -\frac{\partial H}{\partial x_i} \tag{16.8}$$

To determine the values of the singular control E_i^s when one or two of the switching functions in equation 16.6 are nonzero for a finite time period, we need to solve for a singular feedback law (Fraser-Andrews 1989, Volker 1996). In order to obtain the feedback law, we differentiate the switching function with respect to time, as many times as needed (Fraser-Andrews 1989, Volker 1996), substituting equations where appropriate, until the control appears in the expression. For our model, we use $\ddot{\sigma}_i(t) = 0, \dot{\sigma}_i(t) = 0$, and $\sigma_i(t) = 0$ along with $\dot{x}_i$, $\ddot{x}_i$, $\dot{\lambda}_i$, and $\ddot{\lambda}_i$ to obtain a complicated closed-form expression for the singular feedback law.

Given the initial values of the fish stock, $X_i(0)$, and the upper and lower bounds on fishing effort, the full solution to the optimal control path for patches pieces together the following components:

$$E_i \in \{E_i^{min}, E_i^{max}\} \qquad 0 \le t \le t_1, i \in (1,2,3)$$

$$\begin{cases} E_i(t) = E_i^s(t) \\ E_j \in \{E_j^{min}, E_j^{max}\} \end{cases} \qquad t_1 \le t \le t_2, (i,j) \in (1,2,3) \text{ with } i \ne j$$

$$\begin{cases} E_i(t) = E_i^s(t) \\ E_k(t) = E_k^s(t) \\ E_j \in \{E_j^{min}, E_j^{max}\} \end{cases} \qquad t_2 \le t \le t_3, (i,j,k) \in (1,2,3) \text{ with } i \ne j, j \ne k$$

$$E_i(t) = E_i^s(t) \qquad t_3 \le t, \forall i \in (1,2,3)$$

In linear control problems with only one control, the full dynamic solution is found by solving for the switch time, which is the date at which the control switches to the singular arc. In our case, we must find three switch times (t_1, t_2, t_3) and search for them using a shooting algorithm that is described in Sanchirico et al. (2008). We also employ the necessary but not sufficient Generalized Legendre-Clebsch (GLC) condition (Volker 1996) for optimality. We utilize the numerical closed-form singular feedback law along the singular arcs to describe the singular dynamics. We do this by building into our numerical algorithm the symbolic solver capabilities of MATLAB®.

16.3 Optimal spatial-dynamic paths

In designing optimal extraction plans, the regulator needs to trade off not just the economic values associated with standing biomass in the local patch, but also the value associated with the local patch's contribution to productivity of other patches via various dispersal processes. Because the regulator is setting the optimal fishing effort which in turn determines the catch in each patch in each period, he or she will need to trade off catching more fish in patch 1, with the consequences to the rest of the system, namely lower population levels and therefore fewer adults dispersing to patch 2 and/or patch 3. This trade off explicitly accounts for the relative profitability associated with harvesting in the patch, itself a function of bioeconomic parameters associated with each patch as well as with the nature of connectivity between patches.

We describe a set of the results that highlight the interplay between different connectivity structures, such as the closed or independent system, fully-integrated, and source-sink, and the biological and economic characteristics of the patch. In each case, we assume that patch 1's initial biomass density is above its economically optimal steady-state and patch 2 and 3 are considerably lower $((X_1(0), X_2(0), X_3(0)) = (.7, .4, .1))$. Such a condition is likely, for example, if patch 1 is further offshore than patches 2 and 3 and therefore has received less fishing pressure. Our assumption that two out of three stocks need to be rebuilt is consistent with global conditions. For example, in the United States' exclusive economic zone one out every four major fish stocks was classified as an overfished stock in 2006.

We parameterize the system to be consistent with our initial conditions where fishing costs are ordered as $c_1 > c_2 > c_3$ and p_3 is 5 percent higher than the prices in the other patches. The higher price represents circumstances in which the catch might be delivered to market sooner (e.g., fresher) due to the relative proximity of the patch to the port. Catchability coefficients and intrinsic growth rates are equal across the patches. We set $E_i^{min} = 0$ and E_i^{max} is equal to 1.5 times the open-access level of fishing effort. (See Sanchirico and Wilen (2007) for a discussion of open-access in a metapopulation context.) In each analysis, the generalized Legendre-Clebsch condition is satisfied and the trebly singular steady-state solutions are all interior.

We start with the closed or independent system to develop intuition on the optimal dynamics (see Figure 16.2 panel A-C). In this case, our 3 patch result mimics the most rapid approach path (bang-bang control) results found in Spence and Starrett (1975). That is, patch 1 is fished at the maximum rate possible driving down the population density until the steady-state level is reached at which point the effort level switches to the singular path, which is time-invariant. Patch 2 and 3 effort levels are set at their minimums (zero) and, hence, rebuild at varying rates due to different initial conditions, even though the intrinsic growth rates of the populations are identical. If the economic and ecological parameters are homogenous and there is no dispersal, then all three patches end up at the same steady-state. With habitat heterogeneity, as we assume, or asymmetric dispersal processes, such as in the source-sink case, the steady-states are no longer identical but the bang-bang nature of the dynamics is the same.

The time scales presented are determined in the simulation and caution is in order when trying to translate these values into calendar scales, as the parameter definitions are not specific to a unit of time (e.g., growth rate per year, etc). In the closed system, we find that patch 1's effort level switches off E_1^{max} at $t_1* = .7361$ and goes onto its time-invariant singular path. Patch 3, however, is under a moratorium with zero effort optimal until it switches to its singular path at $t_3* = 8.88$. Patch 2 with a higher initial condition has a shorter moratorium than patch 3. Once patch 3 switches off E_3^{min}, the system has reached the trebly-singular steady-state.

At the trebly-singular steady-state, we find that $x_1^* > x_2^* > x_3^*$ and the corresponding optimal effort levels are in the reverse order. The equilibrium biomass levels align with the relative profitability of the patches, where patch one has the highest per unit cost of fishing effort, patch 2 is slightly lower, and patch 3 has both a higher per unit price and the lowest per unit cost of fishing.

Next we introduce connectivity into the system by considering the optimal dynamics for the fully-integrated (See Figure 16.2 panel D-F) and source-sink system (see Figure 16.2 panel G-I), where patch 1 is assumed the source. Some of the key differences between these cases and the independent system are:

1. The singular portion of the effort dynamics is time-varying rather than time-invariant.
2. The biomass in patch 1 is driven to levels below its steady-state levels and then allowed to rebuild with fishing occurring.
3. The system reaches the steady-state in less time with connectivity than without connectivity.
4. In the source-sink system, it is optimal to fish patch 1 harder initially and then to essentially abandon fishing for a period of time to allow the stock to rebuild.

All of these differences are driven by the fact that connectivity creates conditions where the sole-owner needs to consider not just the return from fishing or not in any given patch but also the effect that an action in patch i has on the economic returns in other patches throughout the system. We observe these trade-offs between the fully-integrated and independent system in the following ways:

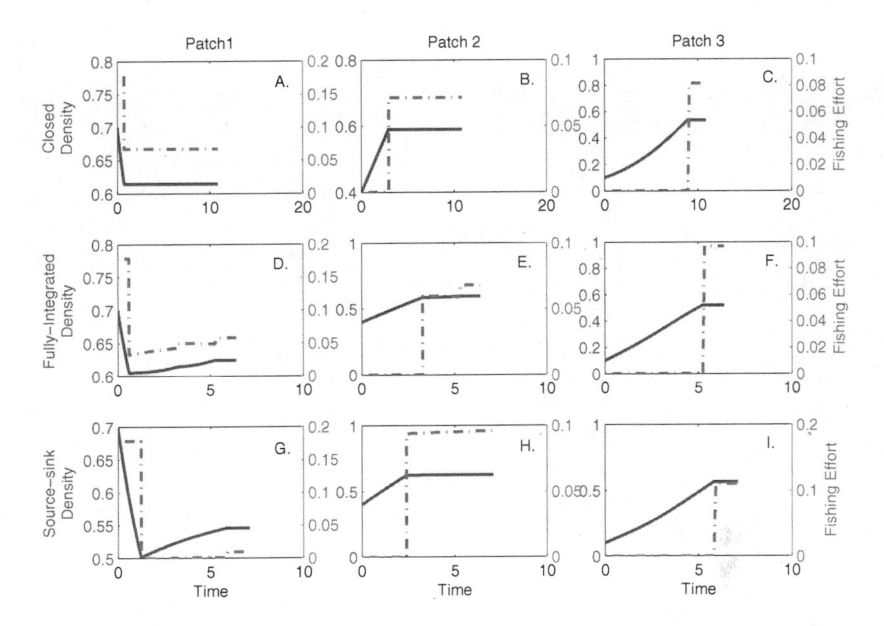

Figure 16.2 Optimal management of a metapopulation for the closed (panels A-C), fully-integrated system (panels D-F) and source-sink system with patch 1 as the source patch (panels G-I). The left y-axis measures the density of the population in the patch (solid lines), and the right y-axis measures the amount of fishing effort (dashed lines). The x-axis is in the time units of the simulation and should not be interpreted in calendar units, such as years. The parameters used in the numerical analysis are: $(c_1, c_2, c_3) = (.48, .42, .3)$, $(p_1, p_2, p_3) = (1, 1, 1.05)$,$(q_1, q_2, q_3) = (1.5, 1.5, 1.5)$, $(r_1, r_2, r_3) = (.26, .26, .26)$, $d_{ij} = b = .0525$ for $i \neq j$ and $d_{ij} = 2b$ for $i = j$, and $\delta = .05$. (See color insert following page 202.)

1. Relative to the independent system, the moratorium in patch 2 is longer in the fully-integrated system. The sole owner holds off fishing in patch 2, which implies that the owner is foregoing economic returns, to let the population rebuild to a higher level that helps to speed up the recovery in patch 3.
2. The sole owner switches off of the maximum effort level in patch 1 sooner in the fully-integrated system relative to the closed system, which implies a higher population that also can feed the recovery of patch 2 and patch 3.
3. The equilibrium biomass level in patch 1 is slightly higher in the fully-integrated system than in the closed system, as the owner holds a higher biomass level as a means to increase the perpetual flow of fish from the high cost patch (patch 1) to the lower cost patches (patches 2 and 3) in the system.

In the source-sink system where the biomass is flowing from patch 1 to patch 2 and from patch 1 to patch 3, we see similar responses in the dynamics but now the results are not as pronounced as in the fully-integrated system. This is because only patch 1 can be used to feed the recovery of the other patches. For example, while the recovery time in patch 3 is less than in the closed system, it is not as fast as in the fully-integrated system (it is only 34 percent faster than the closed system). We also see that the gains from maintaining fishing at the maximum rate in patch 1 are too great to forego in the source-sink, as the switch time is approximately double that in the closed system. The reason for this is that patch 1 is contributing biomass to the other patches independently of the density in the other patches. Therefore, with everything else being equal, the net flows from patch 1 to patch 2 or 3 are greater in the source-sink than in the fully-integrated system. This reduces the need to hold larger populations in the patch. We also find that the moratorium in patch 2 is 26 percent shorter in the source-sink than in the fully-integrated system. The shorter moratorium is because it no longer pays to delay fishing in patch 2, as it is not connected to patch 3 in the system.

16.4 Economic costs when not accounting for resource patchiness

What is the cost of ignoring the spatial characteristics of resources, such as habitat heterogeneity and dispersal processes? To investigate this, we solve for the optimal controls in the case where all of the heterogeneity is averaged out of the system. This might be the case, for example, if the resource is considered to be uniformly distributed throughout a homogenous environment. In particular, we assume that $c_i = \bar{c}$ where $\bar{c}$ is the average of the cost parameters used in the previous section. The same rule is applied for the price difference. We also assume that the initial conditions are equal across the three patches and set to the average of the "true" levels that were employed above $((\hat{X}_1(0), \hat{X}_2(0), \hat{X}_3(0)) = (.4, .4, .4))$. The reduction in the net present value from applying aspatial policies in a spatial setting is the cost associated with ignoring the spatial characteristics.

To calculate the costs, we undertake the following steps:

1. We solve for the optimal solution in the averaged system, denoted $(\hat{x}_1, \hat{x}_2, \hat{x}_3, \hat{E}_1, \hat{E}_2, \hat{E}_3)$ and the optimal switch times $(\hat{t}_1, \hat{t}_2, \hat{t}_3)$.
2. We take $(\hat{E}_1, \hat{E}_2, \hat{E}_3)$ and plug them into equations (16.1), (16.2), and (16.3) with $(X_1(0), X_2(0), X_3(0)) = (.7, .4, .1)$ and the heterogenous parameters to generate a path of biomass levels for each patch i, denoted $(\hat{x}_1, \hat{x}_2, \hat{x}_3)$.
3. Using $(\hat{x}_1^*, \hat{x}_2^*, \hat{x}_3^*, \hat{E}_1, \hat{E}_2, \hat{E}_3)$ and the heterogenous parameters, we evaluate the objective function (16.4), and we denote the net present value level as $\widehat{NPV}$.
4. We calculate the percent difference in the net present value between NPV^* and $\widehat{NPV}$, where NPV^* is the net present value evaluated at $(x_1^*, x_2^*, x_3^*, E_1^*, E_2^*, E_3^*)$

Figure 16.3 illustrates the differences in the patch density levels between the spatial and aspatial policies in the closed, fully-integrated, and source-sink system. That is, we illustrate both (x_1^*, x_2^*, x_3^*) and $(\hat{x}_1, \hat{x}_2, \hat{x}_3)$. The corresponding differences in the effort paths can be discerned from the stock dynamics. The single largest difference comes from the averaging of the different initial conditions, which is a likely outcome

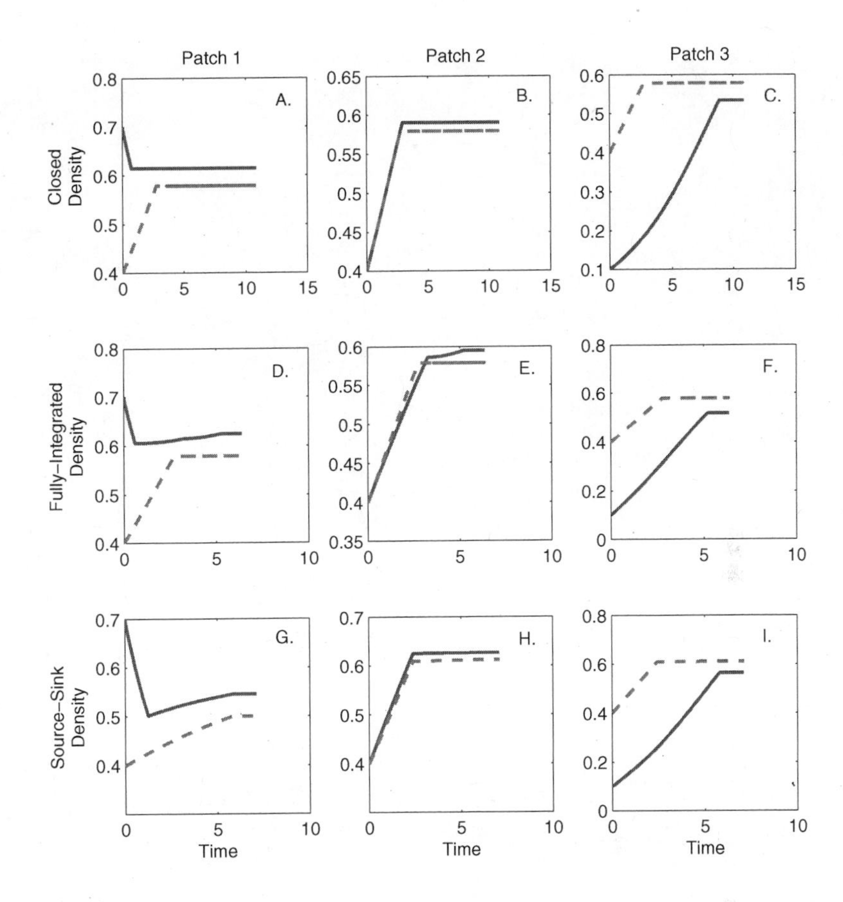

Figure 16.3 Patch density levels for the closed (panels A-C), fully-integrated system (panels D-F) and source-sink system with patch 1 as the source patch (panels G-I) in the spatial (solid line) and aspatial or averaged system (dashed line).The parameters used in the aspatial numerical analysis are: $(c_1, c_2, c_3) = (.4, .4, .4)$, $(p_1, p_2, p_3) = (1.01, 1.01, 1.01)$,$(q_1, q_2, q_3) = (1.5, 1.5, 1.5)$, $(r_1, r_2, r_3) = (.26, .26, .26)$, $d_{ij} = b = .0525$ for $i \neq j$ and $d_{ij} = 2b$ for $i = j$, and $\delta = .05$.

when stock assessments, for example, mask the underlying spatial heterogeneity. We also observe variations in the switch times and steady-state levels that reflect the averaging. Interestingly, the economically sub-optimal case (in the sense that the "true" system is heterogeneous) results in a longer moratorium in patch 1 that speeds up the recovery for patch 3 in both the fully-integrated and source-sink setting.

When the solutions from the aspatial case are run through the "true" metapopulation system, we find more dramatic differences between the spatial and averaged setting. In Figure 16.4 panels A, D, and G, the stock initially grows as a harvest moratorium

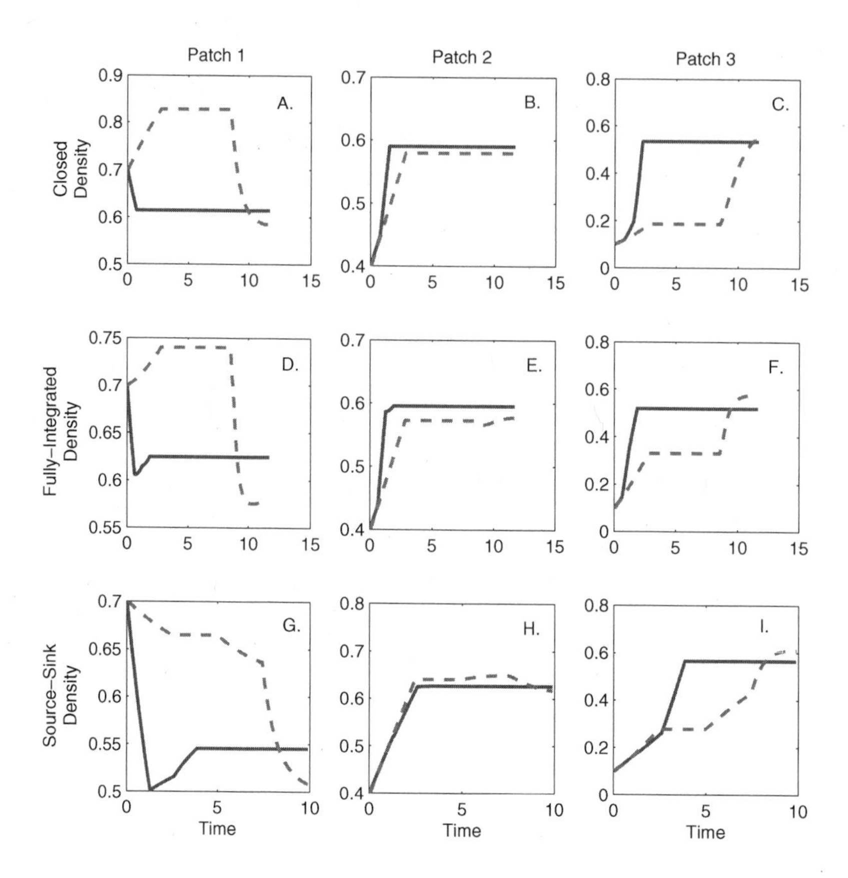

Figure 16.4 Patch density levels for the closed (panels A-C), fully-integrated system (panels D-F) and source-sink system with patch 1 as the source patch (panels G-I) in the spatial system (solid line) and when inserting the aspatial or averaged system optimal fishing effort trajectories into the "true" or spatial configuration (dashed line).

is put in place as opposed to fishing down the stock. We also see much more dramatic changes in the stock levels.

We also observe that there are potentially significant economic costs. In our example, applying the averaged results in a spatially-heterogeneous world results in a 54 percent reduction in the net present value in the closed system, 52 percent reduction in the fully-integrated, and a 46 percent reduction in the source-sink system.

Given that the averaged policy eliminates the initial economic returns, it is likely that a higher discount rate would lead to even higher losses in net present value from the sub-optimal averaging. Indeed this is what we found when we increased the discount rate to 7.5 percent from 5 percent. At the higher discount rate, the costs were 63, 62, and 56 percent for the closed, fully-integrated, and source-sink system, respectively.

For both the 7.5 and 5 percent discount rates, we also investigated the role of the dispersal rate between the patches (we lowered the common dispersal rate by 50 percent). In each case, the losses in net present value essentially remained the same for the fully-integrated system but increased slightly in the source-sink system. In particular, we find that at $\delta = .075$ the losses in the net present value were 59 percent and at $\delta = .05$ they were 49 percent of optimal. One explanation for the higher losses is that with a lower dispersal rate, the ability to tailor the policies to optimize the dispersal flow has greater value in terms of higher economic returns.

Of course, in general the less heterogeneity, in terms of habitat heterogeneity, initial stock levels, and asymmetrical dispersal processes, the lower we expect the losses from sub-optimal system averaging to be, everything else being equal. It is also possible that the implementation of an aspatial policy in a spatial world could lead to localized or global stock collapse. Such a result seems much more likely when conditions exist that lead to low stock levels even in the optimal spatial setting. For example, a combination of very low fishing costs, high prices, and a high discount rate is one possibility. Investigating the conditions under which this might occur seems like a worthwhile topic for further research.

16.5 Conclusion

Modern management of fisheries is mostly carried out under a "whole fisheries" paradigm. Whole fisheries management utilizes uniformly applied instruments over large regions. Better information on the role of space and spatial processes in marine systems will be a catalyst for the creation of new institutions to incorporate this understanding. In particular, it is likely that the future will move toward systems that are more finely delineated spatially and temporally, in ways that incorporate the patchy nature of abundance and the processes that connect various metapopulations (Sanchirico and Wilen 2007).

At this stage of the scientific revolution where spatial characteristics such as dispersal are still very difficult to pin down (Shanks et al. 2003), developing models that can help to predict the value of this new information for management provides a rich opportunity for future research. During this process, surprises will be found that challenge our basic intuition on resource management (Sanchirico and Wilen 2007). For example, we find a discontinuity on the time-varying portion of singular arc that is a function of having more than two controls that are interdependent along with the extreme/singular nature of the control paths. This is not a result that emerges in single or two control problems that are typical in the renewable resource economics literature and are currently being used to guide the design of policies.

In general, interdependent multiple control systems such as this one are a fundamental feature of spatial-dynamic economic-ecological problems, such as invasive species, disease, erosion, marine and freshwater pollution, air pollution, and will be a reoccurring theme as the management of marine systems moves towards ecosystem-based management (Wilen 2007 and Smith et al. 2009).

16.6 Acknowledgments

This research was made possible by funding from NSF Biocomplexity in the Environment Program (OCE-0119976) and EPA Science to Achieve Results (R832223). We thank Conrad Coleman (RFF) for help with the development of the MATLAB® code and an anonymous reviewer for his helpful comments.

16.7 References

G. M. Brown (1970), An optimal program for managing common property resources with congestion externalities, *Journal of Political Economy* **82**: 163-174.

O. Burt and R.G. Cummings (1970), Production and investment in natural resource industries, *American Economic Review* **60**: 576-590.

M. J. Caley, M. H. Carr, M. A. Hixon, T. P. Hughes, G. P. Jones, and B. A. Menge (1996), Recruitment and the population dynamics of open marine populations, *Annual Review of Ecology and Systematics* **27**: 477-500.

R. K. Cowen, K. M. M. Lwiza, S. Sponaugle, C. B. Paris, and D. B. Olson (2000), Connectivity of marine populations: Open or closed? *Science* **287**: 857-859.

R. K. Cowen, C.B. Paris, and A. Srinivasan (2006), Scaling connectivity in marine populations, *Science* **311**: 522-527.

C. W. Clark (1990), *Mathematical Bioeconomics: The Optimal Management of Renewable Resources*, Second Ed., John Wiley & Sons, New York: USA.

C. W. Clark and G. Munro (1975), Economics of fishing and modern capital theory: A simplified approach, *Journal of Environmental Economics and Management* **2**: 92-106.

J.A. Crutchfield and A. Zellner (1962), Economic Aspects of the Pacific Halibut Industry, Fishery Industrial Research, Vol. 1, no. 1, U.S. Department of the Interior, Washington D.C.

P. Cury, C. Roy, R. Mendelssohn, A. Bakun, D. M. Husby, and R. H. Parrish (1995), Moderate is better: Exploring nonlinear climatic effects on the Californian northern anchovy (Engraulis mordax) Canadian special publication of fisheries and aquatic sciences, Ottawa.

G. Fraser-Andrews (1989), Finding candidate singular optimal controls: A state of the art survey, *Journal of Optimization Theory and Applications* **60**: 173-190

A. Hastings and L. W. Botsford (2003), Comparing designs of marine reserves for fisheries and for biodiversity, *Ecological Applications* **13**: S65-S71.

R. Hilborn, T. P. Quinn, D. E. Schindler, and D. E. Rogers (2003), Biocomplexity and fisheries sustainability, *Proceedings of the National Academy of Sciences (USA)* **100**: 6564-6568.

M.L. Kamien and N. L. Schwartz (1991), *Dynamic Optimization: The Calculus of Variations and Optimal Control in Economics and Management*, New York.

J. P. Kritzer and P. F. Sale (2004), Metapopulation ecology in the sea: From Levins' model to marine ecology and fisheries science, *Fish and Fisheries* **5**: 131-140.

S. A. Levin (1992), The problem of pattern and scale in ecology, *Ecology* **73**: 1943-1967.

N. J. Mantua, S. R. Hare, Y. Zhang, J. M. Wallace, and R. C. Francis (1997), A Pacific interdecadal climate oscillation with impacts on salmon production, *Bulletin of the American Meteorological Society* **78**: 1069-1079.

M. Neubert and G.E. Herrera (2008), Triple benefits from spatial resource management, *Theoretical Ecology* **1**: 5-12.

C. G. Plourde (1970), A simple model of replenishable resource exploitation, *American Economic Review* **60**: 518-522.

T. J. Quinn II and R. B. Deriso (1999), *Quantitative Fishery Dynamics*, Oxford, New York.

J. N. Sanchirico and J. E. Wilen (2005), Optimal spatial management of renewable resources: Matching policy scope to ecosystem scale, *J. of Environ. Econom. Management* **50**(1): 23-46.

J. N. Sanchirico and J. E. Wilen (2007), Sustainable use of renewable resources: Implications of spatial-dynamic ecological and economic processes, *International Review of Environmental and Resource Economics* **1**: 367-405.

J. N. Sanchirico, J. E. Wilen, and C. Coleman (2008), Optimal rebuilding of a metapopulation, Manuscript available upon request.

P. Sale, I. Hanski, and J. P. Kritzer (2006), The merging of metapopulation theory and marine ecology: Establishing the historical context, in *"Marine Metapopulations,"* ed. by J. P. Kritzer and P. Sale, Elsevier, New York, pp. 3-28.

A. D. Scott (1955), *Natural Resources: The Economics of Conservation*, University of Toronto, Toronto.

A. Shanks, B. Grantham, and M. Carr (2003), Propagule dispersal distance and the size and spacing of marine reserves, *Ecol. Apps.* **13**(1)Supplment: S159-S169.

M. D. Smith and J. E. Wilen (2003), Economic impacts of marine reserves: The importance of spatial behavior, *Journal of Environmental Economics and Management* **46**(2): 183-206.

M. D. Smith, J. N. Sanchirico, and J. Wilen (2009), The economics of spatial-dynamic processes: applications to renewable resources, *Journal of Environmental Economics and Management* (forthcoming).

V. L. Smith (1968), Economics of production from natural resources, *American Economic Review* **58**: 409-431.

M. Spence and D. Starrett (1975), Most rapid approach paths in accumulation problems, *International Economic Review* **16**(2): 388-403.

M. A. Steele (1997) Population regulation by post-settlement mortality in two temperate reef fishes, *Oecologia* **112**(1): 64-74.

S. E. Swearer, J. S. Shima, M. E. Hellberg, S. R. Thorrold, S.R., G. P. Jones, D. R. Robertson, S. G. Morgan, K. A. Selkoe, G. M. Ruiz, and R. R. Warner (2002), Evidence of self-recruitment in demersal marine populations, *Bull. Mar. Sci.* **70**(1)Supplement: 251-271.

S. R. Thorrold, G. P. Jones, M. E. Hellberg, R. S. Burton, S. E. Swearer, J. E. Neigel, S. G. Morgan, and R. R. Warner (2002), Quantifying larval retention and connectivity in marine populations with artificial and natural markers, *Bull. Mar. Sci.* **70**(1)Supplement: 291-308.

M. Volker (1996), *Singular Optimal Control - The State of the Art*, Laboratory of Technomathematics, Geomathematics Group, University of Kaiserslautern.

R. R. Warner and R. K. Cowen (2002), Local retention of production in marine populations: Evidence, mechanisms, and consequences, *Bull. Mar. Sci.* **70**(1)Supplement: 245-249.

J. E. Wilen (2007), Economics of spatial-dynamic processes, *American Journal of Agricultural Economics* **89**: 1134-1144.

CHAPTER 17

Models of harvesting

Donald B. Olson
University of Miami

Abstract. The explicit dynamics of harvesting either through hunting or fishing or by growing crops is explored with attention to means of formulation of models for these activities. In particular carrying capacity is expressed in these models as set by the area of landscape being used for these activities. In the case of foraging the use of the area involves the encounter rate between prey and the forager. In terms of human use of natural resources the stability of the system involves both the encounter rate and the handling time involved. In terms of agriculture the system involves both the land area devoted to particular crops and the use of capital to promote their growth and marketing. In this case it is the capitalization that controls the stability of the activity. Ecologically it is interesting to consider the case where the capitalization can be borrowed so that the capital pool is negative. In extreme states the system allows negative capital and loss of production across an area, i.e., bankruptcy. The conclusion is that it is important to explicitly model the human activities themselves if an understanding of these activities and their stability is going to be gained.

17.1 Introduction

The ecology of humans (*Homo sapiens*) is a topic that is highly diversified into the fields of demographics, anthropology, and various social science fields such as economics and political science. While there is a large body of mathematical literature associated with the first of these, the other fields have fallen behind in the realm of quantitative theory. Here the goal is to treat humans as another element of the ecosystem and explore their impacts in terms of their use of "natural resources." This involves defining *ecosystem services* or human gain from the global ecosystem in relationship to the workings of the natural ecosystem or *ecosystem function* (Ruhl et al., 2007). While our exploitation of ecosystem services involves a wide range of actions that impact the ecosystem here the focus is on the direct taking (consumption) of resources through the act of *harvesting*. The approach is to formulate models for the act of harvesting in two different ways. The first involves the earliest form of foraging or hunting as manifest today primarily in large scale forestry and fishing respectively. The second considers the later advent in human history of actually controlling the ecosystem by husbandry of domesticated plants and animals. This seems to have first arisen in Asia Minor (Bar-Yosef and Belfer-Cohen, 1989; Bar-Yosef, 1998) but clearly arose separately in southeast Asia and the Americas (see Mithen, 2003). In both cases the analysis will consider competition between resource elements and the structure of human society. In the case of the fisheries example this includes food chain and food web dynamics to which human predation is added. In the agricultural situation the choice in commodities and their interaction with capital (monetary) resources is explored as one example of societal structure.

The models suggested here are simple in terms of the overall dynamics behind these situations. They, however, allow the use of dynamical systems theory to provide a means of understanding more complete models that defy simple analytical treatment and therefore enter the realm of complexity of more massive simulation models. In this sense the models here might be termed intermediate models that stand between simple linear analyses and larger scale simulations of fisheries and agriculture (Olson et al., 2005).

In the evolution of human civilization the trend has been in the direction of small foraging groups that grew into larger aggregated structures (tribes, nations) that controlled larger areas but tended to concentrate living areas into smaller areas such as villages and eventually cities. This trend has drastically changed our ways of using resources. To our current understanding this has occurred at least since the end of the last ice age (Mitthen, 2004). In the last 10,000 years the trend has been for human civilization to evolve as hunter/gatherer groups form denser populations that become dependent on modifications of the local environment for their sustenance (Toynbee, 1976, Mithen, 2004).

The early examples of hunters and gatherer groups involve movement across the landscape and concentration in resource rich areas. While there are some terrestrial societies that still fall into this mold, the major remaining vestige of this behavior occurs in the guise of marine fisheries and forestry today. Starting around 8000 years ago humans began a trend for a more sedentary existence that utilized the ecosystem by specifically encouraging certain animals and plants to dominate the environment. At issue here are the factors that underlie this transition and the increase in the density of human populations, i.e. the growth of the village, town, and city. A hypothesis (Mithen, 2004) is that the transition from foraging to agriculture involved the climate (drought) driven decline in small scale resources of grains on the hill sides of modern Turkey which precipated a shift to the large scale irrigated lands of Mesopotamia. In a similar manner the culture shifted from hunting of game, primarily gazelles, to domestication of other species (sheep and goats; Bar-Yosef, 1998). Again these transitions are not unique in that they also occurred in two locations in Southeast Asia (India and Indo-China) and in the Americas. In the latter case it may have also involved two locations, one in the Andes and one in the area of southern Mexico (see Mithen, 2004). In the case of these two areas there is little extant information on the onset of agriculture or even the plants that led to modern rice and corn. The location of the resulting cultures, however, provides a clue to the operant conditions. One factor in all of these locations is that they appear as ecosystem boundaries and locations that provide relatively small areas for human occupation. While there may be some fundamental flaws in the archeological record, the suggestion is that agriculture and domestication occurred in relatively localized areas and then spread out into surrounding regions. The fact that there are at least three independent locations involved suggests that the process is a natural emergent process in our species and its interactions with the environment. In terms of our model formulation it makes it imperative that space and time be considered in terms of demographic structure.

17.2 Basic model formulation

The mathematical formulation of the problem involves a set of populations N_j that can be defined in terms of population numbers, density (numbers per area), or in terms of the probability of finding an individual in an area. Interactions between populations can take several different forms, Lotka-Volterra (modified logistic) or Holling dynamics, for example, but will in general involve nonlinear interactions with products or other functional forms representing the interaction between populations (see Murray, 1993). Here it is assumed that these populations are dependent on space and time (x, t). In the case where specific entities are distributed across space it is useful to associate an area, $A_j(x, t)$, that defines a population carrying capacity. Here this simply represents the physical space required by the population. In addition to the population dynamics the problem demands consideration of other variables, such as physiological status, age, and at longer time scales genetics (see Olson et al., 2005). The goal here is to provide a model system that can directly consider biogeography. Therefore it is important to carefully define $A_j(x, t)$ and its parameterization in the model framework.

For the current discussion $A_j(x, t)$ will be considered to be a logistic variable. The formulation assumes that the success of populations on an area decline with density due to decreased births and increased natural loss of population in an area. Interactions with other organisms, such as competition for area, can be parameterized with the classical addition of a weighted logistic term. Here the equation for a species density (population per area or probability of encountering the jth species,) N_j, is given by

$$DN_j/Dt = RN_j(1 - N_j/A_j - \Sigma b_{ij}N_i/A_j) - P. \tag{17.1}$$

Here the subscript denotes the species with N_i indicating the competition with other species. The system then involves an equation for each species in the model ecosystem. Here humans can be considered as just another species in the formulation. The derivative with the capital D/Dt is used to denote all of the variables that structure a population across space including physical movement such as movement of animals (walking, swimming), air or current flows (advective drift), and diffusion. Here diffusion is an analog of molecular diffusion, but involves random movements tied to animal behavior or in the case of marine or atmospheric transport of insects by turbulence in these media. The total derivative also includes changes associated with aging, metabolic factors, or genetics. See Olson (2007) for a detailed discussion of the total derivative. The rate constant, R, is discussed in more detail below, but as the controlling rate constant it will include physiological factors in response to environmental variations other than forage and in the case of human harvesting capital (money) available to sustain the harvest. The b_{ij} coefficient indicates the interaction between competing species for the carrying capacity of the system. Again, here we are considering this to be the ecosystem area used by the species. It is possible to reconstruct the ratio of b_{ij}/A_j to express this as a carrying capacity (area niche) for the other species, A_i. While the logistic formulation inherently considers losses, in real systems it is necessary to include a predation term (P). As discussed below this adds a considerable amount of complexity.

In terms of hunting or foraging the P term involves the encounter rate (ε_{ij}) between species within their respective ecosystems and the time h_i it takes for a consumer to handle or harvest its resource (prey or crop). A fairly general formulation based on mass action (see Cosner et al. 1999) leads to a P term of the form

$$P = \varepsilon_{ij} N_j N_i / (1 + \varepsilon_{ij} h_i N_j). \tag{17.2}$$

In most realistic cases harvesting involves collections (village, city, global market) and large aggregations of the harvested species. In this case harvesting is concentrated and the result of the harvest is shared within the community involved. As derived by Cosner et al. (1999) a Beddington-DeAngelis functional response is used in such cases which modifies P to the form

$$P = \varepsilon_{ij} N_j N_i / (1 + \varepsilon_{ij} h_i N_j + s_i N_i). \tag{17.3}$$

Here the additional term basically demands that per capita gain is lower due to sharing of the resource (Cosner et al., 1999).

Before considering the solutions of these systems, it is important to return to rate function R. The maximum harvesting rate is dependent on a number of variables that are completely expressed in the above formulation for the target species except for factors that involve the ability of the harvesters themselves. In general these factors can be put into the P-term or into the total derivative (DN_j/Dt), but to simplify the system and to provide some further understanding of the harvesting process, it is useful to consider the factors involved in putting the harvesting into action. Here the focus is on humans and their use of the ecosystem. The formulation above for hunting or foraging over an area A_j includes the process of encounter (ε_{ij}) and processing of the resource (h_i). While these terms treat the actual return involved with encounters and the time it takes for each encounter to yield a successful result, the model does not consider issues that lead to variations in effort. Effort here is defined in the typical fisheries context as the resource applied to harvesting. In both the marine and terrestrial context R will involve the important social/economic variables where N_i is a human economic measure and the rate processes involve harvesting of the resources.

The rate of utilization of a resource in various systems depends on their organization and the level at which it is harvested. In ecology there is a considerable discussion involved with food-chains (linear trophic systems) versus food-webs (Fig. 17.1). In the figure both systems represent the minimal description of a real system. Mathematically, however, these represent the limit of classical (nonnumerical) analysis. This means that one can calculate steady states (fixed states) and provide an analysis of their stability. While some portions of the conclusions depend on the formulations described above, the major issues involve the formulation of R and its interaction with the P term. In the context of the current discussion R will be posed as a catalyst term governed by Michaelis-Menten or Holling dynamics. The basic form is then $X_i/(K+X_i)$, where X_i is a controlling variable. The rate term (R) then can include explicit influence of the physical climate, $T(x,t)$, or economic variables such as capital (money, C_j) and markets (M_j). The subscripts associate the economic variables with specific populations, but it is assumed that there are transfers between these just as there are interactions between populations. Furthermore, it is assumed that the climate variable impacts the rate coefficients in the population equations and changes in populations invoke variations in capital and respond to market pressure. These interactions will in general also involve nonlinear functions. While at this point the intent is to allow the specific formulations to be open, there are some limitations to the possible forms that can appear. In particular it is assumed that the equations allow positive equilibrium solutions for the N_j populations, that $T(x,t)$ has a set range of values, and that market functions are positive. Capital, C_j, deserves some special attention since unlike most problems in ecology it can be of either sign. While there will be a constraint on the sum, $\Sigma C_j > 0$, individual capital values can be negative, i.e., a particular population (commodity sector) can be in debt. The goal here is to then consider the variations in A_j, M_j, and $T(x,t)$ that lead to positive C_j. The alternative view of the problem is to consider routes to negative C_j and loss of area under occupation, i.e., failure.

A schematic for the general problem is given in Fig. 17.2. The problem specifically includes structure in space, time, and in associated variables such as capital, area, and market. A discussion of the equations for this type of system can be found in Olson (2007) along with an argument that the only feasible manner of handling these problems in simulations involves a Lagrangian or particle following system that tracks individual populations across space and time. Here the focus is on particular behaviors in these equations across space and time. In particular, the stability of populations and capital are of interest. Since the analysis is explicitly dependent on space and time a natural ecosystem based question is the manner in which populations reach limits in the sense of the determinants of population ranges (biogeography) and the nature of extinction, i.e., $N_j \rightarrow 0$. On the side of economics the corresponding state in which $C_j < 0$ and the commodity in question is no longer allowed to occupy area $(A_j = 0)$, is of interest since this involves the failure of a harvesting endeavor or in modern terms bankruptcy. The important question to ask is what conditions lead to disappearance of populations and in the economic state the failure of human enterprises involving abandonment of sites in early times and demise of markets in modern times.

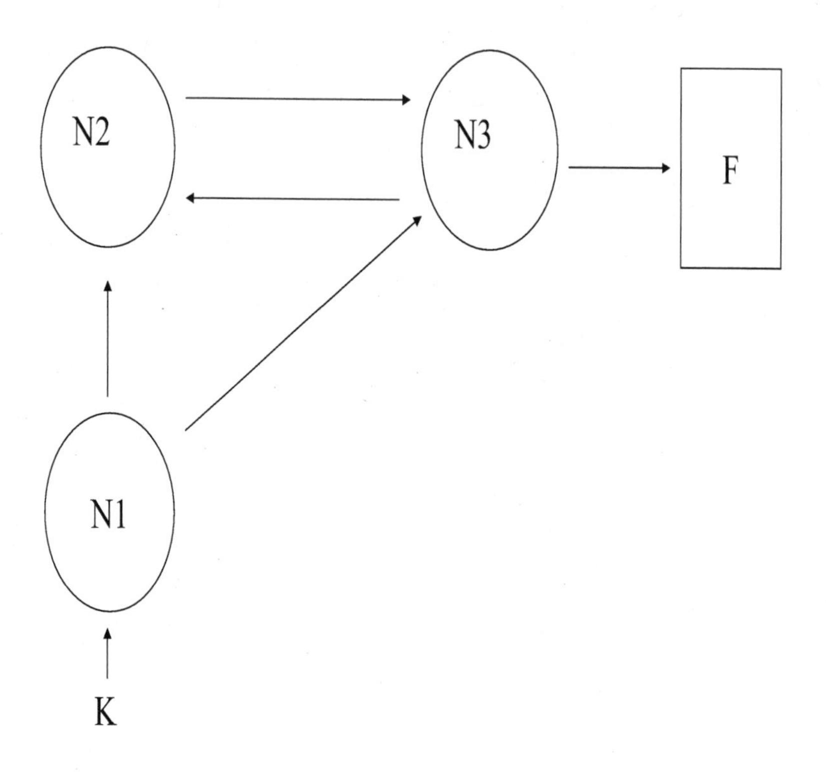

Figure 17.1 A conceptual model including a simple food chain (N_1 to N_2) and foodweb structure (interaction between N_2 and N_3) as well as the impact of fishing (F) on the system. The figure is modified from Olson et al. (2005).

17.3 Explicit examples

There are two types of systems to be explored. These include basic hunting or fishing involving the search for natural populations and their exploitation and the domestication of animals or plants in what can be called "farming." There has been a historical progression from the dominance of the former to the dominance of the latter. This transition has had major influences on the progression of human cultures. Both the geography and investment (effort or funds) involved in the hunting process have evolved markedly with technological advances, while the advances in domestication efforts, again a technological change, have made the more primitive reliance on natu-

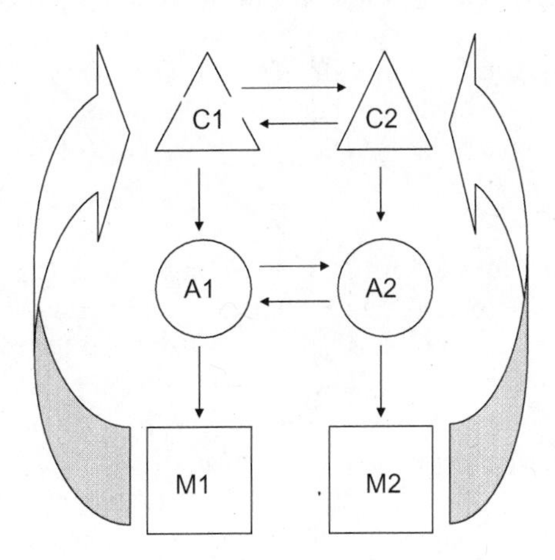

Figure 17.2 A conceptual model of an agricultural activity involves fields of two commodities with areas A_1 and A_2. The fields yield produce to two markets (M_1 and M_2) that in turn return capital to the farming activity (C_1 and C_2). This capital then is used to catalyze the farming activity in these areas. See text for more details.

ral populations less extensive. From the native hunting cultures of the past to modern times the changes in culture have seen a reduction of "hunting" to fisheries and harvesting of old growth forests. There are also some recreational activities that involve the old style of pursuit and harvest albeit with ever more complex technology. In order to understand these transitions it is possible to consider the intermediate model formulation introduced above.

To begin an examination of the interplay between technology and exploited population dynamics consider the case of two types of fishing gear in use today (Olson et al., 2005) applied to an age structured population of tuna. This example, while related to an existent set of pelagic fisheries, is informative as regards the succession of exploitation involved with different technologies imposed on harvest of a common species. The target in this case is yellowfin tuna that is harvested by long-line fisheries involving arrays of hooks or by purse siene nets that harvest younger fish. The conclusion of an analysis including the age structure of the tuna population and the potential fishery yield is that the purse sienes will exclude the less efficient long-line fleet. This conclusion is basically consistent with the distributions of the two types of fishing gear in the ocean today. Failure to consider this replacement of fishing (hunting) effort in context has led to some rather strange views of the ocean ecosystem (Worm et al., 2005) where consideration of a single fishery (long lines) suggests that regions are nonproductive. In actual terms the particular type of fishing has been replaced economically. Therefore the Worm et al. (2005) conclusion that particular areas of the tropical oceans are nonproductive is biased by their use of the Japanese long-line data set. Economically the long-line fisheries have been subplanted in the areas of high potential production by the harvesting of younger fish via the net fisheries that essentially remove the economic viability of the areas involved for long-lining (see Olson et al., 2005).

The same conditions apply to a terrestrial scenario in terms of hunting. Here the outcome of hunting depends not only on the applied rate of harvest, but also on the specific features of the target species (such as age or sex) and the technology employed in the hunt. This is not generally true of agriculture, however. In agriculture the area dependence has some fundamental differences from the hunting situation. Here starting with the women planting grains as the highlands dried, there is a intrinsic occupation of space. A field occupies area in a different way than an area that is foraged or hunted. Active cultivation formally occupies land in a way that hunting and gathering does not. Planting a field makes explicit use of A_i. It also makes a certain expenditure of capital that is hard to quantify in terms of hunting and gathering. In this sense it ties the people to the land. In the same sense it ties wealth, C_i, to a set space of land, A_i, that determines the output of human endeavors. Historically the nature of the catalytic term in C_i and the rate constants changes with technology. In the case of the foraging (fishing) technology and expenditures modify both the encounter rate and the handling time. Since these two terms establish the bifurcation parameter, technology and C_i can shift the stability of systems (Martin and Ruan, 2001). Initial explorations of the agricultural model suggest that in the continuous formulation capital (C_i) can become depleted hence taking the system to the $C_i = 0$ and $A_i = 0$ states. In the discrete system or in a system with time lags the solutions can go past zero to $C_i < 0$ and $A_i = 0$ states. These correspond to bankruptcy for the endeavor.

17.4 Conclusions

In both the case of foraging (hunting and fishing) and in agriculture (growing crops) the nature of the human activities determines the dynamics of the ecosystem. To ignore the dynamics of the human agents in these cases obscures the actual reality and allows consequences of human actions to be mistakenly related to population dynamics on the part of the harvested quantities. In particular, it is important to consider explicitly the impact of economic factors and technological advances on the stability of harvesting regimes.

17.5 Acknowledgments

The author would like to thank A. McCrea-Strub who worked on the fisheries food-webs and chains. The agriculture work gained from the analysis skills of C. Laciana. G. Podesta, and D. Letson were also very helpful in laying out the system and discussions on the capitalization issues. Finally, the author would like to thank the editors for being very patient with the development rate of this manuscript.

17.6 References

O. Bar-Yosef (1998), The Natufian cluture in the Levant, threshold to the origins of agriculture, *Evol. Anthro.* **6**:159-177.

O. Bar-Yosef and A. Belfer-Cohen (1989), The origins of sedentism and farming communities in the Levant, *J. World Prehistory* **3**:477-498.

C. Cosner, D.L. DeAngelis, J.S. Ault and D.B. Olson (1999), Effects of spatial grouping on the functional response of predators, *Theor. Pop. Biol.* **56**:65-75.

A. Martin and S. Ruan (2001), Predator-prey models with delay and prey harvesting, *J. Math. Biol.* **43**:247-267.

S. Mithen (2004), *After the Ice*. Harvard University Press, Cambridge, 622 pp.

D.B. Olson, C. Cosner, S. Cantrell, and A. Hastings (2005), Persistence of fish populations in time and space as a key to sustainable fisheries, *Bull. Mar. Sci.* **76**:213-231.

D.B. Olson (2007), Lagrangian biophysical dynamics, in *Lagrangian Analysis and Prediction in Coastal and Ocean Systems,* ed. by A. Griffa, D. Kirwan, A. Mariano, and T. Ozgokman, Cambridge Uni. Press, Cambridge, 275-348.

J.B. Ruhl, S.E. Kraft, and C.L. Lant (2007), *The Law and Policy of Ecosystem Services.* Island Press, Washington, 345 pp.

A.J. Toynbee (1976), *Mankind and Mother Earth.* Oxford University Press, Oxford, 670 pp.

B. Worm, M. Sandow, A. Oschlies, J.K. Lotze, and R.A. Myers (2005), Global patterns of predator diversity in the open ocean, *Science* **309**:1365-1369.

CHAPTER 18

Spatial optimal control of renewable resource stocks

Guillermo E. Herrera
Bowdoin College

Suzanne Lenhart
University of Tennessee

Abstract. To understand the impact of habitat heterogeneity and other spatial features on the management of fisheries and other resource stocks management, we investigate models which include explicit representations of space. We give the ideas behind some metapopulation models, with interconnected subpopulations in discrete space. These models are based upon ordinary differential equations. We then discuss some recent work focusing on partial differential equation models, in which the resource stock moves continuously in both time and space. The basic underlying ideas of optimal control of partial differential equations are given and illustrated by a harvesting example.

18.1 Introduction

It is well known that some property rights structures give rise to socially undesirable outcomes in the spatially homogeneous case – the marquis example being the complete erosion of pecuniary net benefits (or "rent dissipation") in an open-access fishery; this is Clark's (1976) bionomic equilibrium. In spatially heterogeneous systems, decentralized decisions about resource exploitation often fail to account for spatial variations in the *in situ* value of the resource.

The spatial structure of a renewable natural resource has potentially important implications for the outcome of management. Improved understanding of spatial dynamics can enhance the predictive power of behavioral models, illuminate sources of inefficiency in decentralized resource use, and suggest qualitatively new means of addressing these inefficiencies. By contrast, regulations which are naive to the spatial dynamics of a resource generally produce suboptimal results (Tuck and Possingham 1994, Sanchirico and Wilen 2005).

Spatial heterogeneity and spatial dynamics give rise to variations in *in situ* economic values (or anthropocentric benefits): units of biomass or individual organisms in different locations provide different dynamic contributions to the level of the resource stock, and to the net benefits emerging from the resource over time. It has long been recognized (Gordon, 1954) that when harvesters in decentralized systems have incomplete property rights over these future benefits, they tend to exert not only a suboptimal amount of aggregate harvest effort per unit time, but also a suboptimal spatial distribution of effort. That is, due to divergences between private and social objective functions, the spatial dimension provides another opportunity for harvesters independently pursuing their own interests to "dissipate rents," i.e., to decrease net economic benefits below their potential levels.

Spatial bioeconomic analysis seeks to characterize the outcome of decentralized exploitation of a resource, to compare this to the efficient spatial-dynamic pattern of use which would arise under centralized management or sole ownership, and to assess the relative effectiveness of different regulatory instruments in mitigating spatiotemporal inefficiencies. While optimality is often seen as synonymous with dynamic efficiency, or the maximization of discounted pecuniary net benefits over time, alternative objectives such as equity (the distribution of net benefits across human stakeholders) and sustainability can be assumed. It is not surprising that spatially structured regulations outperform their nonspatial analogs; but adding spatial resolution to regulation is costly, so it is valuable to better understand the contexts in which such resolution is particularly beneficial (Quinn 2003, Walters and Martell 2004).

To include movement of the species involved or a hetergeneous environment, a representation of the spatial dimension is needed (Kareiva et al. 1990). One common approach is to divide the environment in a collection of patches, frequently called metapopulation modeling. The patches can also have specific spatial coordinates with varying movement between them. Including dispersal in a continuous way is frequently done with partial differential equations (PDEs) with diffusion and advection terms (Cantrell and Cosner, 2003). We note that spatial spread can also be represented with dispersal kernels (Lockwood et al. 2002, Kot et al. 1996, Lewis and Van Kirk 1997).

In the next section, we discuss some bioeconomic results from metapopulation models, in which the underlying models are systems of ordinary differential equations (ODEs) and space is included as a discrete feature.

Then the following section gives some background of optimal control results for partial differential equations (PDEs) related to renewable resources. Since the technique of optimal control of PDEs has some different techniques from optimal control of ODEs, we also want to illustrate the differences. An explicit example is worked out in some detail for demonstration of these techniques.

18.2 ODE models with spatial components

The majority of bioeconomic analyses that incorporate spatial dynamics are based on metapopulation models (see pp. S60-S61 of Gerber et al. 2003). The dynamics of these systems are governed by ordinary differential equations, so the dynamic optimization of their exploitation yields to conventional optimal control techniques. It is dynamically efficient to remove those units of biomass, and only those, for which the current net benefits of extraction meet or exceed the *in situ* value, represented by the co-state, or adjoint variable, in the optimality system associated with the Hamiltonian.

Sanchirico and Wilen brought the issue of spatial structure in renewable resources into the bioeconomic mainstream with a series of models assuming logistic growth within patches and density-dependent dispersal between patches. Their initial contribution (1999) was not optimal control *per se*, but instead showed that persistent patterns of spatial exploitation emerge in a spatially heterogeneous system in the context of open-access, where harvest effort is assumed to increase in all patches until the point where equilibrium profits are driven to zero.

Maintaining their assumption of rent dissipation in all harvested areas, Sanchirico and Wilen (2001) showed that a no-take reserve can in some cases increase both standing stock and the level of harvest (if not profit). They then showed (2002) that no-take reserves can simultaneously increase biomass and the equilibrium value of harvest quotas (a proxy for the economic rents emerging from the resource).

Gordon (1954) recognized the potential for misallocation of harvest effort in space in a simple model of two spatially isolated fishing grounds. One of his conclusions is that the specification of the spatial distribution of harvest effort may yield significant benefits. Tuck and Possingham (2000) used coupled difference equations to model the harvested and closed local populations of a single-species, two-patch metapopulation. They consider the problem of optimally exploiting the single species local population that is connected by dispersing larvae to an unharvested second local population. They apply dynamic optimization techniques to derive the optimal equilibrium escapement for the harvested stock. They also consider how a reserve affects both yield and spawning stock abundance when compared to policies that ignore the spatial structure of the metapopulation. Closed areas (reserves) are found to decrease gross harvest levels only slightly, and to have positive net benefits in terms of both stock abundance and economic rents.

Brown and Roughgarden (1997) were among the first to explicitly demonstrate that a closed area can be part of a dynamically efficient solution. They applied optimal control to a system with larval pool dynamics and space-constrained settlement (recruitment) and found it to be optimal to harvest at only one patch, i.e., to induce and maintain an equilibrium outcome consisting of a source-sink dynamic. Herrera (2007) also used a common larval pool model to show that different patterns of closures (alternating between areas, toggling on and off in one area, etc.) can be efficient over time, given lower costs of enforcing closures relative to positive harvest quotas.

18.3 PDE models

The metapopulation formulation is a simplification of spatial structure and dynamics. While it may be a realistic depiction of some systems - for example reefs separated by inhospitable habitat - it is generally more realistic to think of a resource stock as diffusing across a continuous habitat. Such a formulation gives rise to a richer set of outcomes and potentially regulatory approaches, but such systems require a different set of optimal control methods.

In contrast to these discrete space models, Neubert (2003) considered a resource existing in a continuous, finite one-dimensional spatial domain. The resource grows logistically and diffuses continuously. He solved for the spatial distribution of fishing effort that maximizes the yield at the steady state in which no reserves are imposed *a priori.* After rescaling the variables, the model is:

$$-u_{xx} = u(1-u) - h(x)u, \text{ on } 0 < x < l$$
$$u(0) = u(l) = 0.$$

The assumption of equilibrium turns the problem of optimally controlling a PDE system into one of controlling a coupled ODE system. Using Pontryagin's Maximum Principle (Pontryagin et al. 1962) with x as the underlying variable, he showed that no-take marine reserves are always part of an optimal harvest designed to maximize yield. Also he found that the size, number, and location of the optimal reserves depend on a dimensionless length parameter. For small values of this parameter, the maximum yield is obtained by placing a large reserve in the center of the habitat. For large values of this parameter, the optimal harvesting strategy is a spatial "chattering control" with infinite sequences of reserves alternating with areas of intense fishing. Such a chattering strategy would be impossible to actually implement due to the difficulty of monitoring the reserves. In this model, the population is zero on the boundary of the region, which means that the boundary is absorbing and individuals who encounter the boundary die. Such a 'lethal' boundary may be due to a hostile region surrounding the habitat.

Neubert and Herrera (2007) extended Neubert's (2003) yield maximization framework by including harvest costs. This allowed for a nondegenerate characterization of the open-access equilibrium (open-access inevitably leads to resource extinction in the absence of harvest costs), and a comparison of the equilibrium stock and effort distributions emerging from open-access and optimal spatial regulation. The optimal solution in the positive cost case is qualitatively similar to the zero-cost case (i.e., reserves play a role, and effort is largely focused near the habitat boundaries), though the chattering phenomenon vanishes. Another result of this paper is that, under certain circumstances characterized in the paper, spatially optimal exploitation employs a greater amount of harvest effort in aggregate than is brought to bear under open-access. Because reduction in industry participation is an important political-economic impediment to the implementation of fisheries regulations, the existence of this "employment benefit" is a potentially important result.

When choosing a PDE to model a scenario, one must consider the types of diffusion, advection, and growth terms to include. But boundary conditions can have an essential effect. As mentioned above, those results are for zero boundary conditions (called Dirichlet conditions). If the individuals encountering the boundary are reflected back and thus do not leave the domain, then 'no flux' boundary conditions, $\frac{\partial u}{\partial n} = 0$ (called Neumann conditions), would be valid and will give different results. If the flux across the boundary is proportional to the population at the boundary, then Robin conditions would be used, $\frac{\partial u}{\partial n} + bu = 0$.

Optimal control techniques for PDEs are just beginning to be applied to resource problems. There has been some work done on harvesting problems from a mathematical viewpoint using PDEs. Leung and Stojanovic (1993) studied the optimal harvesting control of a biological species, whose growth is governed by the diffusive Volterra-Lotka equation. The species concentration satisfies a steady-state equation with no-flux (Neumann) boundary condition. The optimal control criterion is to maximize profit which is the difference between economic revenue and cost. Leung (1995) also studied the corresponding optimal control problem for steady-state prey-predator diffusive Volterra-Lotka systems and obtained similar results to the single species case. Cañada et al. (1998) and Montero (2000) studied an optimal control problem for a nonlinear elliptic equation of the Lotka-Volterra type with Dirichlet boundary condition. The conditions for the optimality system and uniqueness of the optimal control depend on the eigenvalues of the Laplacian operator. These papers emphasize the mathematical analysis and do not give economic interpretations of the results nor numerical examples. Existence and characterization results for an optimal control are given.

Kurata and Shi (2007) studied a reaction-diffusion model with logistic growth, constant effort harvesting (depending only on space and not on time), and Dirichlet boundary conditions. By minimizing an intrinsic biological energy function, which is different from the yield, they obtained an optimal spatial harvesting strategy which would benefit the population the most. They found out a nonharvesting zone should be designed. On the other hand, in the zone which allows harvesting, the effort should be put at the maximum value. Their objective function involves the gradient, the square, and the cube of the population and does not seem to have a biological interpretation.

Ding and Lenhart (2009) extended Neubert's (2003) work to a multidimensional domain, i.e., considering an optimal fishery harvesting problem using a spatially explicit model with a semilinear elliptic PDE, Dirichlet boundary conditions, and logistic population growth. They considered two objective functions: maximizing the yield and minimizing the cost or the variation in the fishing effort (control). Minimizing variation was considered to avoid the 'chattering' effect in Neubert's results. The optimal control when minimizing the variation is characterized by a variational inequality instead of the usual algebraic characterization, which involves the solutions of an optimality system of nonlinear elliptic partial differential equations. Some interesting conclusions were found. If one only wants to maximize yield, then a reserve is part of the optimal harvesting strategy. The problem of maximizing yield only with Neumann boundary condition gives a simple optimal control, a singular case. When considering the problem with cost in the objective function, the optimal benefit increases when domain size increases, but the structure of the reserve was preserved. There are few very control results known about the effect of changing domain size (Montero, 2001).

Lenhart and Bhat (1993) considered a harvesting problem in a parabolic PDE with logistic growth in which the population was considered to be a nuisance population. The goal was to minimize the damage due to the population and the cost of harvesting. The numerical illustration used 10 years of beaver data from some counties in New York and showed the optimal harvest level as a function of space and time.

We note that Brock and Xepapadeas (2005) treated an optimal harvesting problem for a population modeled by a parabolic PDE. The necessary conditions are given without including bounds on the controls. In the infinite time horizon case, an approximation with a linear PDE and quadratic objective function is used to show the existence of a Turing space of diffusive instability, which leads to the emergence of a spatial pattern in the optimal state.

18.3.1 Techniques for optimal control of PDEs

Note that in the multidimensional PDE case, one cannot use Pontryagin's Maximum Principle, thus some further analysis is needed to justify the necessary conditions. J.-L. Lions (1971) laid the foundation of the basic ideas of optimal control of PDEs in the 1970's. There is no complete generalization of Pontryagin's Maximum Principle to PDEs, but the book by Li and Yong (1995) deals with corresponding "maximum principle" type results.

Now we give a brief sketch of the technique of optimal control of PDEs in the parabolic system case. Choosing the underlying solution space for the states is a crucial feature for optimal control of PDEs. Classical solutions (solutions with all the derivatives occurring in the PDE being continuous) will not exist for most nonlinear PDE problems. Deciding in what "weak" sense to solve the PDEs is essential. We refer to Evans (1998) and Friedman (1982) for the rigorous definitions of Sobolev spaces and weak derivatives and give only an informal treatment. For most parabolic PDE control problems, the appropriate solution space is

$$L^2([0,T]; H_0^1(\Omega)),$$

where Ω is the spatial domain. The control set frequently consists of the Lebesgue integrable functions, which have specified upper and lower bounds.

The general idea starts with a PDE with state solution w and control u. Take A to be a parabolic partial differential operator with appropriate initial conditions (IC) and boundary conditions (BC),

$$Aw = f(w,u) \text{ in } \Omega \times [0,T], \text{ along with } BC, IC, \tag{18.1}$$

assuming the underlying variables are x for space and t for time. We are treating problems with space and time variables, but one could treat steady state problems with only spatial variables.

The objective function represents the goal of the problem; here we write our function in an integral form. We seek to find the optimal control u^* in an appropriate control set to maximize our goal

$$J(u^*) = \sup_u J(u), \tag{18.2}$$

with objective function

$$J(u) = \int_0^T \int_\Omega g(x, y, t, w(x, y, t), u(x, y, t))\, dxdy\, dt. \tag{18.3}$$

After specifying a control set and a solution space for the states, one can usually obtain the existence of a state solution given a control. Namely, for a given control u, there exists a state solution $w = w(u)$, showing the dependence of w on u.

Proving the existence of an optimal control in the PDE case requires *a priori* estimates of the norms of the states in the solution space to justify convergence. Thus obtaining the existence results for optimal controls in the PDE case is different than the ODE case. These estimates give the existence of a minimizing sequence u_n of controls where

$$\lim_{n\to\infty} J(u_n) = \sup_u J(u). \tag{18.4}$$

In the appropriate weak sense, this usually gives

$$u_n \rightharpoonup u^* \quad \text{in} \quad L^2(\Omega \times [0,T]), \quad w_n = w(u_n) \rightharpoonup w^* \quad \text{in the solution space,} \tag{18.5}$$

for some u^* and w^*. One must show $w^* = w(u^*)$, which means that w^* is the state corresponding to control u^*. We must also show that u^* is an optimal control, control, i.e.,

$$J(u^*) = \sup_u J(u). \tag{18.6}$$

To derive the necessary conditions, we need to differentiate the objective function with respect to the control, namely, differentiate the map

$$u \longmapsto J(u). \tag{18.7}$$

Since $w = w(u)$ contributes to $J(u)$, we must also differentiate the map

$$u \longmapsto w(u). \tag{18.8}$$

This step is another main difference in deriving the necessary conditions. Such a step is not required in applying Pontryagin's Maximum Principle to a system of ODEs.

The map $u \mapsto w(u)$ is weakly differentiable in the directional derivative sense (Gateaux):

$$\lim_{\epsilon\to 0} \frac{w(u+\epsilon l) - w(u)}{\epsilon} = \psi, \tag{18.9}$$

where l is the variation function. The function ψ is called the *sensitivity* of the state with respect to the control.

We use the operator in the sensitivity PDE to find the operator in the adjoint PDE. *A priori* estimates of the difference quotients in the norm of the solution space give the existence of the limit function ψ and that it solves a PDE, which is a linearized version of state PDE

$$L\psi = F(w, l, u) \text{ with appropriate } BC, IC. \tag{18.10}$$

Note that the linear operator L comes from linearizing the state PDE operator A. The 'adjoint' operator (Conway, 1990) of the L operator is used to find the operator to be used in adjoint PDE.

We use the adjoint λ and the sensitivity ψ PDEs to simplify the derivative of the map $u \to J(u)$ and to obtain the explicit characterization of the optimal control in terms of the state and the adjoint. We will illustrate how to explicitly find the sensivity and adjoint PDEs as well as how to derive the characterization of the optimal control in the example below. The state and adjoint equations together with the control characterization, is called the *optimality system.* We refer the reader to the book by Lenhart and Workman (2007) for a discussion of uniqueness of the optimal control and more detail about numerical solutions of the optimality system.

18.3.2 Illustrative PDE example

In the following example, we will concentrate on the calculations of the sensitivity equation, adjoint equation, and the characterization of the optimal control. We do not treat the details of proving existence of the optimal control here. We will also assume the appropriate difference quotients converge to the sensitivity function, which would need to be proven in a fully justified solution. See Evans (1998) for details on deriving such needed estimates. We also refer the reader to the references (Fister, 1997, Lenhart and Bhat, 1993, Lenhart et al. 1999) to see examples with such details.

We consider the problem of harvesting in a diffusing population. In Joshi et al. (2009), a general parabolic equation for the stock density is treated with the justification of the corresponding optimal control analysis. Here we illustrate this technique with an simple equation

$$\begin{aligned} w_t - (w_{xx} + w_{yy}) &= w(1-w) - uw \text{ in } \Omega \times (0,T), \\ w(x,y,t) &= 0 \text{ on } \partial\Omega \times (0,T) \quad \text{(boundary condition)}, \\ w(x,y,0) &= w_0(x,y) \geq 0 \text{ on } \Omega, t = 0 \quad \text{(initial condition)}, \end{aligned}$$

where Ω is an open, connected subset of $\Re^2$ and $\partial\Omega$ is the boundary of Ω. The state $w(x,y,t)$ is the density of the population and the harvesting control is $u(x,y,t)$. Note the state equation has logistic growth $w(1-w)$. Note that the initial population, $w_0(x,y)$, is not identically zero. The "profit" objective function is

$$J(u) = \int_0^T \int_\Omega e^{-\delta t}(pu(x,y,t)w(x,y,t) - Bu(x,y,t)^2)\, dxdy\, dt, \tag{18.11}$$

which is a discounted "revenue less cost" stream. With p representing the price of harvesting population, puw represents the revenue from the harvested amount uw. We use a quadratic cost for the harvesting effort with a weight coefficient B. At first, we consider the case of a positive constant B, and then we discuss the case of $B = 0$. In the $B = 0$ case, the problem is linear in the control and only the yield is being maximized. The coefficient $e^{-\delta t}$ is a discount term with $0 \leq \delta < 1$. For convenience, we now take the price to be $p = 1$.

A main point of interest is if "marine reserves" are part of the optimal solution, and if so, where they should be placed, that is, the regions of no harvesting, where $u^*(x, y, t) = 0$. We seek to find u^* such that

$$J(u^*) = \max_u J(u), \tag{18.12}$$

where the maximization is over all measurable controls with $0 \leq u(x, y, t) \leq M < 1$. Under this set-up, we note that any state solution will satisfy

$$w(x, y, t) > 0 \quad \text{on} \quad \Omega \times (0, T), \tag{18.13}$$

by the Maximum Principle for parabolic equations (Evans, 1998).

First, we differentiate the $u \rightarrow w$ map. Given a control u, consider another control $u^\epsilon = u + \epsilon l$, where l is a variation function and $\epsilon > 0$. Let $w = w(u)$ and $w^\epsilon = w(u^\epsilon)$ be the corresponding states. The state PDEs corresponding to controls, u and u^ϵ, are

$$w_t - (w_{xx} + w_{yy}) = w(1 - w) - uw$$

$$w_t^\epsilon - (w_{xx}^\epsilon - w_{yy}^\epsilon) = w^\epsilon(1 - w^\epsilon) - u^\epsilon w^\epsilon.$$

We form the difference quotient

$$\frac{w^\epsilon - w}{\epsilon}, \tag{18.14}$$

and find the corresponding PDE satified by the difference quotients

$$\left(\frac{w^\epsilon - w}{\epsilon}\right)_t - \left(\frac{w^\epsilon - w}{\epsilon}\right)_{xx} - \left(\frac{w^\epsilon - w}{\epsilon}\right)_{yy} = \frac{w^\epsilon - w}{\epsilon}(1 - u) - \left(\frac{(w^\epsilon)^2 - w^2}{\epsilon}\right) - lw^\epsilon. \tag{18.15}$$

A priori estimates of the states and those quotients will justify that as $\epsilon \rightarrow 0$, $w^\epsilon \rightarrow w$ and

$$\frac{w^\epsilon - w}{\epsilon} \rightarrow \psi. \tag{18.16}$$

As for the nonlinear term, note that

$$\frac{(w^\epsilon)^2 - w^2}{\epsilon} = (w^\epsilon + w)\frac{w^\epsilon - w}{\epsilon} \rightarrow 2w\psi. \tag{18.17}$$

The corresponding derivative quotients will converge and then the resulting PDE for ψ is

$$\psi_t - \psi_{xx} - \psi_{yy} = \psi - 2w\psi - u\psi - lw \text{ on } \Omega \times (0, T),$$

$$\psi = 0 \text{ on } \partial\Omega \times (0, T),$$

$$\psi = 0 \text{ on } \{t = 0\}.$$

Given an optimal control u^* and the corresponding state w^*, we rewrite the sensitivity PDE as

$$L\psi = -lw^*, \text{ where } L\psi = \psi_t - \psi_{xx} - \psi_{yy} - \psi + 2w^*\psi + u^*\psi.$$

Now we discuss the process of finding the adjoint equation. The basic idea of the L^* operator in the adjoint PDE is

$$\int_0^T \int_\Omega e^{-\delta t}\lambda L\psi \, dxdy\, dt = \int_0^T \int_\Omega e^{-\delta t}\psi(L^*\lambda + \delta\lambda)\, dxdy\, dt. \tag{18.18}$$

To see the specific terms of L^*, use integration by parts to see

$$\int_0^T \int_\Omega e^{-\delta t}\lambda\psi_t \, dxdy\, dt = \int_0^T \int_\Omega -e^{-\delta t}(-\delta\lambda + \lambda_t)\psi \, dxdy\, dt. \tag{18.19}$$

The boundary terms on $\Omega \times \{T\}$ and $\Omega \times \{0\}$ vanish due to λ and ψ being zero on the top and the bottom of our domain, respectively. The term with δ comes from the discount term in the objective function. Next notice by integrating by parts twice

$$\int_0^T \int_\Omega e^{-\delta t}\lambda\psi_{xx} dxdy\, dt = \int_0^T \int_\Omega e^{-\delta t}\lambda_{xx}\psi \, dxdy\, dt \tag{18.20}$$

since λ and ψ are zero on $\partial\Omega \times (0, T)$. The linear terms of L go directly in L^* as the same types of terms. Our operator L^* and the adjoint PDE are

$$L^*\lambda = -\lambda_t - \lambda_{xx} - \lambda_{yy} - \lambda + 2w^*\lambda + u^*\lambda$$

$$\text{adjoint PDE } L^*\lambda + \delta\lambda = u^* \text{ on } \Omega \times (0, T)$$

$$\lambda = 0 \text{ on } \partial\Omega \times (0, T)$$

$$\lambda = 0 \text{ on } \Omega \times \{t = T\}.$$

The nonhomogeneous term u on the right hand side (RHS) in the adjoint PDE comes from

$$\frac{\partial(\text{integrand of J})}{\partial(\text{state})} = \frac{\partial(uw)}{\partial w} = u \tag{18.21}$$

where we use the integrand of J without the discount factor $e^{-\delta t}$, which came into play in the integration by parts above.

Next, we use the sensitivity and adjoint functions in the differentiation of the map $u \to J(u)$. At the optimal control u^*, the quotient is nonpositive since $J(u^*)$ is the maximum value, i.e.,

$$0 \geq \lim_{\epsilon \to 0^+} \frac{J(u^* + \epsilon l) - J(u^*)}{\epsilon}. \tag{18.22}$$

Rewriting the adjoint equation as $L^*\lambda + \delta\lambda = u^*$, this limit simplifies to

$$\begin{aligned} 0 \geq \lim_{\epsilon \to 0^+} & \int_0^T \int_\Omega e^{-\delta t}\frac{1}{\epsilon}((u^* + \epsilon l)w^\epsilon - u^*w^* - (B(u^* + \epsilon l)^2 - B(u^*)^2))\, dxdy\, dt \\ &= \int_0^T \int_\Omega e^{-\delta t}[u^*\psi + lw^* - 2Bu^*l]\, dxdy\, dt \end{aligned}$$

$$= \int_0^T \int_\Omega e^{-\delta t}(\psi(L^*\lambda + \delta\lambda) + lw^* - 2Bu^*l)\, dxdy\, dt$$
$$= \int_0^T \int_\Omega e^{-\delta t}(\lambda L\psi + lw^* - 2Bu^*l)\, dxdy\, dt$$
$$= \int_0^T \int_\Omega e^{-\delta t}(-\lambda lw^* + lw^* - 2Bu^*l)\, dxdy\, dt$$
$$= \int_0^T \int_\Omega e^{-\delta t} l(w^*(1-\lambda) - 2Bu^*)\, dxdy\, dt,$$

using that the RHS of the ψ PDE is $-lw^*$.

On the set $\{(x, y, t) : 0 < u^*(x, y, t) < M\}$, the variation l can have any sign. Thus on this set, in the case that $B \neq 0$, the rest of the integrand must be zero, so that

$$u^* = \frac{w^*(1-\lambda)}{2B}. \tag{18.23}$$

By taking the upper and lower bounds into account, we obtain

$$u^* = \min\left(M, \max\left(\frac{w^*(1-\lambda)}{2B}, 0\right)\right). \tag{18.24}$$

This completes the analysis in the case of positive cost constant B.

However, the $B = 0$ case is also important. In this case, we are maximizing yield only. When $B = 0$, the problem is linear in the control u. The argument above goes through with $B = 0$ until the end, before we solve for u^*. The quotient calculation becomes

$$0 \geq \int_0^T \int_\Omega e^{-\delta t} lw^*(1-\lambda)\, dxdy\, dt.$$

On the set $\{(x, y, t) : 0 < u^*(x, y, t) < M\}$, the variation l can have any sign. Since $e^{\delta t}$ and w^* are positive, the inequality above implies $\lambda = 1$ on this set. This case is called singular because the integrand of the objective function drops out on this set. Suppose $\lambda = 1$ on some set of positive measure. By looking at the adjoint PDE, and noting the derivatives of λ are 0, we can solve for the state

$$w^* = \frac{1-\delta}{2}. \tag{18.25}$$

Now use this constant for w in the state equation and solve for the optimal control

$$u^* = \frac{1+\delta}{2}. \tag{18.26}$$

On the set $\{(x, y, t) : u^*(x, y, t) = M\}$, the variation l must be nonpositive and this corresponds with

$$\lambda(x, y, t) < 1.$$

By a similar argument, we treat the case of the optimal control at lower bound, and then we conclude

$$u^*(x, t) = 0 \quad \text{if } \lambda(x, y, t) > 1$$

$$= \frac{1+\delta}{2} \quad \text{if } \lambda(x,y,t) = 1$$
$$= M \quad \text{if } \lambda(x,y,t) < 1.$$

A forward-backward sweep iteration method was used to solve this problem. Each sweep was done by a finite difference scheme. Note that the state PDE has an initial condition while the adjoint PDE has a final time condition, so both PDEs cannot be solved forward in time together. Starting with a guess for the control, the state PDE is solved forward in time and then these state values are used to solve the adjoint PDE backward in time. The control is updated using the calculated state and adjoint values. Then forward and backward sweeps are done again and are followed by a control update. This method continues until successive iterative values are close.

We illustrate one numerical example of this bang-bang case in Figure 18.1. Note that $M = .9$, $\delta = 0$, and initial condition $.5sin(\pi x/4)sin(\pi y/4)$ were used in this figure with space (x, y) and time bounds of 0 and 4. The singular case does not occur in this numerical example. In this example, plotted at time $t = 2$, there is a region in the center of the spatial domain with no harvest, which would be considered a marine reserve, as in Neubert (2003). We see that the region of no harvest varies in space as time changes. Joshi et al. (2009) includes the analyses that vary the initial conditions, the discount rate, and the time horizon. Including an advection term is also treated. We note that a marine reserve was always present in the optimal solution in their various scenarios.

We also note that our illustration of optimal control of a PDE is in the case of finite time horizon. One can also consider the case of infinite time horizon, where the results could differ from the finite time horizon depending on the particular PDEs involved.

18.4 Conclusions

Surveying some of the literature shows the transition from metapopulation models with space as a discrete feature to PDE models with space as a continuous feature. The results obtained vary with the types of model and the objective function.

In this paper, we illustrated the basic ideas of optimal control applied to PDE bioeconomic models and hope that the reader can see possibilities for future work. We showed a simple PDE example, but these techniques can be extended to more general PDEs with other features like advection, variable diffusion coefficients, different boundary conditions, and more general growth coefficients. Even terms in the boundary conditions can be taken as controls (Lenhart et al., 1999). Control theory for PDEs with age-structure has recently been developed, but the techniques are a bit different than those given here due to the different solution space and estimates in the age-structure case (Fister and Lenhart, 2004 and 2006). Optimal harvesting of resources has also been investigated in integrodifference models by Gaff et al. (2007) and Joshi et al. (2006 and 2007), which are discrete in time and continuous in space.

If decisions and policy changes occur only at discrete times, models with a mixture of discrete and continous features or with discrete time and space would be valuable to consider.

Optimal control has been applied to a variety of systems with interacting populations, especially competition, cooperative, and predator-prey models (Lenhart et al. 1997, Fister 1997). But very little has been done on applications to managing resources in the system case. Considering the management of multiple resources which may interact in some way would be interesting.

18.5 Acknowledgments

Both authors would like to acknowledge the support of the National Science Foundation grant DMS-0532378.

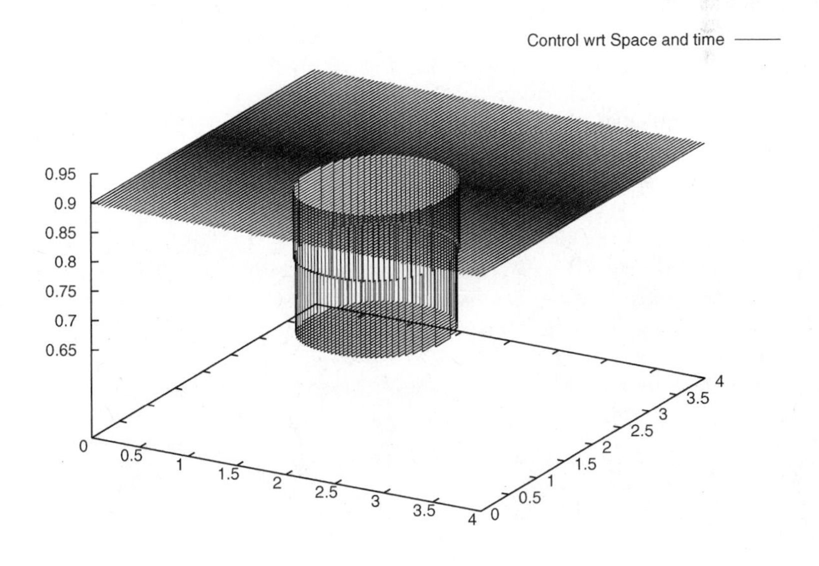

Figure 18.1 Proportion to be harvested.

18.6 References

W. Brock and A. Xepapadeas (2005), Optimal control and spatial heterogeneity: Pattern formation in economic-ecological models, *Fondazione Eni Enrico Mattei* 9605: 1-44.

G. M. Brown and J. Roughgarden (1997), A metapopulation model with private property and a common pool, *Ecological Economics* 30(1):293-299.

A. Cañada, J. L. Gámez and J. A. Montero (1998), Study of an optimal control problem for diffusive nonlinear elliptic equations of logistic type, *SIAM J. Control Optim.* 36(4): 1171-1189.

R. S. Cantrell and C. Cosner (2003), *Spatial Ecology via Reaction-Diffusion Equations*, Wiley, New Jersey.

C. W. Clark (1976), *Mathematical Bioeconomics: The Optimal Management of Renewable Resources*, Wiley, New York.

W. Ding and S. Lenhart (2009), Optimal harvesting of a spatially explicit fishery model, *Natural Resource Modeling* (to appear).

L. C. Evans (1998), *Partial Differential Equations*, American Mathematical Society, Providence.

K. R. Fister (1997), Optimal control of harvesting in a predator-prey parabolic system, *Houston J. Math.* 23: 341-55.

K. R. Fister and S. Lenhart (2004), Optimal control of a competitive system with age structure, *J. Math. Anal. Appl.* 291: 526-537.

K. R. Fister and S. Lenhart (2006), Optimal harvesting in an age-structured predator-prey model, *Appl. Math. Optim.* 54: 1-15.

A. Friedman (1982), *Foundations of Modern Analysis*, Dover, New York.

H. Gaff, H. R. Joshi, and S. Lenhart (2007), Optimal harvesting during an invasion of a sublethal plant pathogen, *Environment and Development Economics Journal* 12: 673-686.

L. R. Gerber, W. Botsford, A. Hastings, H. P. Possingham, S. D. Gaines, S. R. Palumbi, and S. J. Andelman (2003), Popuation models for marine reserve design: A retrospective and prospective synthesis, *Ecological Appliations* 13: S47-S64.

H. S. Gordon (1954), The economic theory of a common property resource: the fishery, *Journal of Political Economy* 62: 124-142.

G. E. Herrera (2007), Dynamic use of closures and imperfectly enforced quotas in a metapopulation, *American Journal of Agricultural Economics* 89(1): 176-189.

H. R. Joshi, G. E. Herrera, S. Lenhart, and M. G. Neubert (2009), Optimal dynamic harvest of a diffusive renewable resource, *Natural Resource Modeling* (to appear).

H. R. Joshi, S. Lenhart, and H. Gaff (2006), Optimal harvesting in an integro-difference population model, *Optimal Control Applications and Methods* 27: 61-75.

H. R. Joshi, S. Lenhart, H. Lou, and H. Gaff (2007), Harvesting control in an integrodifference population model with concave growth term, *Nonlinear Analysis: Hybrid Systems* 1(3): 417-429.

P. Karieva, A. Mullen, and R. Southwood (1990), Population dynamics in spatially complex environments: Theory and data (and discussion), *Philosophical Transactions: Biological Sicences* 330(1257): 175-190.

M. Kot, M. A. Lewis and P. van den Driessche (1996), Dispersal data and the spread of invading organisms, *Ecology* 77: 2027-2042.

K. Kurata and J. Shi (2008), Optimal spatial harvesting strategy and symmetry-breaking, *Appl. Math. Optim.* 58: 89-110.

S. Lenhart and M. Bhat (1993), Application of distributed parameter control model in wildlife damage management, *Math. Meth. Appl. Sci.* 2: 423-39.

S. Lenhart, M. Liang, and V. Protopopescu (1999), Optimal control of the effects of the boundary habitat hostility, *Math. Meth. Appl. Sci.* 22: 1061-1077.

S. Lenhart, V. Protopopescu, and A. Szpiro (1997), Optimal control for competing coalitions, *Nonlinear Analysis: Theory, Methods & Applications* 28: 1411-1428.

S. Lenhart and J. T. Workman (2007), *Optimal Control Applied to Biological Models*, Chapman and Hall/CRC, Boca Raton, FL.

A. W. Leung (1995), Optimal harvesting-coefficient control of steady-state prey-predator diffusive Volterra-Lotka systems, *Appl. Math. Optim.* 31: 219-241.

A. W. Leung and S. Stojanovic (1993), Optimal control for elliptic Volterra-Lotka type equations, *J. Math. Anal. Appl.* 173: 603-619.

M. A. Lewis and R. W. Van Kirk (1997), Integrodifference models for persistence in fragmented habitats, *Bull. Math. Biol.* 59: 107-137.

X. Li and J. Yong (1995), *Optimal Control Theory for Infinite Dimensional Systems*, Birkhauser, Boston.

J. L. Lions (1971), *Optimal Control of Systems Governed by Partial Differential Equations*, Springer-Verlag, New York.

D. R. Lockwood, A. Hastings, and L. W. Botsford (2002), The effects of dispersal patterns on marine reserves: Does the tail wag the dog? *Theoretical Population Biology* 61: 297-309.

J. A. Montero (2000), A uniqueness result for an optimal control problem on a diffusive elliptic Volterra-Lotka type equation, *J. Math. Anal. Appl.* 243: 13-31.

J. A. Montero (2001), A study of the profitability for an optimal control problem when the size of the domain changes, *Natural Resource Modeling* 14(1): 139-146.

M. G. Neubert (2003), Marine reserves and optimal harvesting, *Ecology Letters* 6: 843-849.

M. G. Neubert and G. E. Herrera (2007), Triple benefits from spatial resource management, *Theoretical Ecology* DOI 10.1007/s12080-007-0009-6.

L. S. Pontryagin, V. G. Boltyanskii, R. V. Gamkrelize, and E. F. Mishchenoko (1962), *The Mathematical Theory of Optimal Processes*, Wiley.

T. J. Quinn II (2003), Ruminations on the development and future of population dynamics models in fisheries, *Natural Resource Modeling* 16(4): 341-392.

J. N. Sanchirico and J. E. Wilen (1999), Bioeconomics of spatial exploitation in a patchy environment, *Journal of Environmental Economics and Management* 37(2): 129-150.

J. N. Sanchirico and J. E. Wilen (2001), A bioeconomic model of marine reserve creation, *Journal of Environmental Economics and Management* 42(3): 257-276.

J. N. Sanchirico and J. E. Wilen (2002), The impacts of marine reserves on limited-entry fisheries, *Natural Resource Modeling* 15(3): 380-400.

J. N. Sanchirico and J. E. Wilen (2005), Managing renewable resource use with market-based instruments: Matching policy scope to ecosystem, *Journal of Environmental Economics and Management* 50(1): 23-46.

G. N. Tuck and H. P. Possingham (2000), Marine protected areas for spatially structured exploited stocks, *Marine Ecology Progress Series* 192: 89-101.

G. N. Tuck and H. P. Possingham (1994), Optimal harvesting strategies for a metapopulation, *Bull. Math. Biol.* 56(1): 107-127.

Index